INTRODUCTION TO LINEAR PROGRAMMING

INTRODUCTION TO LINEAR PROGRAMMING

JAY E. STRUM

Quantitative Methods Department
Graduate School of Business Administration
New York University

hd

HOLDEN-DAY, INC.

San Francisco, Cambridge, London, Amsterdam

500 Sansome Street, San Francisco, California 94111

Library of Congress Catalog Card Number: 76–188125
ISBN: 0-8162-8536-5

Printed in the United States of America

To Joseph H. Taggart, who has profoundly influenced many lives, among which the author's

PREFACE

Linear programming is a relatively new, very important branch of applied mathematics. It is desirable, and possible, to offer the quantitatively oriented student an introduction to linear programming as early as the freshman year, conjointly with an introduction to calculus. With this aim in mind, linear programming is developed in these pages with a minimum of prerequisites—in fact, the only real prerequisite is the ability to read Chapter 1, because little is assumed and the progression to new ideas is gradual.

Students of applied mathematics should understand theory thoroughly and, as a rule, should not be treated like automatic computing machines. In this text an attempt has been made to render the theory understandable through example and counterexample without emphasizing formal proofs and, at the same time, to motivate the reader to study further—as any introduction should.

The problem in teaching linear programming is largely one of choosing a frame of reference from which to work, that is, of deciding whether to adopt a geometric, algebraic, or "complementarity" approach, and, if algebraic, whether to emphasize row equation or column vector. Here emphasis is given to row equations, but the other approaches are not neglected. The instructor should note that (1) rigid choice of tableau format has been purposely avoided, (2) in this text, sensitivity analysis precedes and motivates duality, (3) although matrix algebra is neither assumed nor used, an elementary example of column generation is given in the treatment of the cutting-stock problem in Chapter 9, and (4) if desired, matrix methods can be used to render more concise the material on multiplier theory and postoptimality analysis in Chapters 7 and 8, respectively.

I deeply acknowledge indebtedness to the following reviewers: M. Chen, H. Dwyer, P. Nielsen, and A. Schiff, all former students of mine; D. Aulicino, lecturer at the City University of New York; Professor J. Owen, New York University; Professor M. Eisner, Cornell University; Dr. A. J. Hoffman, IBM Research, Yorktown Heights, New York. For encouragement given me over the years, let me here thank Dean C. Clark, Graduate School of Business, the University of Kansas. Lastly, I should like to express my debt to E. Millman, Senior Editor at Holden-Day, Inc., for the expert red-penciling and other help given me.

J. Strum

New York
Summer 1970

CONTENTS

1

THE CORNER-POINT THEOREM

This textbook is written with the reader in mind. Minimal prerequisites are assumed and, as needed, additional concepts and techniques are carefully developed. This introductory chapter touches upon certain aspects of "linear programming," not systematically but rather in an exploratory way. Certain ideas are introduced but their significance is not emphasized—in fact, nowhere in Chapter 1 is there a formal definition of "linear programming." In Chapter 2, however, we reexamine the ideas introduced in Chapter 1 in a more systematic, more algebraic way.

It is assumed that the reader knows or can verify that an equation of the first degree, say $3x + y = 48$, represents a straight line when graphed in the usual coordinate system. He should be familiar with the inequality notation and know that "x is greater than or equal to 3" can be symbolized in two ways, either $x \geq 3$ or, equivalently, $3 \leq x$. He should also realize that the inequalities

$$6x + 6y \leq 420$$

and

$$x + y \leq 70$$

are *equivalent*, in the sense that any pair of values of x and y which satisfies one of these inequalities also satisfies the other. Lastly, he should understand or be able to demonstrate by substitution that

$$x + y \leq 70$$

and

$$-x - y \geq -70$$

are equivalent inequalities.

The following problems give some idea of the subject matter of linear programming.

PROBLEM 1.1 (ALLOCATION OF RESOURCES TO PRODUCTION). A manufacturer produces two different products, A and B. Each of these products must be processed by two machines, I and II. To manufacture one unit of A it is necessary to use $\frac{3}{4}$ hour of time on machine I and $\frac{1}{4}$ hour on machine II. One unit of product B requires $\frac{1}{4}$ hour on machine I and $\frac{1}{3}$ hour on machine II. Machine I may not operate for more than 12 hours per day, while machine II is limited to 10 hours of operation per day. Each unit of A contributes \$5 to profit, and each unit of B contributes \$6.

How should the manufacturer allocate his facilities for production; that is, what amount of each product should be produced daily in order to maximize profit? We summarize the data in the following table:

	Product A	*Product B*	*Maximum machine-hours available per day*
Machine I	$\frac{3}{4}$ hr	$\frac{1}{4}$ hr	12 hr
Machine II	$\frac{1}{4}$ hr	$\frac{1}{3}$ hr	10 hr
Profit contribution per unit	\$5	\$6	

Note that the rows focus attention on machines, while the columns focus attention on products. For example, the *Product A* column tells us that to produce 1 unit of product A we must use $\frac{3}{4}$ hour on machine I and $\frac{1}{4}$ hour on machine II. The *Product B* column is interpreted similarly.

Since the profit contribution of product B is larger than that of product A, it might seem that we ought to produce only B. How many units of B could we produce daily? By the restrictions placed on the machines, no more than 30 units of B can be produced. This result follows from an examination of column B. Every unit of B must spend $\frac{1}{4}$ hour on machine I and $\frac{1}{3}$ hour on machine II. Therefore, after producing the thirtieth unit of B, machine I will have run $(30)(\frac{1}{4})$ hours, and machine II will have run $(30)(\frac{1}{3})$ hours. Even though the available time on the first machine has not been exhausted, the available running time on the second machine would be completely utilized.

This production program of 0 units of A and 30 units of B will lead to \$180 of profit. What is surprising is the fact that this program will not maximize profit. Consider a second program in which 8 units of A and 24 units of B are to be produced daily. Such a program is technically possible, for it requires $(\frac{3}{4})(8) + (\frac{1}{4})(24) = 12$ hours of time on machine I and $(\frac{1}{4})(8) + (\frac{1}{3})(24) = 10$ hours of time on machine II. Furthermore, the net profit for this program is $(5)(8) + (6)(24) = \$184$. Thus the second production program is more profitable than the first, even though it produces fewer units of the seemingly more profitable product B. Why is this so? Perhaps it is because our first program allowed too much *idle time* on machine I. In producing 30 units of B, machine I is used for only $7\frac{1}{2}$ hours, but this machine could have been running for an additional $4\frac{1}{2}$ hours.

We now demonstrate that the second program is the most profitable program possible. Let x equal the number of units of product A and y equal the number of units of product B that are to be produced daily. Daily profit, in dollars, can then be denoted by

$$P = 5x + 6y$$

Using the notation of mathematics, our production problem is reduced to finding values of x and y which maximize $P = 5x + 6y$, subject to the inequality *constraints*

$$x \geq 0 \qquad (1)$$

$$y \geq 0 \qquad (2)$$

$$\tfrac{3}{4}x + \tfrac{1}{4}y \leq 12 \qquad (3)$$

$$\tfrac{1}{4}x + \tfrac{1}{3}y \leq 10 \qquad (4)$$

Constraints (1) and (2) express the fact that it is impossible to manufacture a negative amount of any product, while constraints (3) and (4) indicate that machines I and II are limited to 12 and 10 hours of use per day, respectively, no matter how those hours are allocated to production of products A and B.

Note that our problem has two parts: an *objective function* to be maximized and inequalities (1) through (4) which describe the constraints on our problem. These constraints determine the *region of feasibility*, which can be defined as the totality of points (x, y) representing technically feasible production programs, that is, values of x and y which jointly satisfy all the constraints. To illustrate, the program $(x, y) = (4, 8)$, which produces $x = 4$ units of A and $y = 8$ units of B, represents a feasible program, as does $(1, 1)$. The reader should check these values to see that the constraints are satisfied. The program $(16, 4)$ is one which is not feasible, since $(\frac{3}{4})(16) + (\frac{1}{4})(4)$ is greater than 12, so that constraint (3) is violated.

We can avoid fractions in inequalities (3) and (4). If we multiply both sides of inequality (3) by 4 and both sides of inequality (4) by 12, we obtain the inequalities

$$3x + y \leq 48 \qquad (3')$$

$$3x + 4y \leq 120 \qquad (4')$$

Any numbers x and y which satisfy inequality (3) also satisfy inequality (3′), and conversely. This is also true for inequalities (4) and (4′). Being creatures of habit, we would work with (3′) and (4′) to avoid fractions. (Of course, an electronic computer might have different habits.)

We now exhibit graphically the region of feasibility determined by (1), (2), (3′), and (4′). Constraint (1), $x \geq 0$, corresponds to all the points on and to

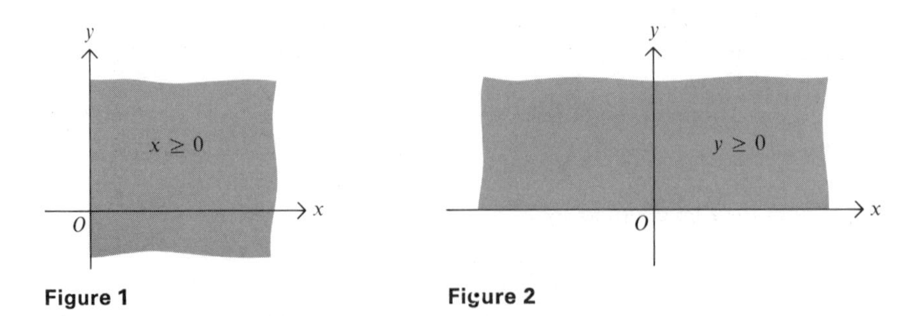

Figure 1

Figure 2

the right of the line $x = 0$ (that is, of the y axis) (see Figure 1). Constraint (2), $y \geq 0$, limits the region of feasibility to points lying on and above the x axis (see Figure 2). Constraints (1) and (2) hold simultaneously in the doubly shaded region in Figure 3, which is the "intersection" of the shaded regions in Figures 1 and 2.

In order to plot constraint (3′) we first sketch the associated line $3x + y = 48$ and then decide on which side of this line the inequality $3x + y < 48$ holds. Figure 4 shows this.

The best way to think about inequality (3′) and the associated equality $3x + y = 48$ is to realize that the equality represents the *boundary*, or *limiting*

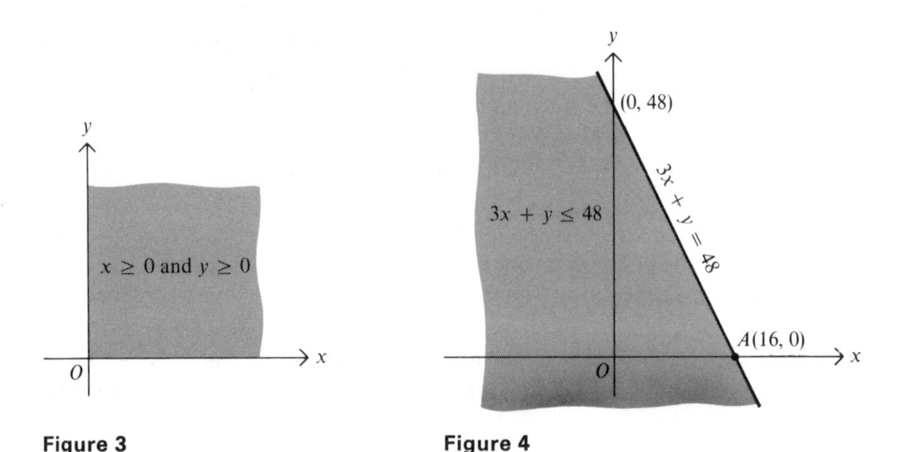

Figure 3

Figure 4

line, beyond which the inequality will cease to hold. Exactly which side of the line represents the "beyond" side can be easily determined by testing a point or two. For example, at the point ($x = 10, y = 10$) the value of $3x + y$ is only 40, and hence the side of the line $3x + y = 48$ containing ($x = 10, y = 10$) is the region where the inequality $3x + y < 48$ holds. Similarly, points where both x and y are very large, say ($x = 100, y = 100$), must be on the side of the line where the inequality $3x + y > 48$ holds.

The set of points where constraints (1), (2), and (3′) are satisfied simultaneously is the shaded triangular region in Figure 5. It is the intersection of the shaded regions of Figures 3 and 4.

If we now plot constraint (4′), namely, $3x + 4y \leq 120$, and determine where its shaded region agrees with the others, then the aforementioned triangle is cut down to quadrilateral $OABC$ (see Figure 6). The region of feasible production programs is now completely represented by all points in the interior,

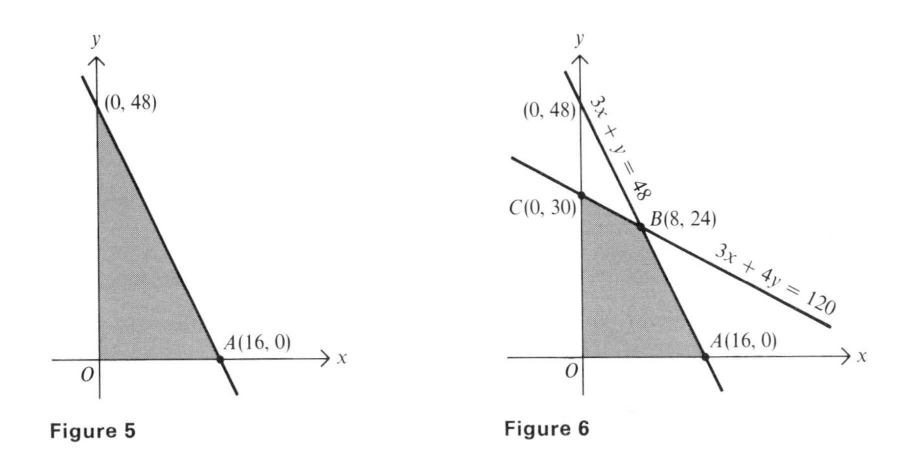

Figure 5

Figure 6

or on the boundary, of $OABC$. Point O has the coordinates (0, 0), point A the coordinates (16, 0), point C the coordinates (0, 30). In order to determine the coordinates of B, note that this point lies at the intersection of the limiting lines whose equations constitute the system

$$\begin{aligned} 3x + y &= 48 \\ 3x + 4y &= 120 \end{aligned}$$

This system has the solution $x = 8$, $y = 24$. (The reader should verify this. In a later chapter we study systems of simultaneous linear equations.)

Motivated by Figure 6, we call the points O, A, B, and C *corners*. We can now list all the corners of the region of feasibility and the values of the profit function at these corners:

Corner	*Value of* $P = 5x + 6y$
$O(0, 0)$	0
$A(16, 0)$	80
$B(8, 24)$	184
$C(0, 30)$	180

Of course, A is not the only point where $P = 80$, nor is C the only point where

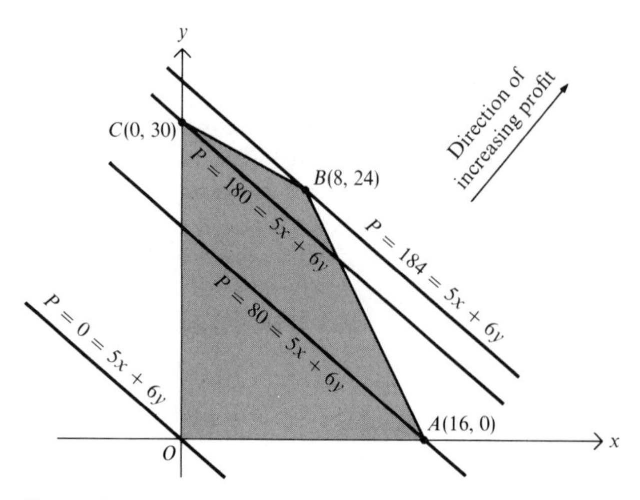

Figure 7

$P = 180$. In Figure 7 the lines representing the profit schedules (objective functions)

$$
\begin{aligned}
5x + 6y &= 0 \\
5x + 6y &= 80 \\
5x + 6y &= 184 \\
5x + 6y &= 180
\end{aligned}
$$

have been superimposed on the quadrilateral of Figure 6. These schedules correspond to the values of the profit function at the corners O, A, B, and C, respectively. To draw any such limiting line we need only find two of its points. For instance, to draw the line whose equation is

$$5x + 6y = 180$$

note that if $x = 0$ in this equation, then $y = 30$, and hence the point (0, 30) lies on the line. Similarly, the point (36, 0) lies on this line. Plotting these points and drawing the line they determine gives us the graph of $5x + 6y = 180$.

Suppose k represents an arbitrary constant. Clearly, the function $P = 5x + 6y$ is constant all along the line $5x + 6y = k$. For this reason any equation of the form

$$5x + 6y = k$$

is often called an *isoprofit line*, an *isoquant*, or simply a *line of constant profit*. By varying k we obtain a family of parallel lines like the four we graphed in Figure 7.

If we wish to maximize P we must move the line $5x + 6y = 0$ parallel to itself, up and to the right, as far from the origin as the feasible region permits. It is easy to do this graphically by placing a straightedge along $5x + 6y = 0$

and then "sliding it parallel to itself" up and across the region of feasibility in the direction of increasing P, until further movement carries us beyond the feasible region. The reader can actually see that the largest profit is attainable at the *corner* $B(8, 24)$ of the feasible region, where $P = 184$.

PROBLEM 1.2. Suppose that in Problem 1.1, 15 more minutes of time become available on machine I. All other things being equal, we should be able to make at least as much net profit as before, for the new region of feasibility contains the region of Problem 1.1. Mathematically, the new problem is to maximize the objective function

$$P = 5x + 6y$$

subject to the trivial constraints $x \geq 0, y \geq 0$ and the nontrivial constraints

$$\tfrac{3}{4}x + \tfrac{1}{4}y \leq 12\tfrac{1}{4}$$
$$\tfrac{1}{4}x + \tfrac{1}{3}y \leq 10$$

Clearing fractions, Problem 1.2 can be rewritten as follows:

$$\begin{aligned} &\text{Maximize} \quad P = 5x + 6y \\ &\text{where} \quad x \geq 0, \quad y \geq 0 \\ &\text{and} \quad 3x + y \leq 49 \qquad (1) \\ &\qquad\quad 3x + 4y \leq 120 \qquad (2) \end{aligned}$$

This is the formulation we henceforth call "Problem 1.2."

Figure 8 shows the feasible region and corners of Problem 1.2. The reader should compare Figures 7 and 8 very carefully. These figures differ only in line AB, which is somewhat more "up and to the right" in Figure 8 than in Figure 7, corresponding to the availability of 49 rather than 48 quarter-hours on machine I.

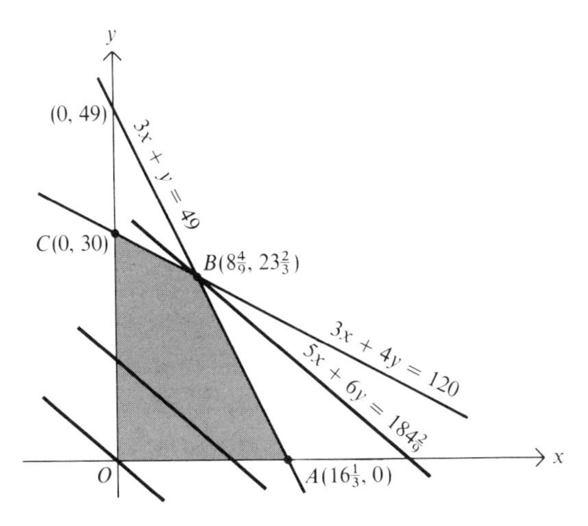

Figure 8

The following table gives the feasible corners and corresponding values of P:

Corner	$P = 5x + 6y$
$O(0, 0)$	$P = 0$
$A(\frac{49}{3}, 0)$	$P = \frac{245}{3} = 81\frac{2}{3}$
$B(8\frac{4}{9}, 23\frac{2}{3})$	$P = 184\frac{2}{9}$
$C(0, 30)$	$P = 180$

Maximum profit is achieved at corner B, where $x = 8\frac{4}{9}$, $y = 23\frac{2}{3}$, and $P = 184\frac{2}{9}$.

The optimal solution to this problem involves fractions. Unless otherwise stated, the variables in a linear-programming problem are assumed to be any numbers, not necessarily whole numbers. Sometimes a linear-programming problem will lead to a solution involving meaningless fractions which can be discarded without substantially affecting the solution. In other problems, however, discarding fractions can lead to nonfeasible programs or to large changes in the objective function.

Note that optimal profit in Problem 1.2 is greater by $\frac{2}{9}$ than that in Problem 1.1, which suggests that we should consider $\frac{2}{9}$ to be the *implicit* or *incremental value* of having 1 more unit of time available for machine I, that is, 15 more minutes.

The two problems above illustrate the key theorem of linear programming, the so-called *fundamental corner-point theorem*, which may be stated as follows: *If a linear-programming problem possesses an optimal solution then there is a corner point of the region of feasibility at which the objective function achieves its optimal value (although this same optimal value may also be achieved at other points*). In Chapter 2 we shall reexamine the corner-point theorem in greater detail. The reader can be certain that all problems presented in this chapter do possess an optimal solution.

PROBLEM 1.3. A company is planning production for a certain period of time, say a week. It produces two products, A and B, each of which requires three types of processing. The length of time for processing each unit is shown in the following table:

	Product A	*Product B*	*Maximum process capacity per week*
Process I	6 hr/unit	6 hr/unit	420 hr
Process II	3 hr/unit	6 hr/unit	300 hr
Process III	4 hr/unit	2 hr/unit	240 hr
Profit contribution per unit	\$0.30	\$0.20	

How much of each product should the company produce per week in order to maximize profit?

To solve this problem, let x denote the amount of product A and y the amount of product B to be produced per week. Our problem is to maximize

$$P = 0.3x + 0.2y$$

subject to the following constraints:

$$\begin{aligned} x \quad\;\; &\geq 0 \\ y &\geq 0 \\ 6x + 6y &\leq 420 \\ 3x + 6y &\leq 300 \\ 4x + 2y &\leq 240 \end{aligned}$$

Instead of working with the inequalities stipulated above, we can work with the simplified but equivalent inequalities

$$x \geq 0 \qquad (1)$$
$$y \geq 0 \qquad (2)$$
$$x + y \leq 70 \qquad (3)$$
$$x + 2y \leq 100 \qquad (4)$$
$$2x + y \leq 120 \qquad (5)$$

The region of feasibility in this problem can be found by sketching the lines associated with the above inequalities. We list the corresponding equations of these lines below, giving them the same numbers as the associated inequalities, so that no confusion will arise.

$$x = 0 \qquad (1)$$
$$y = 0 \qquad (2)$$
$$x + y = 70 \qquad (3)$$
$$x + 2y = 100 \qquad (4)$$
$$2x + y = 120 \qquad (5)$$

Any method of sketching will do. For instance to graph

$$x + 2y = 100$$

we need only realize that this line passes through the points (0, 50) and (100, 0). But these points make it hard for us to scale the x and y axes. Fortunately, an accurate graph is not what is needed, but rather a rough sketch of how the relevant lines intersect and an *accurate labeling of their intersections* (thanks to the corner-point theorem). It will often be convenient to use the symbol $[E1, E2] = (x_0, y_0)$ to abbreviate the fact that the point (x_0, y_0) is the intersection of the lines, or equations, labeled (1) and (2), and so on.

Figure 9 gives the desired picture. The region of feasibility is the interior of $ORSTU$, where $O = [E1, E2]$, $R = [E2, E5] = (60, 0)$, $U = [E1, E4] = (0, 50)$,

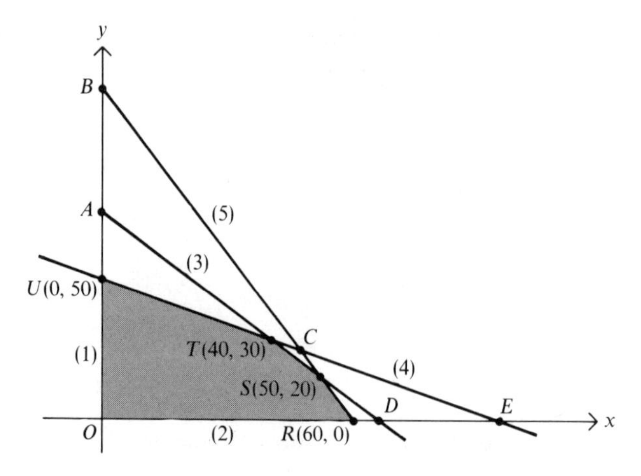

Figure 9

$S = [E3, E5] = (50, 20)$, and $T = [E3, E4] = (40, 30)$. For example, we obtain point S by solving the system of two equations in two unknowns:

$$x + y = 70 \tag{3}$$

$$2x + y = 120 \tag{5}$$

By the fundamental corner-point theorem, we should evaluate the objective function at each corner of the region of feasibility:

Feasible corner	*Value of $P = 0.3x + 0.2y$*
$O(0, 0)$	0.0
$R(60, 0)$	18.0
$S(50, 20)$	19.0
$T(40, 30)$	18.0
$U(0, 50)$	10.0

Thus, the maximum value of P is 19.0, attained at the feasible corner $S(50, 20)$. In other words, the maximum profit is \$19, attainable by producing 50 units of product A and 20 units of product B.

In the above analysis we located the corners of the region of feasibility by rough graphing. Instead, we could have reasoned as follows. We "know" that our problem reduces to finding the true corners of the region of feasibility determined by the straight lines

$$x = 0 \tag{1}$$

$$y = 0 \tag{2}$$

$$x + y = 70 \tag{3}$$

$$x + 2y = 100 \tag{4}$$

$$2x + y = 120 \tag{5}$$

But all candidates for corners can be found by solving simultaneously each system of two equations chosen from the five above. These candidates are:

Corner	*Coordinates*	
corner 1 = [E1, E2]	$x = 0,$	$y = 0$
corner 2 = [E1, E3]	$x = 0,$	$y = 70$
corner 3 = [E1, E4]	$x = 0,$	$y = 50$
corner 4 = [E1, E5]	$x = 0,$	$y = 120$
corner 5 = [E2, E3]	$x = 70,$	$y = 0$
corner 6 = [E2, E4]	$x = 100,$	$y = 0$
corner 7 = [E2, E5]	$x = 60,$	$y = 0$
corner 8 = [E3, E4]	$x = 40,$	$y = 30$
corner 9 = [E3, E5]	$x = 50,$	$y = 20$
corner 10 = [E4, E5]	$x = \frac{140}{3},$	$y = \frac{80}{3}$

For example, corner 8, or ($x = 40$, $y = 30$), is the solution to the system of equations

$$x + y = 70 \quad (3)$$

$$x + 2y = 100 \quad (4)$$

Of these ten corners only five are *feasible corners*. They are corners 1, 3, 7, 8, and 9. Corners 2 and 4 are not part of the feasible region, because they do not satisfy *inequality* (4), namely, $x + 2y \leq 100$; corner 5 does not satisfy inequality (5), and so on. (The reader should check these, and all future contentions, carefully.) This can be seen in Figure 9.

The remainder of the solution is the same. By evaluating the objective function at each feasible corner, the maximum of P is found.

PROBLEM 1.4 (A DIET PROBLEM). The problem is to extract certain minimum amounts of three special "nutrients," or "vitamins," from two raw materials, X and Y. These raw materials cost \$3 and \$5 per 100 pounds, respectively. Each of the raw materials contains the three nutrients, but in differing amounts, as shown in the following table:

	Amount of nutrient in 100 lb of		
	Material X	*Material Y*	*Minimum requirements*
Nutrient 1	1 lb	3 lb	9 lb
Nutrient 2	1 lb	1 lb	5 lb
Nutrient 3	2 lb	1 lb	6 lb
Cost per 100 lb of material	\$3	\$5	

The objective is to obtain at least 9 pounds of nutrient 1, 5 pounds of nutrient 2, and 6 pounds of nutrient 3 at minimum cost.

If we let x be the number of 100-pound units of material X we employ, and y the number of 100-pound units of material Y, then, in mathematical terms, our problem is to

$$\begin{aligned} &\text{Minimize} \quad C = 3x + 5y \\ &\text{where} \quad x \geq 0, \quad y \geq 0 \\ &\text{and} \quad x + 3y \geq 9 \\ &\qquad\quad x + y \geq 5 \\ &\qquad\quad 2x + y \geq 6 \end{aligned}$$

Figure 10 is a sketch of the region of feasibility. Note that the origin O is always on the wrong side of our nontrivial limiting lines. The region of feasibility

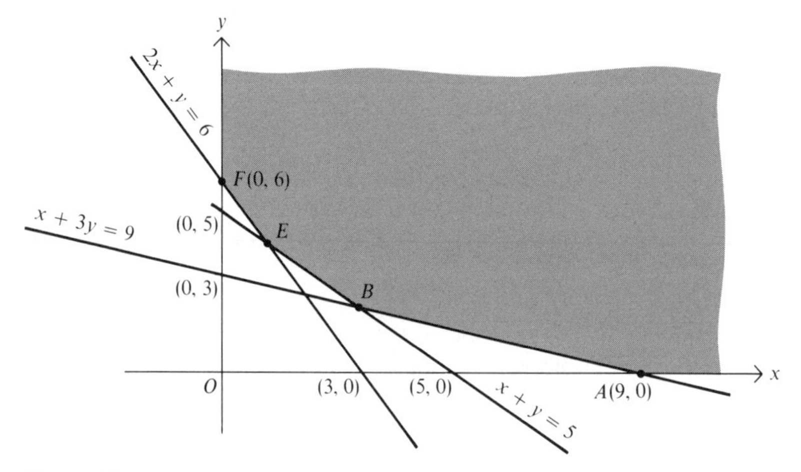

Figure 10

is the shaded area having A, B, E, and F as corner points and extending indefinitely up and to the right. The coordinates of A and F were found directly when we sketched in the limiting lines on which they lie: $A = (9, 0)$ and $F = (0, 6)$. Point B is given by the solution to the system

$$\begin{aligned} x + y &= 5 \\ x + 3y &= 9 \end{aligned}$$

It is easy to see that $B = (3, 2)$. Similarly, point E is given by the solution to

$$\begin{aligned} 2x + y &= 6 \\ x + y &= 5 \end{aligned}$$

Hence $E = (1, 4)$.

Note that the region of feasibility is *unbounded* in the sense that it contains points where x or y can be arbitrarily large. Therefore, no finite point in the region could make C maximum. But a moment's thought, or a plot of isocost lines moving toward the origin as cost falls, should convince the reader that only corners A, B, E, or F could give the minimum of C.

The following table lists these corners and the corresponding values of the objective function:

Corner	*Coordinates*	*Value of* $C = 3x + 5y$
A	(9, 0)	27
B	(3, 2)	19
E	(1, 4)	23
F	(0, 6)	30

Hence the minimum of C is attained at $B(3, 2)$. Thus, we should buy 300 pounds of raw material X and 200 pounds of raw material Y. With this plan we are able to meet the minimum requirement of nutrients 1 and 2 exactly, and we get 200 pounds more than needed of nutrient 3, as the reader can easily verify. The results of optimization often seem opposed to intuition. We can only reduce the excess of 200 pounds of nutrient 3 by increasing cost.

We can work Problem 1.4 without appealing to a graph by drawing up a list of all potential corners. To this end, let us label the limiting equations as below:

$$x = 0 \quad (1)$$

$$y = 0 \quad (2)$$

$$x + 3y = 9 \quad (3)$$

$$x + y = 5 \quad (4)$$

$$2x + y = 6 \quad (5)$$

All corners are listed in the following table:

Corner	*Coordinates*
corner 1 = [E1, E2]	$x = 0, \quad y = 0$
corner 2 = [E1, E3]	$x = 0, \quad y = 3$
corner 3 = [E1, E4]	$x = 0, \quad y = 5$
corner 4 = [E1, E5]	$x = 0, \quad y = 6$
corner 5 = [E2, E3]	$x = 9, \quad y = 0$
corner 6 = [E2, E4]	$x = 5, \quad y = 0$
corner 7 = [E2, E5]	$x = 3, \quad y = 0$
corner 8 = [E3, E4]	$x = 3, \quad y = 2$
corner 9 = [E3, E5]	$x = \frac{9}{5}, \quad y = \frac{12}{5}$
corner 10 = [E4, E5]	$x = 1, \quad y = 4$

The following corners are not feasible, because each violates at least one constraint:

corner 1: $x = y = 0$ does not satisfy $x + y \geq 5$.

corner 2: $x = 0, y = 3$ does not satisfy $x + y \geq 5$.

corner 3: $x = 0, y = 5$ does not satisfy $2x + y \geq 6$.

corner 6: $x = 5, y = 0$ does not satisfy $x + 3y \geq 9$.

corner 7: $x = 3, y = 0$ does not satisfy $x + y \geq 5$.

corner 9: $x = \frac{9}{5}, y = \frac{12}{5}$ does not satisfy $x + y \geq 5$.

The remaining corners constitute the same set of feasible corners as previously obtained graphically.

Exercises

Many of the following problems are worked out in the next chapter. We suggest that each problem be solved by graphical means. As a further exercise, the reader should then attempt to solve the problems without the use of graphs.

1. Find the maxima and minima of the function $F = 3x - 2y$, subject to the constraints $0 \leq x \leq 5$; $0 \leq y \leq 6$.

2. Maximize $P = x + y$, subject to the constraints $x \geq 0$; $y \geq 0$; $y \leq 5$.

3. Minimize $C = x + y$, subject to the constraints $x \geq 0$; $x \leq 5$; $y \geq 0$; $y \leq 5$.

4. Maximize $P = x + y$, subject to the constraints $x \geq 0$; $y \geq 0$; $x + y \leq 10$.

5. Minimize $C = 7x + 3y$, subject to the constraints $x \geq 0$; $y \geq 0$; $x \leq 10$; $y \leq 20$; $x + y \leq 15$; $x + y \geq 5$.

6. Minimize $I = x + y$, subject to the constraints $x \geq 0$; $y \geq 0$; $3x + 10y \geq 20$; $5x + 2y \geq 20$; $5x + 6y \geq 40$.

7. The XYZ Company transforms its own house brand of base metal into gold and silver. The transmutation process consists of two steps, superheating and supercooling. To produce 1 unit of gold, heat must be applied for 1 hour and cold for 2 hours, while to produce 1 unit of silver, heat must be applied for 2 hours and cold for 1 hour. The heating phase of the operation can be performed for a maximum of 10 hours per day, the freezing operation for a maximum of 5 hours per day.

The company makes a profit of \$45 per unit of gold and \$30 per unit of silver. The demand for these metals is unlimited. How many units of each precious metal should be produced daily so as to achieve the highest profit?

8. Make up a "situation" which results in the following problem:

$$\begin{aligned} &\text{Maximize} \quad P = 2x + 3y \\ &\text{where} \quad x \geq 0, \quad y \geq 0 \\ &\text{and} \quad 3x + y \leq 24 \\ &\qquad\quad 4x + 2y \leq 60 \end{aligned}$$

9. Bags of plant food are to be produced, containing a mixture of bone meal and processed vegetable matter. Each bag must contain at least 6 ounces of nitrogen and 3 ounces of phosphorus. Each pound of bone meal contains 2 ounces of nitrogen and 4 ounces of phosphorus; each pound of vegetable matter contains 3 ounces of nitrogen and 2 ounces of phosphorus. Bone meal costs 112¢ per pound and processed vegetable matter costs 94¢ per pound. How many pounds of bone meal and vegetable matter should be used to produce a bag of plant food at least cost?

10. Make up a "situation" which results in the following problem:

$$\begin{aligned} &\text{Minimize} \quad C = 2u + 5v \\ &\text{where} \quad u \geq 0, \quad v \geq 0 \\ &\text{and} \quad 3u + 4v \geq 2 \\ &\qquad\quad u + 2v \geq 3 \end{aligned}$$

2

THE CORNER-POINT THEOREM PURSUED

This chapter is also devoted to the corner-point theorem. We begin by developing, more carefully than previously, the geometry of two-dimensional linear-programming problems; then the corner-point theorem will be extended to problems involving three variables. Our purpose is twofold: (1) to show the need for replacing geometrical arguments with algebraic ones; and (2) to show some of the power—as well as many of the limitations—of the corner-point theorem. The chapter is divided into three sections: Section 2.1 is a study of some relevant questions of two-dimensional geometry; in Section 2.2 we solve a series of two-dimensional linear-programming problems; and in Section 2.3 we turn to three-dimensional problems.

For brevity, we will henceforth refer to linear-programming problems as *linear programs.*

2.1 SOME GEOMETRY OF TWO-DIMENSIONAL SPACE

As we saw in Chapter 1, a linear program in two variables is composed of two parts:

(1) An objective function to be optimized, that is, a profit function to be maximized or a cost function to be minimized

(2) A region of feasibility, its graph bounded by straight lines

We now study both objective function and region of feasibility in greater detail.

DEFINITION 1. *A linear function F* (in two variables) is an expression of the form $F = ax + by + c$, where a, b, and c are arbitrary constants, positive or negative.

For example, $F = \frac{1}{2}x - 7y + 1000$ is a linear function in x and y. Linear functions with constant term $c = 0$ are so important that they have received a special name.

DEFINITION 2. *A homogeneous linear form H* (in two variables) is a linear function whose constant term is 0.

For example, the profit function $P = 5x + 6y$ is a homogeneous form.

We stress that a linear form is a *function*, or *schedule*, mathematically meaningful for all real numbers x and y. To illustrate, the linear form

$$H = 3x + y$$

has a well-defined value at all points in the cartesian plane. If $x = 1$ and $y = 6$, then $H = 9$; if $x = 0.5$ and $y = 3.5$, then $H = 5$; if $x = 3$ and $y = 4$, then $H = 13$, etc.

In order to study the behavior of linear functions we need the following result from analytic geometry: Let $A(x_1, y_1)$ and $B(x_2, y_2)$ be two points in the two-dimensional plane, and let $M(x_M, y_M)$ denote the midpoint of the line segment joining A and B. Then the coordinates of M are *averages* of those of A and B; that is,

$$x_M = \frac{x_1 + x_2}{2}, \quad y_M = \frac{y_1 + y_2}{2}$$

Figure 11 depicts these relationships in the case where $A = (1, 6)$, $B = (3, 4)$, and $M = (2, 5)$.

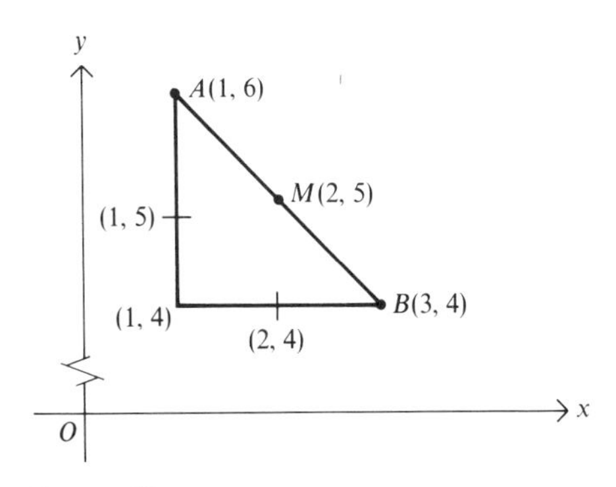

Figure 11

The following theorem is fundamental:

THEOREM 2.1. If a linear function F assumes the value F_A at the point A and the value F_B at the point B, then the value F_M assumed by F at the midpoint M

of the segment AB is

$$F_M = \frac{F_A + F_B}{2}$$

Before we prove Theorem 2.1, consider the following numerical illustration: Suppose that $F = 3x + 4y + 10$, and that $A = (1, 5)$ and $B = (3, 7)$. Then $F_A = (3)(1) + (4)(5) + 10 = 33$, and $F_B = (3)(3) + (4)(7) + 10 = 47$. The midpoint M of the segment AB is then the point $[(1 + 3)/2, (5 + 7)/2] = (2, 6)$. What is the value F_M taken on by F at M? It is $F_M = (3)(2) + (4)(6) + 10 = 40$. But $(F_A + F_B)/2 = (33 + 47)/2 = 40$, and so Theorem 2.1 holds in the present case.

To prove Theorem 2.1, let $F = cx + dy + k$ denote the function in question, and let $A(x_1, y_1)$ and $B(x_2, y_2)$ be two points in the plane. Then, by definition, the coordinates of the midpoint M of the segment AB are

$$M = \left(\frac{x_1 + x_2}{2}, \frac{y_1 + y_2}{2}\right)$$

Furthermore, again by definition, the values of F at the points A, B, and M are, respectively,

$$F_A = cx_1 + dy_1 + k$$

$$F_B = cx_2 + dy_2 + k$$

and

$$\begin{aligned} F_M &= cx_M + dy_M + k \\ &= c\left(\frac{x_1 + x_2}{2}\right) + d\left(\frac{y_1 + y_2}{2}\right) + k \end{aligned}$$

By simple arithmetic,

$$\begin{aligned} F_M &= \frac{cx_1 + cx_2 + dy_1 + dy_2 + 2k}{2} \\ &= \frac{cx_1 + dy_1 + k}{2} + \frac{cx_2 + dy_2 + k}{2} \end{aligned}$$

But $(cx_1 + dy_1 + k)$ equals F_A and $(cx_2 + dy_2 + k)$ equals F_B. Hence

$$F_M = \frac{F_A + F_B}{2}$$

as contended.

We mention in passing that if the linear function

$$F = ax + by + c$$

takes on the same value k at the two points P_1 and P_2, then F assumes the value k everywhere on the line through P_1 and P_2. This result may appear new, but actually we assumed it throughout Chapter 1. In fact, the equation of the line through P_1 and P_2 must be $F = k$, that is,

$$F = ax + by + c = k$$

or

$$ax + by + (c - k) = 0$$

To illustrate, suppose we are studying the linear function

$$F = 5x + 6y + 2$$

At the point $A(2, 0)$ the value of F is 12, and at the point $B(0, \frac{5}{3})$ the value of F is also 12. Hence the equation of the line through A and B must be

$$F = 5x + 6y + 2 = 12$$

or

$$5x + 6y - 10 = 0$$

(see Figure 12).

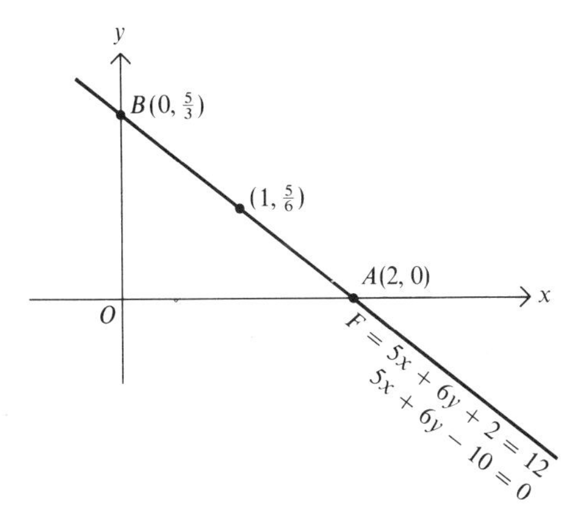

Figure 12

The following definitions and theorems are needed to study the region of feasibility of a linear program.

DEFINITION 3. A set of points S in the plane of analytic geometry is said to be *convex* if and only if, for each pair of points P_1 and P_2 belonging to S, the entire line segment P_1P_2 also belongs to S.

For some examples of convex sets in the plane see Figure 13. Note that sets (d) and (e) in Figure 13 are *not* convex because it is possible to find two points in each of these sets, P_1 and P_2 as shown, such that the segment P_1P_2 does not lie entirely in the set.

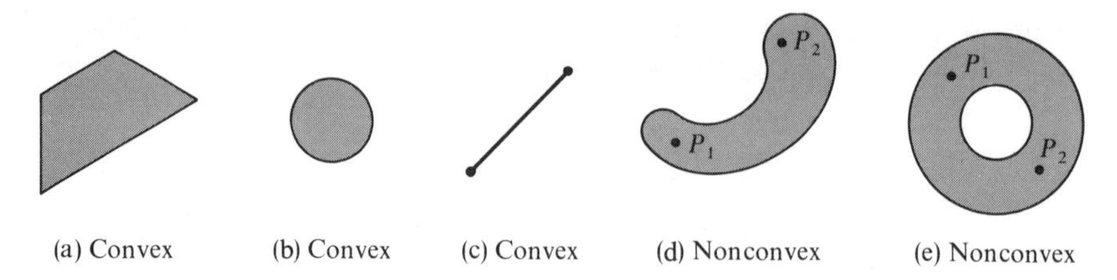

Figure 13

DEFINITION 4. The *intersection* of a finite collection of sets $S_1, S_2, \ldots, S_n$ is the set of points common to all sets in the collection.

DEFINITION 5. If a set contains no points, it is *empty* and is referred to as the *empty set* or *null set.*

For example, the set of points representing the intersection of the set $x \geq 7$ and the set $2x \leq 10$ is empty. By convention, we agree to regard as convex any set containing strictly fewer than two points. Thus the empty set is convex.

THEOREM 2.2. The intersection of a finite number of convex sets is convex.

Proof. Consider a finite collection of convex sets and suppose their intersection contains at least two points. Let A and B be any two points in the intersection. Then A and B must be contained in *each* convex set of the given collection (because, by definition, the intersection of a collection of sets contains all points common to these sets). Therefore, each of these sets contains the entire segment AB. This means that the segment AB is contained in the intersection of these convex sets, and hence the intersection is convex.

For example, suppose that S_1 is the set of points in the plane where $x \geq 0$, and S_2 the set of points in the plane where $y \geq 0$. The intersection of these two sets is the convex set of points which constitute the first quadrant, including the positive x and y axes and the origin.

THEOREM 2.3. Given the linear function

$$F = ax + by + c$$

the totality of points such that $F = 0$ constitutes a straight line. This line divides the plane into three regions: the line itself; one side of the line, where $F > 0$; the other side of the line, where $F < 0$. Each of these three regions is convex. Furthermore, the regions defined by $F \geq 0$ and $F \leq 0$ are also convex.

We will not prove this result. (The proof is not difficult, but our main interest lies in the use and understanding of theorems, rather than in their demonstration.) To illustrate Theorem 2.3, consider the function

$$F = 3x + 6y - 12$$

The set of points where $F = 0$ is the straight line labeled $F = 0$ in Figure 14. This line is clearly a convex set. Likewise, the set of points strictly above the

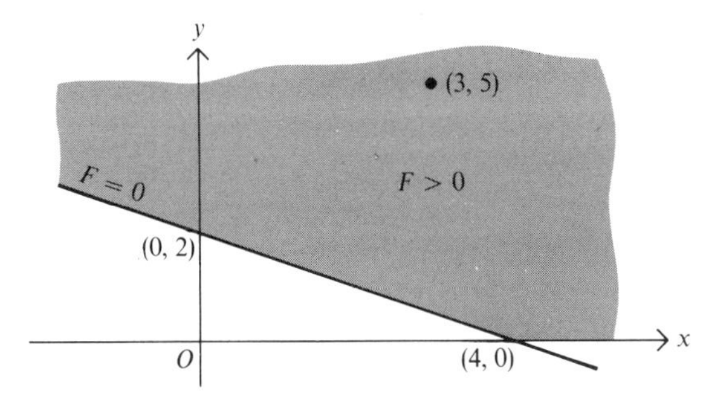

Figure 14

line is convex, and so is the set of points strictly below the line. The half-plane above the line is the totality of points where $F > 0$, while that below the line represents the totality of points where $F < 0$. At the point (3, 5) the value of $F = 3x + 6y - 12$ is $(3)(3) + (6)(5) - 12 = 27 > 0$. At the origin O (0, 0), the value of F is -12, that is, $F < 0$. *A good rule for deciding where* $F > 0$ *or* $F < 0$ *is to examine the behavior of F at the origin.* In the above example, $F = -12$ at O and we can conclude that the side of the line containing O is the negative side. (If F is 0 at the origin, test with some other point.)

In the following example we give another method for determining the region defined by a given inequality. This method can be used to construct a proof of Theorem 2.3.

Example. Sketch the set of points in the plane where $x + y \geq 5$.

First draw the line $x + y = 5$. A simple way to do this is to locate two points on the line, for example, $A(5, 0)$ and $B(0, 5)$, and then draw the unique line on which they lie. This is the line all of whose points satisfy the condition

$$x + y = 5$$

Consider a typical point P on this line, for example, $P(4, 1)$ as shown in Figure 15. A point like P', directly above P, must have a y coordinate strictly

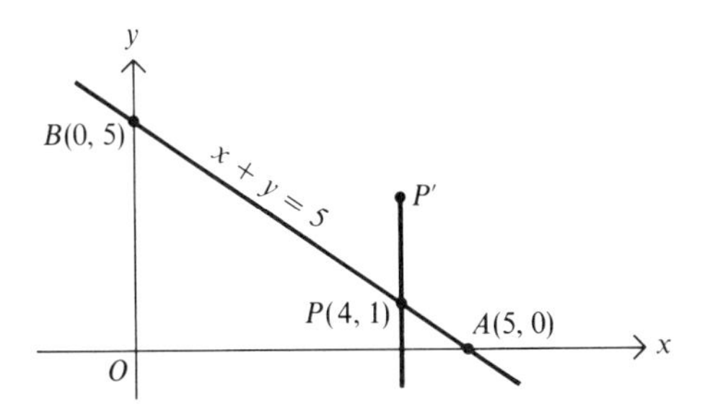

Figure 15

greater than the y coordinate at P. But at P the coordinate y satisfies the equation $x + y = 5$, or, equivalently,

$$y = 5 - x$$

and hence at P'

$$y > 5 - x$$

This holds true for any points above the line in question. In other words, the "top side" of the line $y = 5 - x$ represents the locus of all points where $y > 5 - x$. But to say that $y > 5 - x$ is equivalent to saying that $x + y > 5$. The totality of points where $x + y \geq 5$ is accordingly the set of all points *on and above* the line through A and B.

As suggested earlier, we come to the same conclusion simply by examining the value of the linear form $F = x + y$ at the origin $O(0, 0)$. At O the value of $F = x + y$ is 0, which is less than five. Thus, the side of the line $x + y = 5$ containing the origin is the side where $x + y < 5$. The other side must therefore correspond to the inequality $x + y > 5$.

The number of constraints in a linear program will always be finite. The region of feasibility which they determine is necessarily convex, for the following reasons:

(1) The region of feasibility is the intersection of a finite number of sets, each of which corresponds to the appropriate side of the limiting line specified by one of the constraints, limiting line included.

(2) The set of points to one side of a line is always convex, by Theorem 2.3.

(3) The intersection of a finite number of convex sets is convex, by Theorem 2.2.

To illustrate, consider the region determined by the usual trivial constraints on x and y, that is, $x \geq 0$ and $y \geq 0$, as well as by the following nontrivial constraints

$$3x + y \leq 48 \tag{1}$$

$$3x + 4y \leq 120 \tag{2}$$

$$3x + y \geq 36 \tag{3}$$

The reader may recall that constraints (1) and (2) occurred in the product-mix problem of Chapter 1. The region determined by $x \geq 0$, $y \geq 0$, and constraints (1) and (2) is the convex quadrilateral sketched in Figure 6. Constraint (3) cuts away part of this region. To determine the points which satisfy this constraint we graph the limiting equation $3x + y = 36$. The origin $O(0, 0)$ is a point where $3x + y \leq 36$. Hence the region defined by $3x + y \geq 36$ is the side of $3x + y = 36$ which does not contain the origin, that is, the side above and to the right of the line, but with the line included. The region determined by all the constraints is shown in Figure 16. It is the set of points lying on the boundary of, or interior to, the quadrilateral $FABG$, which is clearly convex.

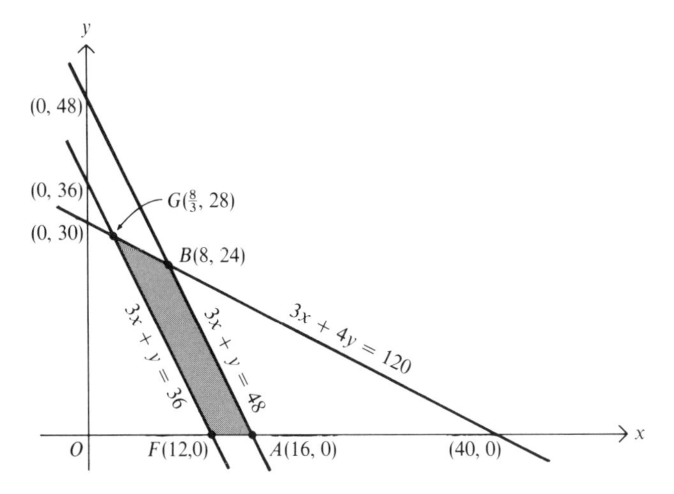

Figure 16

We stress two points:

(1) A constraint like $x \geq 0$ is "trivial" only from the viewpoint of its algebraic structure. Such constraints are very important and could be referred to as the "nonnegativity" constraints—we use the word "trivial' in order to refer to the other constraints as the "nontrivial" constraints.

(2) *The constraints in a linear program always allow for equality.* They are of the $\geq$ type or the $\leq$ type and never strictly of the $>$ or $<$ type. If equalities were not allowed, not every program would have an optimum point. To illustrate, consider the program

$$\begin{aligned} &\text{Maximize} \quad P = x + y \\ &\text{subject to} \quad x \geq 0, \quad y \geq 0 \\ &\text{and} \quad x < 5, \quad y < 6 \end{aligned}$$

According to these constraints the point (5, 6) is not feasible. This means that no point of the region of feasibility can maximize P, because we can find feasible points arbitrarily close to (5, 6), points at which the value of P gets larger and larger. *In other words, no feasible point deserves to be called the optimum point.* For example, consider the value of P at the following points:

$$\begin{aligned} x &= 4.8, & y &= 5.8 \\ x &= 4.9, & y &= 5.9 \\ x &= 4.99, & y &= 5.99 \\ x &= 4.999, & y &= 5.999 \\ &\cdots \end{aligned}$$

Exercises

1. Sketch the set of points in the plane corresponding to each of the following systems of constraints:

(a) $0 \le x \le 5$, $0 \le y \le 7$

(b) $-1 \le x \le 4$, $-3 \le y \le 2$

(c) $x \ge 0$, $y \ge 0$, $x + y \le 10$

(d) $x \ge 0$, $y \ge 0$, $2x + y \ge 6$

(e) $x \ge 0$, $y \ge 0$, $x - y \le 10$

(f) $x \ge 0$, $y \ge 0$, $2x - y \ge 6$

2. Verify that $3x + y \le 10$ and $-3x - y \ge -10$ are equivalent constraints by showing that they describe the same set of points. Are $ax + by \le c$ and $-ax - by \ge -c$ equivalent constraints?

3. (a) Find the corner points of the region determined by the set of constraints

$$\begin{aligned} 3x - 2y &\le 12 \\ x - y &\ge 0 \\ x &\le 4 \end{aligned}$$

(b) Consider the following three linear forms:

$$\begin{aligned} f &= x + y \\ g &= 2x - y \\ h &= x - 2y \end{aligned}$$

Show that, over the region defined in (a) above: (i) f takes on a maximum but no minimum value; (ii) g takes on a maximum but no minimum; (iii) h takes on a minimum but no maximum.

4. *True or false?*

(a) If $p < q$, then $\frac{1}{2}(p + q) < q$.

(b) If $p \le q$, then $\frac{1}{2}(p + q) < q$.

(c) If $p < q$, then $\frac{1}{2}(p + q) > p$.

5. In each of the following sets of inequalities at least one is superfluous. Find one such superfluous inequality in each set.

(a)
$$\begin{aligned} x_1 &\ge 4 \\ x_2 &\ge -4 \\ x_1 + x_2 &\ge 0 \end{aligned}$$

(b)
$$\begin{aligned} x_1 &\ge -1 \\ -x_2 &\le 2 \\ x_1 + x_2 &\le 3 \\ -x_1 - x_2 &\ge 0 \end{aligned}$$

6. Without solving, compare the optimum values of P in the following problems:

(a) Maximize $P = 5x + 6y$
where $x \geq 0$, $y \geq 0$
and $3x + y \leq 48$

(b) Maximize $P = 5x + 6y$
where $x \geq 0$, $y \geq 0$
and $3x + y \leq 48$
$3x + 4y \leq 120$

7. Compare the following definitions of convexity to that given in the text:

(a) A set K is convex if and only if the line segment joining each pair of points P and Q in K contains the midpoint M of the segment PQ.

(b) A set is convex if a "tangent" at any point of its perimeter leaves the whole set to one side of the tangent.

(c) A set is convex if with every pair of points P and Q on its perimeter it contains the whole line segment which joins them.

8.

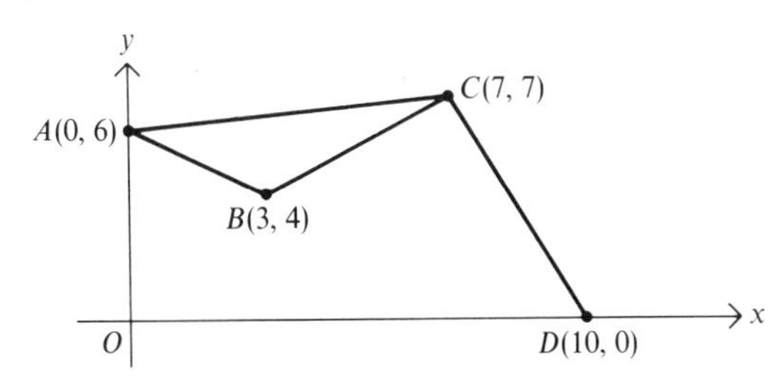

(a) In the diagram, consider the region having O, A, B, C, and D as consecutive corners. Is this region convex? Let $P = x + y$. What can we say about increases and decreases in P as we go from O to A, A to B, B to C, C to D, D to O?

(b) Consider the region having O, A, C, and D as consecutive corners. Is this region convex? What can we say about changes in $P = x + y$ as we go from O to A, A to C, C to D, D to O?

9. Given the points $P_1(-2, 3)$ and $P_2(4, 5)$ and the linear form $H = 5x - 8y$. Determine the values assumed by H at P_1 and P_2 and at the midpoint of the segment joining P_1 and P_2. Where on P_1P_2 does the linear form assume its maximum? Its minimum?

10. *True or false?* The set of points satisfying two linear inequalities in two variables has either no corner points or precisely one corner point.

2.2 LINEAR-PROGRAMMING PROBLEMS

We now consider several two-dimensional problems, each of which illustrates some aspect of the general n-variable linear-programming problem. We proceed from properly behaved examples to less well-behaved ones.

In Chapter 1 we used sliding isoquants to show that the corner-point theorem is valid in two-dimensional space. Problem 2.1 below suggests another way of establishing this result.

PROBLEM 2.1. Find the maximum and minimum of $F = 3x - 2y$

$$\text{subject to } x \geq 0, \quad y \geq 0$$

$$\begin{aligned} \text{and } x \quad & \leq 5 \\ y & \leq 6 \\ x + y & \leq 9 \end{aligned}$$

The region of feasibility of Problem 2.1 is shown in Figure 17. It is obviously convex. Theorem 2.1 can be used to guide our search of this convex

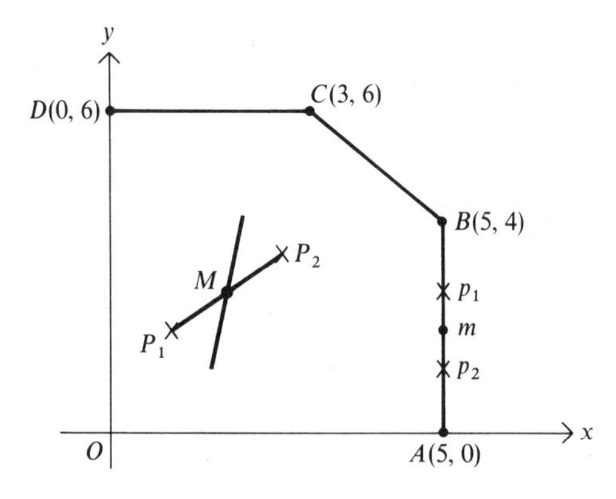

Figure 17

region for those points where F takes on its optimal, or *extreme*, values, that is, its maximum and minimum.

Consider any point not on the boundary of the region, such as point M in Figure 17. It is clear that there exists a segment P_1P_2 lying completely within the region and having M as its midpoint. We contend that one of the values assumed by the objective function F at the endpoints P_1 and P_2 is at least as optimal as the value assumed at M. From Theorem 2.1 we know that the value assumed at M is the average of the values assumed at P_1 and P_2. *Since the average of any two numbers is never larger or smaller than both the given numbers*, the search for a point where F assumes its optimal value can be reduced to the examination of points on the boundary of the region of feasibility.

Now let m be a point on the boundary but not at a corner, as shown in Figure 17. Again we see that m can be made the center of a segment like p_1p_2, all of whose points are feasible. But the argument at M also holds at m. Hence our search can be reduced to the examination of the corner points of the region of feasibility. In a phrase, linear programming is a subject of extremes—*the extreme values of the objective function are assumed at the "extreme," that is, corner, points of the region of feasibility.*

The value assumed by F at the corners of the region are:

Corner	*Value of* $F = 3x - 2y$
$O(0, 0)$	0
$A(5, 0)$	15
$B(5, 4)$	7
$C(3, 6)$	-3
$D(0, 6)$	-12

The maximum value of F is 15, achieved at the corner $A(5, 0)$, and the minimum value of F is -12, achieved at $D(0, 6)$.

The following problem shows that not every point where optimality is achieved need be a corner point. However, the only way this can occur is for the objective function to assume its optimum value at some point on a line segment joining two optimal corners.

PROBLEM 2.2. Maximize $P = x + y$

where $x \geq 0, \quad y \geq 0$

and $x + y \leq 10$

The region of feasibility is the shaded area in Figure 18. The corners of this region are: $O(0, 0)$, $A(0, 10)$, and $B(10, 0)$. At both A and B, the function P

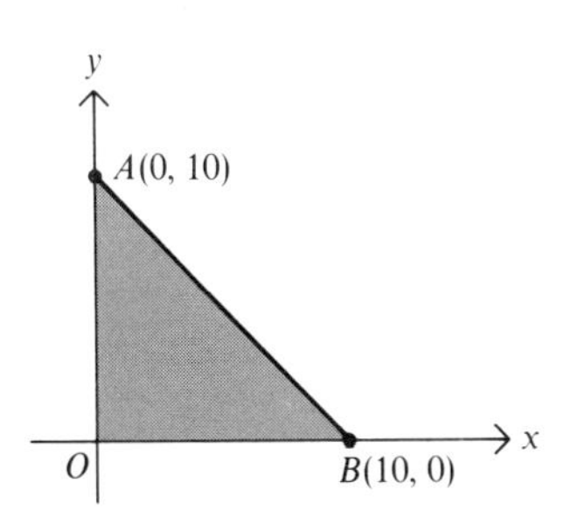

Figure 18

assumes the value 10. Hence the optimum value of P is 10, and it is assumed at every point on the segment AB.

The existence of multiple optima should always be pointed out, because knowledge of alternative optima may allow the decision-maker to meet other objectives. The next problem emphasizes this.

PROBLEM 2.3. Maximize $P = 6x + 2y + 77$

subject to the usual nonnegativity constraints on x and y, and the following nontrivial constraints:

$$3x + y \leq 48 \quad (1)$$

$$3x + 4y \leq 120 \quad (2)$$

$$3x + y \geq 36 \quad (3)$$

Note that the maximum of P occurs at those values (x_0, y_0) which maximize the homogeneous part of P, namely, $H = 6x + 2y$. The constant part of P, that is, 77, is best thought of as a "fixed return," having no effect on the point where P is maximized.

The region of feasibility for Problem 2.3 has already been drawn in Figure 16. It is the set of points within and on the boundary of the quadrilateral $FABG$. Accordingly, to solve Problem 2.3 we need only determine the value of $H = 6x + 2y$ at the corners of the region of feasibility. These feasible corners, their coordinates, and the corresponding values of H are:

Corner	*Value of* $H = 6x + 2y$
$F(12, 0)$	72
$A(16, 0)$	96
$B(8, 24)$	96
$G(\frac{8}{3}, 28)$	72

What conclusions can be drawn? The maximum value of H is 96 and hence the maximum value of P is $96 + 77 = 173$. Furthermore, this maximum is obtained at both corners A and B, so that all points on the segment AB represent optimal solutions.

Probelm 2.4 poses a somewhat different situation. The techniques discussed so far are sufficient to solve it.

PROBLEM 2.4 (A TRANSPORTATION PROBLEM). Suppose a firm produces certain products and stores them in warehouses S_1 and S_2 situated in two different towns. The firm receives orders from customers living in three other towns, D_1, D_2, and D_3. (We can think of S as signifying supply and D demand.) The unit costs of shipping from each warehouse to each customer are shown in the table below, as well as the capacities of the warehouses and the requirements of the customers. Thus it costs \$4 per unit to ship from S_1 to D_1, \$5 per unit to ship from S_2 to D_3, and so on.

	D_1	D_2	D_3	*Capacity*
S_1	\$4	\$2	\$1	15
S_2	\$1	\$3	\$5	25
Requirements	10	20	10	(40)

The margins of this table are to be read as follows: Supply depot 1 has a capacity of 15, supply depot 2 a capacity of 25, customer 1 has a demand of 10, customer 2 a demand of 20, customer 3 a demand of 10. (In this problem, the total capacity and the total demand are each equal to 40. In general, whenever total capacity equals total demand, the transportation problem will be called *balanced.*)

The objective is to find a shipping program which meets all requirements at least cost. At first glance it appears that Problem 2.4 involves finding six unknowns, namely, the amounts to be shipped from each of two supply depots to each of three customers. Closer scrutiny shows that really only two unknowns are involved. Let x represent the amount shipped from S_1 to customer 1, and y the amount shipped from S_1 to customer 2. We can then write all other amounts shipped in terms of x and y, as follows:

	D_1	D_2	D_3
S_1	x	y	$15 - x - y$
S_2	$10 - x$	$20 - y$	$-5 + x + y$

Since all amounts shipped must be nonnegative, our problem is subject to the following constraints, one for each cell in the table above:

$$x \geq 0, \quad y \geq 0, \quad x + y \leq 15$$
$$x \leq 10, \quad y \leq 20, \quad x + y \geq 5$$

The total cost of shipping is

$$C = 4x + 1(10 - x) + 2y + 3(20 - y) + 1(15 - x - y) + 5(-5 + x + y)$$
$$= 7x + 3y + 60$$

The objective is to minimize this function, subject to the above constraints. The graphical solution to this problem follows.

Figure 19 exhibits the region of feasibility. As usual, we construct the

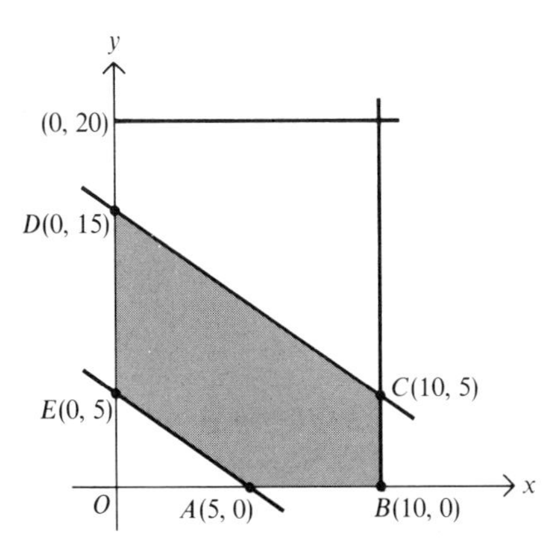

Figure 19

region by drawing the limiting line associated with each constraint and then shading the relevant side of each line. Because several of the limiting lines are parallel, we can read off the feasible corners directly from the graph. They are (5, 0), (10, 0), (10, 5), (0, 15), (0, 5). In the following table we evaluate the homogeneous form $C' = 7x + 3y$ at each feasible corner:

Corner	*Value of* $C' = 7x + 3y$
$A(5, 0)$	35
$B(10, 0)$	70
$C(10, 5)$	85
$D(0, 15)$	45
$E(0, 5)$	15

Corner E is the optimal corner. At this corner $C' = 15$ and hence $C = 15 + 60 = 75$. Since $x = 0$ and $y = 5$ at corner E, we use these values to fill in the optimal plan of shipment, as given in the schedule below.

	D_1	D_2	D_3
S_1	0	5	10
S_2	10	15	0

Note that all quantities shipped are integers, that is, whole numbers. One of the (unlisted) restrictions of a transportation problem is that the amounts shipped be integers. In solving this problem as a linear program, we simply ignored this condition. Fortunately, at the optimum corner, x and y were integers. This will always be the case in transportation problems where the capacities and requirements are given as integers.

The following problem is a "diet" or blending problem in disguise. The reader will soon see that types of investment are like different foods and that investment goals are like nutritional requirements.

PROBLEM 2.5 (A PORTFOLIO PROBLEM). Two stocks have been analyzed as to immediate growth, long-term growth, and anticipated dividend rate. The results of the analysis are shown in the following table:

	Stock A	*Stock B*
Anticipated growth in market value next year, per dollar invested	\$0.045	\$0.15
Anticipated growth in market value in next 10 years, per dollar invested	\$2.50	\$1.00
Anticipated dividend rate	2.5%	3%

What is the smallest amount we should invest in stocks A and B so as to meet the following investment goals:

(1) At least \$300 appreciation over the next year
(2) At least \$10,000 appreciation over the next decade
(3) At least \$200 in income per year

To formulate this problem mathematically, let x denote the amount in dollars invested in stock A, and y the amount invested in stock B. The objective is to minimize

$$I = x + y$$

subject to $x \geq 0$, $y \geq 0$, and

$$\begin{aligned} 0.045x + 0.15y &\geq 300 \\ 2.50\ x + 1.00y &\geq 10{,}000 \\ 0.025x + 0.03y &\geq 200 \end{aligned}$$

The first of the nontrivial inequalities expresses the condition that next year's anticipated growth should amount to at least \$300; the second constraint expresses the condition that the long-run growth should amount to at least \$10,000; the third inequality stipulates that yearly dividend income be greater than or equal to \$200.

Because of the relative smallness of the coefficients in the inequalities, let us change the meanings of x and y. Assume that the variables x and y still represent amounts invested, but measured now in \$1000 bills. The constraints become

$$\begin{aligned} 0.045x + 0.15y &\geq 0.3 \\ 2.5\ x + y &\geq 10 \\ 0.025x + 0.03y &\geq 0.2 \end{aligned}$$

or, equivalently,

$$\begin{aligned} 4.5x + 15y &\geq 30 \\ 2.5x + y &\geq 10 \\ 2.5x + 3y &\geq 20 \end{aligned}$$

After clearing fractions, we have the equivalent constraints

$$\begin{aligned} 3x + 10y &\geq 20 \\ 5x + 2y &\geq 20 \\ 5x + 6y &\geq 40 \end{aligned}$$

We now show how to solve Problem 2.5 without the aid of any graph. In order not to omit any corner from our list, we first label the lines associated with the inequalities of our problem, as follows:

$$x = 0 \tag{1}$$
$$y = 0 \tag{2}$$
$$3x + 10y = 20 \tag{3}$$
$$5x + 2y = 20 \tag{4}$$
$$5x + 6y = 40 \tag{5}$$

In the table below all corners are listed. As in Chapter 1, we use the symbol [E1, E2], for example, to denote the system consisting of the two equations labeled (1) and (2). The value of the objective function is given only at feasible corners.

System	Solution	Feasible?	Value of $I = x + y$
1. [E1, E2]	$x = 0, \quad y = 0$	No	
2. [E1, E3]	$x = 0, \quad y = 2$	No	
3. [E1, E4]	$x = 0, \quad y = 10$	Yes	10
4. [E1, E5]	$x = 0, \quad y = \frac{20}{3}$	No	
5. [E2, E3]	$x = \frac{20}{3}, \quad y = 0$	No	
6. [E2, E4]	$x = 4, \quad y = 0$	No	
7. [E2, E5]	$x = 8, \quad y = 0$	Yes	8
8. [E3, E4]	$x = \frac{40}{11}, \quad y = \frac{10}{11}$	No	
9. [E3, E5]	$x = \frac{70}{8}, \quad y = -\frac{5}{8}$	No	
10. [E4, E5]	$x = 2, \quad y = 5$	Yes	7

We find the solutions of these systems in the usual way. To illustrate, solution 10 is the intersection of lines (4) and (5), found by solving the set of equations

$$5x + 2y = 20 \tag{4}$$

$$5x + 6y = 40 \tag{5}$$

Only three of these ten corners are feasible. Evidently, solution 9 must be rejected because y is negative. Solutions 1 and 6 must be rejected because they fail to satisfy the inequality $3x + 10y \geq 20$. Solutions 2 and 4 are both infeasible because they do not satisfy the constraint $5x + 2y \geq 20$. Solutions 5 and 8 fail to satisfy the inequality $5x + 6y \geq 40$. Only corners 3, 7, and 10 are feasible. The minimum of I is attained at corner 10, where $x = 2$ and $y = 5$. This

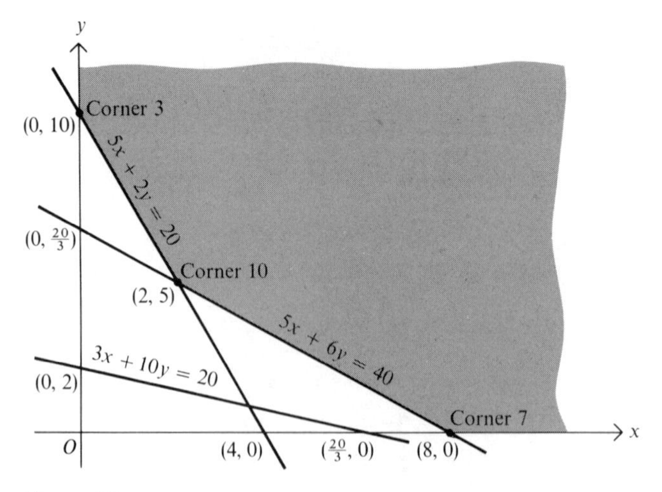

Figure 20

means that the cheapest way of satisfying the portfolio objectives is to invest \$2000 in stock A and \$5000 in stock B.

Although we have solved Problem 2.5 without using geometry, for the sake of completeness the appropriate region of feasibility is shown shaded in Figure 20.

We wish to emphasize the following defect in the corner-point method: Suppose the objective of Problem 2.5 were to *maximize* $P = x + y$ over the same region of feasibility. If we were to evaluate P at the feasible corners without having a picture of the region of feasibility, *or without thinking further*, we could be misled and not realize that P can be made infinitely large.

In the next situation, one of the constraints is an equality. This reduces the dimension of the problem by 1.

PROBLEM 2.6. Maximize $P = 3x + 5y$
subject to $x \geq 0, \quad y \geq 0$
and $2x + y = 10$

This problem differs from all others we have previously encountered in that the nontrivial constraint is an equality rather than an inequality. Note that every equality can be replaced by two inequalities. For example, $2x + y = 10$ is equivalent to the inequalities $2x + y \geq 10$ and $2x + y \leq 10$.

As shown in Figure 21, the region of feasibility for Problem 2.6 encloses

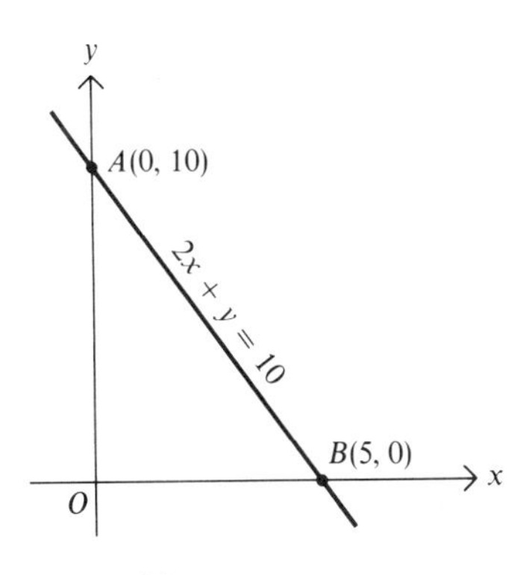

Figure 21

no area. It is the one-dimensional segment AB of the line $2x + y = 10$, with points A and B included. The corners of this convex region are A and B. and the maximum of P will occur at either A or B:

Corner	*Value of* $P = 3x + 5y$
$A(0, 10)$	50
$B(5, 0)$	15

The objective function attains its maximum value of 50 at (0, 10).

Problem 2.6 can also be handled algebraically as follows: Eliminate one of the variables, x or y, from the given equality constraint and thereby reduce Problem 2.6 to a study of the one remaining variable. For example, we can solve for y in terms of x and carry the relation thereby found back to the profit function P. Thus, if

$$2x + y = 10$$

then

$$y = 10 - 2x$$

and

$$\begin{aligned} P &= 3x + 5y \\ &= 3x + 5(10 - 2x) \\ &= 50 - 7x \end{aligned}$$

Our original problem is therefore equivalent to maximizing

$$P = 50 - 7x$$

subject to the constraints $x \geq 0$, $y \geq 0$. But since

$$y = 10 - 2x$$

the constraint $y \geq 0$ is equivalent to the constraint $x \leq 5$. Thus Problem 2.6 can be reduced to the following one-variable problem:

$$\begin{aligned} &\text{Maximize} \quad P = 50 - 7x \\ &\text{where} \quad x \geq 0, \quad x \leq 5 \end{aligned}$$

The maximum of P occurs when $x = 0$.

We digress for a moment to point out the mnemonic value of the one-dimensional linear-programming problem, namely,

$$\begin{aligned} &\text{Maximize (or minimize)} \quad H = kx \\ &\text{where} \quad x \geq a_1, \quad x \geq a_2, \ldots, x \geq a_m \\ &\qquad\qquad x \leq b_1, \ldots, x \leq b_n \end{aligned}$$

Suppose that the largest of these a's is a_m and that the smallest of the b's is b_1. Then the region of feasibility can be described as the set S of all x's such that

$$a_m \leq x \leq b_1$$

Note that if $a_m > b_1$, the set S is empty; if $a_m = b_1$, S contains only one point; if $a_m < b_1$, then S is an interval on the line, endpoints included. In all cases, S is convex.

Suppose we are in the third case. Then clearly H takes on its maximum or minimum at one of the corners of the domain of feasibility, namely, $x = a_m$ or $x = b_1$, depending on the sign of k. (In addition, if H takes on the value d at both the where corner $x = a_m$ and the corner where $x = b_1$, then $H = d$ everywhere between a_m and b_1, because in this case both k and d must be 0.)

The next two problems lead to unusual feasibility regions.

PROBLEM 2.7. Maximize $P = x + y$

subject to $x \geq 0, \quad y \geq 0$

and $x + y \leq 10$

$2x + y \geq 30$

Note in Figure 22 that the region of feasibility is empty. This can be demonstrated algebraically: The inequality $x + y \leq 10$ implies that $2x + 2y \leq 20$.

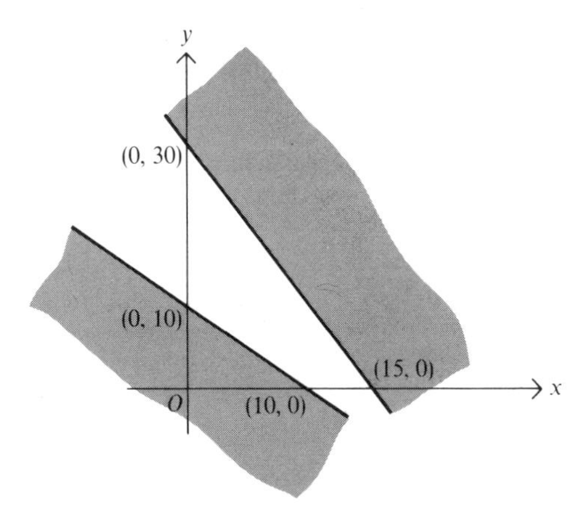

Figure 22

This in turn implies that $2x + y \leq 20$, which contradicts the constraint $2x + y \geq 30$. (Where in this proof did we use the nonnegativity constraints $x \geq 0$ and $y \geq 0$?)

PROBLEM 2.8. Maximize $P = x + y$

subject to $x \geq 0, \quad y \geq 0$

and $y \leq 5$

The region of feasibility is shown in Figure 23. In this example, we can make x, and hence P, larger than any given number. No finite point of the region deserves to be called the optimum point, and hence we say that P does not possess a finite optimum. Observe that if we had proceeded to examine only all

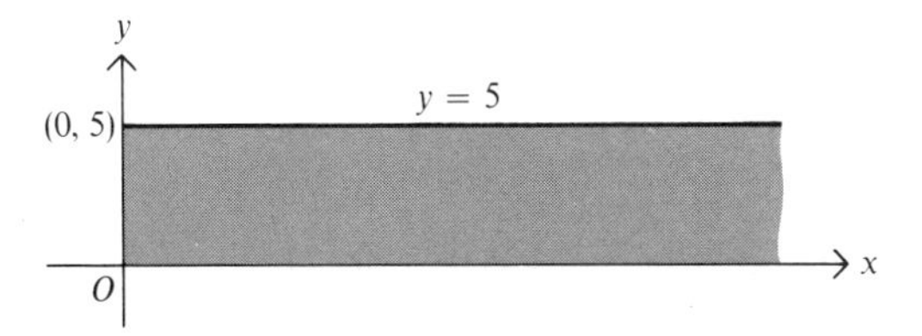

Figure 23

feasible corners without drawing Figure 23 we might have missed the fact that P is unbounded.

The next example shows that an unbounded region of feasibility does not necessarily imply an unbounded objective function.

PROBLEM 2.9. Maximize $P = -x + \frac{1}{2}y$
subject to $x \geq 0, \quad y \geq 0$
and $-x + y \leq 2$

In this example the region of feasibility extends to the right and up, indefinitely, as shown in Figure 24. Nevertheless, P assumes its maximum value of 1 at only one point, the corner where $x = 0, y = 2$.

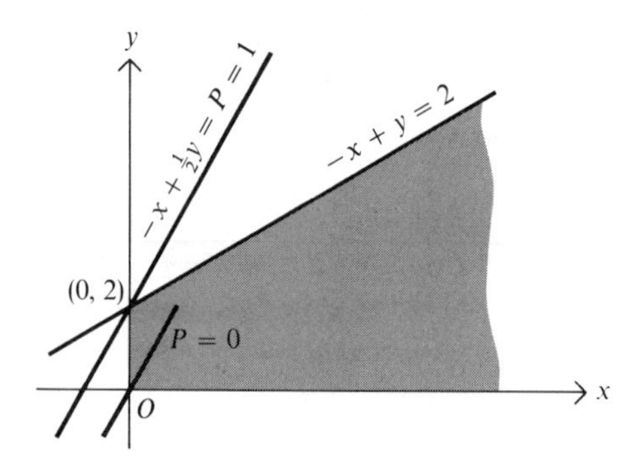

Figure 24

PROBLEM 2.10. Maximize $P = -x + y$
subject to $x \geq 0, \quad y \geq 0$
and $-x + y \leq 2$

The region of feasibility is the same as in the previous problem. The maximum value of P is 2. This value is assumed at the corner where $x = 0$, $y = 2$ *and* at every point on the boundary line $-x + y = 2$.

Exercises

Solve each of the following problems by algebraic means; then check by drawing the appropriate graph.

1. Maximize $5x + 26y$, subject to $x \geq 0, y \geq 0$; and $3x + y \leq 48$, $3x + 4y \leq 120$.
2. Maximize $15x + 6y$, subject to $x \geq 0, y \geq 0$; and $3x + y \leq 48$, $3x + 4y \leq 120$.
3. Maximize $5x + 6y$, subject to $x \geq 0, y \geq 0$; and $3x + y \leq 48$, $3x + 4y \leq 123$.

4. Maximize $5x + 6y$, subject to $x \geq 0$, $y \geq 0$; and $3x + y \leq 48$, $3x + 4y \leq 124$.

5. Minimize $3x_1 + 2x_2$, subject to $x_1 \geq 0$, $x_2 \geq 0$; and $x_1 + x_2 \leq 8$, $6x_1 + 4x_2 \geq 12$.

6. Minimize $3x_1 + 2x_2$, subject to $x_1 \geq 0$, $x_2 \geq 0$; and $x_1 + x_2 \leq 8$, $6x_1 + 4x_2 \geq 12$, $5x_1 + 8x_2 = 20$.

7. Maximize $2x + y$, subject to $x + 2y \leq 4$, $x - y \geq -2$, $x \leq 2$. (*Note:* Do not assume that x and y are nonnegative.)

8. The California Juice Company has two basic juices available, apple and grape. These juices are blended into a mixture and sold as fruit punch. The company has decided to make only two types of punch, A and B, each of which is sold in 5-gallon bottles. A bottle of A contributes \$2 to profit and overhead; a bottle of B contributes \$3 to profit and overhead.

Punch A requires 4 gallons of apple juice and 1 gallon of grape juice per bottle; punch B requires 2 gallons of apple juice and 3 gallons of grape juice per bottle. Suppose the company has 130 gallons of apple juice and 240 gallons of grape juice available. What is the optimal production plan for the current period?

9. In Exercise 8 above, by how much would profit change if:

(a) 150 gallons of apple juice were available (all other resources being held constant)?

(b) 250 gallons of apple juice were available (all other resources held constant)?

(c) 300 gallons of grape juice were available (all other resources held constant)?

10. Draft quotas for a certain month have been set as follows: Boston, 2000; New York, 2000; Dallas, 500. Vacancies in training cycles exist at various basic-training centers, as follows: Fort A, 3000; Fort B, 1500. Transportation rates per man have been fixed by contract. They are given in the following table:

	Destination	
Origin	*Fort A*	*Fort B*
Boston	\$30	\$90
New York	\$25	\$80
Dallas	\$40	\$40

How should the draftees be allocated to each of the training camps in order to minimize the cost of transportation? To maximize the cost of transportation?

11. Maximize $P = 7x + 9999$, subject to $x \geq 0$; and $3x \leq 18$, $x \leq 5$, $2x \geq 4$.

12. *True or false?* If the value of the objective function is unbounded, then every variable can be made arbitrarily large.

13. (a) Explain "linearity."

(b) If $F = 3x + 4y + 5z + 10$, does the value of F double when the coordinates of the point (x, y, z) are doubled?

2.3 LINEAR PROGRAMS IN THREE VARIABLES

The following programs all involve three variables and illustrate some of the difficulties which arise when we attempt to study such problems by the corner-point method. Our purpose is to demonstrate the power of this method—in theory at least—and to point out why an even more powerful method must be sought.

PROBLEM 2.11. Maximize $P = 4x + 3y + 7z$

$$\text{subject to} \quad x \geq 0, \quad y \geq 0, \quad z \geq 0$$

$$\text{and} \quad x + 3y + 2z \leq 120$$

$$2x + y + 3z \leq 120$$

Every triplet of numbers can be regarded as a point in three-dimensional space. Listed below are some definitions and theorems concerning this space. They are the analogs of the key concepts and results of our study of two-dimensional space and are needed to solve such programs as Problem 2.11

(1) Let $P_1 = (x_1, y_1, z_1)$ and $P_2 = (x_2, y_2, z_2)$ represent two points in space. Then the midpoint P_M of the line segment joining these points has coordinates which are the averages of the coordinates of P_1 and P_2. More precisely,

$$P_M = [\tfrac{1}{2}(x_1 + x_2), \tfrac{1}{2}(y_1 + y_2), \tfrac{1}{2}(z_1 + z_2)]$$

(2) Let $H = ax + by + cz$ represent the homogeneous linear form in three variables. If H_1 is the value of H at the point P_1, and H_2 the value of H at the point P_2, then the value assumed by H at P_M is $\tfrac{1}{2}(H_1 + H_2)$.

(3) By definition, the segment P_1P_2 joining the points $P_1(x_1, y_1, z_1)$ and $P_2(x_2, y_2, z_2)$ is the set of all points $P(x, y, z)$ whose coordinates can be expressed in terms of those of P_1 and P_2 by the equations

$$x = (1 - \lambda)x_1 + \lambda x_2$$

$$y = (1 - \lambda)y_1 + \lambda y_2$$

$$z = (1 - \lambda)z_1 + \lambda z_2$$

where $0 \leq \lambda \leq 1$.

Each value of λ between 0 and 1 defines a different point P on the segment P_1P_2. For example, $\lambda = \tfrac{1}{3}$ defines a point P one-third of the way from P_1 toward P_2; in general, $\lambda = \lambda_0$ defines a point P which lies the fraction λ_0 of the way from P_1 toward P_2, as shown in Figure 25.

(4) A set of points in space is said to be *convex* if, whenever it contains two points P_1 and P_2, it also contains the entire segment P_1P_2. All sets with fewer than two points are regarded as convex. The intersection of a finite number of convex sets is convex.

(5) (a) The set of points which satisfies the equation

$$ax + by + cz = d$$

is a *plane* in space. All planes are convex.

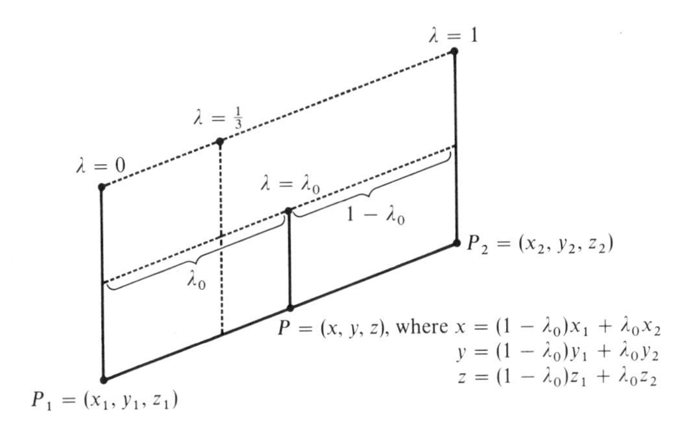

Figure 25

(b) The set of points which satisfies the inequality

$$ax + by + cz < d$$

is said to constitute an *open half-space*. All open half-spaces are convex.

(c) The set of points which satisfies the inequality

$$ax + by + cz \leq d$$

is said to constitute a *closed half-space*. All closed half-spaces are convex.

The closed half-space in (c) is made up of the points in the open half-space in (b) as well as the points on the plane in (a).

(6) The region of feasibility of a linear program is the intersection of a finite number of closed half-spaces, by definition. Hence it is always convex.

(7) A set S of points in space is said to be *bounded* if it can be "packed into a box of definite size," that is, if we can find constants r_1, r_2, s_1, s_2, t_1, t_2, such that, if $P(x, y, z)$ is any point in S, then

$$r_1 \leq x \leq r_2$$
$$s_1 \leq y \leq s_2$$
$$t_1 \leq z \leq t_2$$

(8) A *corner*, or *extreme point*, of a convex region is a point which does not lie on any segment P_1P_2 joining two distinct points, P_1 and P_2, in the region. In particular, if P_1 and P_2 are in the region and P is the midpoint of the segment P_1P_2, then P cannot be a corner point.

It is not hard to show that whenever the region of feasibility of a linear program is *nonempty* and *bounded*, the objective function will attain its optimum at least once over the region. Furthermore, to find a point where optimality is achieved, we need only examine all feasible corners. This is the general corner-point theorem, *restricted to nonempty, bounded regions of feasibility*.

But how can we determine the feasible corners, especially when we cannot draw three-dimensional pictures? *It is a demonstrable fact that feasible corners can be determined algebraically by proceeding as follows:* List all the limiting equations corresponding to the given constraints, form all possible systems of three of these equations taken at a time, and solve each system. Henceforth we call such a system *fundamental.* Whenever the solution of a *fundamental system* is not uniquely defined (in other words, if the system of equations possesses no solution or else more than one solution), we simply drop it from consideration. The only fundamental systems we need consider are those which possess a unique solution. We call such a unique solution a *basic solution.* Furthermore, the only basic solutions of interest are those which satisfy all the given constraints; these are called *basic feasible solutions*. They correspond to the feasible corners, or extreme points, of the region of feasibility, as will be demonstrated in Chapter 10. We shall often use the phrase *potential corner* as a synonym for *fundamental system*.

The three-dimensional corner-point theorem should be intuitively obvious. Assume that all variables are nonnegative; that is, the origin is below and to the left of the region of feasibility. For definiteness, suppose that the isoquant plane $H = 0$ slopes downward, from left to right. The reader can visualize that, as this plane is moved away from the origin, parallel to its previous position and across the bounded, convex region of feasibility, there will be some point on the outer surface of the region where optimality is attained. If the optimal value is unique, this point will be a corner. (The only way optimality can be achieved at a noncorner point is for it to be achieved at two corners and hence all along the segment joining these corners.)

The corner-point theorem can be applied to Problem 2.11 because the region is obviously nonempty and bounded—the first nontrivial constraint limits y to a maximum of 40, whereas the second nontrivial constraint limits x to a maximum of 60 and z to a maximum of 40.

Now label the limiting planes or equations corresponding to the constraints in this problem, as follows:

$$x + 3y + 2z = 120 \tag{1}$$

$$2x + y + 3z = 120 \tag{2}$$

$$x = 0 \tag{3}$$

$$y = 0 \tag{4}$$

$$z = 0 \tag{5}$$

As usual, [E1, E2, E3] denotes the fundamental system formed by three equations (1), (2), and (3), and so on. In this problem, every such fundamental system possesses a unique solution. The table of solutions is

Fundamental system	*Basic solution*	$P = 4x + 3y + 7z$
1. [E1, E2, E3]	$x = 0, \quad y = \frac{120}{7}, \quad z = \frac{240}{7}$	$\frac{2040}{7} = 291\frac{3}{7}$
2. [E1, E2, E4]	$x = -120, \quad y = 0, \quad z = 120$	Infeasible
3. [E1, E2, E5]	$x = 48, \quad y = 24, \quad z = 0$	264
4. [E1, E3, E4]	$x = 0, \quad y = 0, \quad z = 60$	Infeasible
5. [E1, E3, E5]	$x = 0, \quad y = 40, \quad z = 0$	120
6. [E1, E4, E5]	$x = 120, \quad y = 0, \quad z = 0$	Infeasible
7. [E2, E3, E4]	$x = 0, \quad y = 0, \quad z = 40$	280
8. [E2, E3, E5]	$x = 0, \quad y = 120, \quad z = 0$	Infeasible
9. [E2, E4, E5]	$x = 60, \quad y = 0, \quad z = 0$	240
10. [E3, E4, E5]	$x = 0, \quad y = 0, \quad z = 0$	0

To illustrate the calculations, solution 1 for [E1, E2, E3] was found by solving the system

$$x + 3y + 2z = 120 \tag{1}$$

$$2x + y + 3z = 120 \tag{2}$$

$$x = 0 \tag{3}$$

Of the ten basic solutions, four do not satisfy every constraint—for example, in solution 2, x is negative, contradicting the constraint $x \geq 0$; in solution 6, $x = 120$, contradicting the constraint $2x + y + 3z \leq 120$, and so on. From the table we see that P achieves its maximum of $291\frac{3}{7}$ at corner 1, where $x = 0$, $y = \frac{120}{7}$, $z = \frac{240}{7}$. At this optimal point, one of the original variables is 0. This could have been predicted in advance. Of the five limiting equations, only Eqs. (1) and (2) are nontrivial. To determine a corner we must solve every possible system formed by choosing three of the original five equations, so every such system must contain at least one of the three trivial equations, (3), (4), or (5).

PROBLEM 2.12. Maximize $P = 4x + 3y + 7z$

subject to $x \geq 0, \quad y \geq 0, \quad z \geq 0$

and

$$x + 3y + 2z = 120 \tag{1}$$

$$2x + y + 3z \leq 120 \tag{2}$$

Problem 2.12 is almost an exact restatement of Problem 2.11, except that it is more restrictive. Constraint (1) is an *equality* here, whereas, in Problem 2.11, the analogous constraint was an inequality of the "less than or equal to" type. Since Problem 2.12 is more restrictive than Problem 2.11, its maximum cannot be greater than that of Problem 2.11.

Below, we give two methods for solving Problem 2.12.

Method 1. Use the equality to *eliminate* completely one of the unknowns, thereby reducing the problem to a two-dimensional program.

We can solve Eq. (1) for any of the unknowns in terms of the others. It is simplest to solve for x, obtaining

$$x = 120 - 3y - 2z \tag{3}$$

Substituting this expression for x into inequality (2), we obtain

$$2(120 - 3y - 2z) + y + 3z \leq 120$$

or

$$-5y - z \leq -120$$
$$5y + z \geq 120$$

By substituting Eq. (3) into the objective function we obtain

$$\begin{aligned} P &= 4(120 - 3y - 2z) + 3y + 7z \\ &= -9y - z + 480 \end{aligned}$$

Finally, since x is nonnegative, Eq. (3) is equivalent to the inequality

$$3y + 2z \leq 120$$

Problem 2.12 has now been completely reduced to a two-dimensional problem in y and z, namely, Problem 2.12′.

PROBLEM 2.12′. Maximize $P' = -9y - z + 480$
subject to $y \geq 0, \quad z \geq 0$
and $3y + 2z \leq 120$
$5y + z \geq 120$

This is a simple two-dimensional problem. The region of feasibility is the shaded triangle ACE in Figure 26.

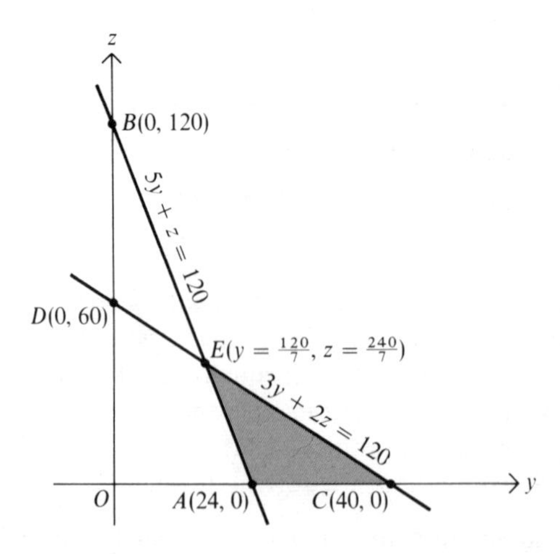

Figure 26

Let us label the lines associated with the inequalities in Problem 2.12′ as follows:

$$y \qquad = 0 \qquad (1)$$

$$z = 0 \qquad (2)$$

$$3y + 2z = 120 \qquad (3)$$

$$5y + z = 120 \qquad (4)$$

As the reader can verify, each fundamental system possesses a unique solution. We tabulate the values of P' at all feasible corners:

Fundamental system	*Basic solution*	*Value of* $P' = -9y - z + 480$
1. [E1, E2]	$y = 0, \quad z = 0$	Infeasible
2. [E1, E3]	$y = 0, \quad z = 60$	Infeasible
3. [E1, E4]	$y = 0, \quad z = 120$	Infeasible
4. [E2, E3]	$y = 40, \quad z = 0$	120
5. [E2, E4]	$y = 24, \quad z = 0$	264
6. [E3, E4]	$y = \frac{120}{7}, \quad z = \frac{240}{7}$	$291\frac{3}{7}$

Thus the optimal solution to Problem 2.12′ is attained at corner 6, where $y = \frac{120}{7}$, $z = \frac{240}{7}$, and $P' = 291\frac{3}{7}$.

To find the optimal solution to Problem 2.12, note that if we substitute $y = \frac{120}{7}$ and $z = \frac{240}{7}$ into Eq. (3) of Problem 2.12, namely, $x = 120 - 3y - 2z$, then it follows that $x = 0$. The optimum point of Problem 2.12 is therefore $x = 0$, $y = \frac{120}{7}$, $z = \frac{240}{7}$. And, of course, at this point, $P = P' = 291\frac{3}{7}$.

Method 2. We keep the problem in three-dimensional space, where a feasible point must satisfy each of the following constraints.

$$x \qquad \geq 0 \qquad (1)$$

$$y \qquad \geq 0 \qquad (2)$$

$$z \geq 0 \qquad (3)$$

$$x + 3y + 2z = 120 \qquad (4)$$

$$2x + y + 3z \leq 120 \qquad (5)$$

At first glance it would seem that we ought to examine all the systems that can be formed by three of the five limiting equations. But Eq. (4) must hold at any corner because it is an *equality*, and so we need only examine those fundamental systems which include Eq. (4). There are six possibilities, all of which happen to be basic solutions, as follows:

Fundamental system	*Basic solution*	*Value of* $P = 4x + 3y + 7z$
1. [E1, E2, E4]	$x = 0, \quad y = 0, \quad z = 60$	Infeasible
2. [E1, E3, E4]	$x = 0, \quad y = 40, \quad z = 0$	120
3. [E1, E4, E5]	$x = 0, \quad y = \frac{120}{7}, \quad z = \frac{240}{7}$	$291\frac{3}{7}$
4. [E2, E3, E4]	$x = 120, \quad y = 0, \quad z = 0$	Infeasible
5. [E2, E4, E5]	$x = -120, \quad y = 0, \quad z = 120$	Infeasible
6. [E3, E4, E5]	$x = 48, \quad y = 24, \quad z = 0$	264

The results of this analysis agree with those found by Method 1.

The reader should note that the region of feasibility of Problem 2.12 is two dimensional, that is, a plane in three-dimensional space.

PROBLEM 2.13. Maximize $P = x + y + z$

subject to $x \geq 0, \quad y \geq 0, \quad z \geq 0$

and
$$x + y - z \leq 10$$
$$y \leq 5$$

We label the limiting equations corresponding to the above inequalities as follows:

$$x = 0 \tag{1}$$
$$y = 0 \tag{2}$$
$$z = 0 \tag{3}$$
$$y = 5 \tag{4}$$
$$x + y - z = 10 \tag{5}$$

Not every system formed by taking three of these equations possesses a solution. For example, any system containing both Eqs. (2) and (4) is necessarily contradictory because $y = 0$ and $y = 5$ are incompatible.

Listed below are all possible systems of three equations which can be formed from Eqs. (1) to (5). These are the potential corners of Problem 2.13.

Fundamental system	*Basic solution*	*Value of* $P = x + y + z$
1. [E1, E2, E3]	$x = 0, \quad y = 0, \quad z = 0$	0
2. [E1, E2, E4]	Nonexistent	
3. [E1, E2, E5]	$x = 0, \quad y = 0, \quad z = -10$	Infeasible
4. [E1, E3, E4]	$x = 0, \quad y = 5, \quad z = 0$	5
5. [E1, E3, E5]	$x = 0, \quad y = 10, \quad z = 0$	Infeasible
6. [E1, E4, E5]	$x = 0, \quad y = 5, \quad z = -5$	Infeasible
7. [E2, E3, E4]	Nonexistent	
8. [E2, E3, E5]	$x = 10, \quad y = 0, \quad z = 0$	10
9. [E2, E4, E5]	Nonexistent	
10. [E3, E4, E5]	$x = 5, \quad y = 5, \quad z = 0$	10

Examination of this table suggests that the maximum value of P is 10 and that it occurs at both corners 8 and 10. But, in reality, there is no point of the region which can be said to correspond to the optimal value of P. This is because both x and z can be made arbitrarily large, and hence so can P. The only constraint that could limit the size of x or z is

$$x + y - z \leq 10$$

Notice, however, that the coefficient of z is negative. This permits us to make both x and z arbitrarily large, provided that we make sure that $x - z$ does not exceed 10. For example, $(x = 100{,}010, y = 0, z = 100{,}000)$ is a feasible point at which $P = 200{,}010$.

In a problem containing many variables and equations it is not at all an easy matter to detect unboundedness. Problem 2.13 exhibits some difficulties that can be encountered when we attempt to apply the corner-point theorem to an unbounded region of feasibility. It shows clearly that the search for the optimal value of a linear function over an unbounded region cannot always be reduced to the examination of all feasible corners. Several things can happen, depending on the nature of the problem under study. Sometimes the unboundedness of the region will play no role, and we can actually reduce the problem to a study of corners. To illustrate, suppose we seek to

$$\begin{aligned} &\text{Minimize} \quad C = 3x + 4y + 5z \\ &\text{subject to} \quad x \geq 0, \quad y \geq 0, \quad z \geq 0 \\ &\text{and} \quad 5x + 6y + 3z \geq 45 \\ &\qquad\; 5x + 10y + 5z \geq 100 \\ &\qquad\; 3x + 4y + 6z \geq 120 \end{aligned}$$

The region of feasibility is unbounded upward because any point whose coordinates are very large is clearly feasible. But our objective is to *minimize C*. For the isocost planes $3x + 4y + 5z = C_0$, it is easy to see that some corner of the region of feasibility near the origin will be optimal.

Not every unbounded problem can be handled so easily. If we are unaware that the region of feasibility is unbounded, the study of all feasible corners can lead us into serious error. What we need is a simple method for detecting unboundedness—or else a new technique for solving linear-programming problems, a technique powerful enough to reveal the existence of an unbounded solution.

The reader might feel that we are being overacademic in stressing unboundedness. After all, why should a practical problem ever be unbounded? Unfortunately, practical problems are often formulated improperly so that they lead to unbounded regions, and it is desirable to have a method of solution which will detect unboundedness whenever it exists. Even for problems where the region of feasibility is bounded and nonempty, the corner-point method is hardly practicable. As the number of variables and equations in a linear program increases, the number of potential corners, that is, fundamental systems, increases much more rapidly. And, of course, we would still have to determine whether or not each such system possesses a unique solution.

The following problem shows that a feasible solution of a fundamental system need not be a corner point.

PROBLEM 2.14. Maximize $P = x_1 + 3x_2 + 5x_3$

where $x_1 \geq 0, \quad x_2 \geq 0, \quad x_3 \geq 0$

and
$$4x_1 + 6x_2 + 12x_3 \leq 22 \tag{4}$$
$$6x_1 + 9x_2 + 18x_3 \leq 33 \tag{5}$$
$$2x_1 + 3x_2 + 6x_3 \leq 11 \tag{6}$$

We denote the limiting lines associated with these constraints as follows:

$$x_1 = 0 \tag{1}$$
$$x_2 = 0 \tag{2}$$
$$x_3 = 0 \tag{3}$$
$$4x_1 + 6x_2 + 12x_3 = 22 \tag{4}$$
$$6x_1 + 9x_2 + 18x_3 = 33 \tag{5}$$
$$2x_1 + 3x_2 + 6x_3 = 11 \tag{6}$$

It is not difficult to see that there are 20 fundamental systems associated with this program. Let us examine the system [E4, E5, E6]: Clearly, this system has the feasible solution $x_1 = x_2 = x_3 = 1$. But this point is not a basic feasible solution, that is, a corner point, because it is not the *unique* solution of a fundamental system. Observe that the system [E4, E5, E6] possesses many solutions, for example,

$$x_1 = 0, \quad x_2 = 0, \quad x_3 = \tfrac{11}{6} \tag{a}$$
$$x_1 = 0, \quad x_2 = \tfrac{11}{3}, \quad x_3 = 0 \tag{b}$$
$$x_1 = \tfrac{11}{2}, \quad x_2 = 0, \quad x_3 = 0 \tag{c}$$
$$x_1 = 0, \quad x_2 = \tfrac{11}{6}, \quad x_3 = \tfrac{11}{12} \tag{d}$$
$$x_1 = 2, \quad x_2 = \tfrac{1}{6}, \quad x_3 = \tfrac{13}{12} \tag{e}$$
$$x_1 = 1, \quad x_2 = 1, \quad x_3 = 1 \tag{f}$$

The reader can check that the point $x_1 = 1$, $x_2 = 1$, $x_3 = 1$, that is, solution (f), is the average of solutions (d) and (e). This shows that solution (f) is not a corner point, since it is the midpoint of the segment having solutions (d) and (e) as endpoints (see Figure 27).

Figure 27 seems to suggest that solutions (a), (b), and (c) are corner points. This is true. To see why, suppose solution (a) is the average of two other feasible solutions. A moment's thought shows that these two other solutions must both have $x_1 = x_2 = 0$, for otherwise solution (a) could never be their average. But if $x_1 = x_2 = 0$, then x_3 necessarily equals $\tfrac{11}{6}$; and hence these

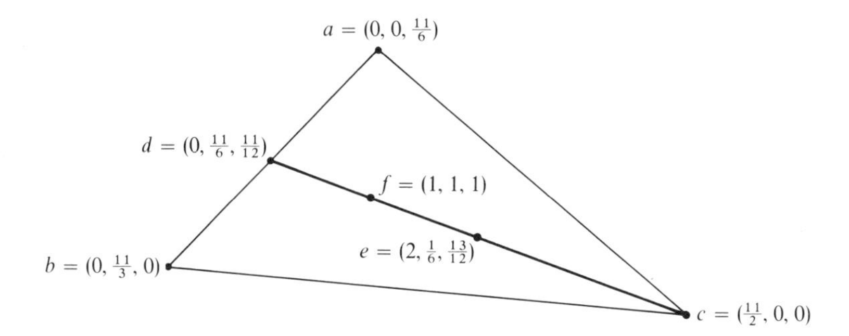

Figure 27

"other" solutions must really be the same as solution (a). By similar reasoning the reader can easily show that solutions (b) and (c) represent corner points.

But is this not a paradox? Did we not state earlier that, in examining only fundamental systems with uniquely defined feasible solutions, we would, in effect, consider all the feasible corner points? If we disregard system [E4, E5, E6], will we not be disregarding the corners corresponding to the preceding solutions (a), (b), and (c)? The answer is no, because the points in question are the unique solutions to fundamental systems other than [E4, E5, E6]. To illustrate, the point ($x_1 = 0$, $x_2 = 0$, $x_3 = \frac{11}{6}$) is the unique solution of system [E1, E2, E4] or system [E1, E2, E5] or system [E1, E2, E6]; the point ($x_1 = 0$, $x_2 = \frac{11}{3}$, $x_3 = 0$) is the unique solution of system [E1, E3, E4], and so on. *We stress that a point is a corner point (basic solution) only if some (perhaps one, perhaps more) fundamental system has the point as its unique solution.*

For a better understanding of Problem 2.14 and some of the difficulties it poses, the reader is encouraged to work the following Exercises 9, 10, and 11 with care.

Exercises

1.

$$\begin{aligned} &\text{Maximize} \quad 2x + 4y + 3z \\ &\text{subject to} \quad x \geq 0, \quad y \geq 0, \quad z \geq 0 \\ &\text{and} \quad x + 2y \leq 4 \\ &\qquad y + 3z \leq 6 \\ &\qquad 2x + y + 2z \leq 10 \end{aligned}$$

2. A wholesaler has 9600 square feet of space available and \$5000 he can spend for products A, B, and C. Product A costs \$4 per unit and needs 4 square feet of space, product B costs \$10 per unit and needs 8 square feet of space, while product C costs \$5 per unit and needs 6 square feet of space. Moreover, only 500 units of A are available to the wholesaler. Assuming the wholesaler expects to make a profit of \$1 on each unit of A, \$3 on each unit of B, and \$2 on each unit of C, how many units should he buy and stock to maximize his profit? [*Hint:* Let a, b, and c represent the units of products

A, B, and C stocked. Then the problem has three nontrivial constraints:

$$4a + 8b + 6c \leq 9600 \qquad (1)$$

$$4a + 10b + 5c \leq 5000 \qquad (2)$$

$$a \leq 500 \qquad (3)$$

Constraint (1) refers to space utilized, constraint (2) to money utilized, and constraint (3) expresses the limitation on units of A available.]

3. Find the feasible corner points of the region given by:

$$\begin{aligned} x_1 + x_2 + x_3 &\leq 3 \\ 2x_1 + 2x_2 + x_3 &\leq 5 \\ x_1 - x_2 &\leq 0 \\ x_1 &\geq 0 \\ x_2 &\geq 0 \\ x_3 &\geq 0 \end{aligned}$$

4. A manufacturer can produce, on a given machine operating 45 hours per week, three different products P_1, P_2, and P_3. Item P_1 brings a net revenue of \$4 per unit, item P_2, \$12, and item P_3, \$3. The hourly output of the machine is 50 units of P_1 alone, or 25 units of P_2 alone, or 75 units of P_3 alone. Potential sales are limited to 1000 units of P_1, 500 units of P_2, and 1500 units of P_3. How should production time be allocated in order to maximize revenue?

5. Maximize $P = 5x + 6y + 10z$

subject to $x \geq 0, \quad y \geq 0, \quad z \geq 0$

and
$$\begin{aligned} x + 2y + 3z &\leq 20 \\ 2x + y + z &\leq 30 \\ 3x + 2y + 3z &\geq 60 \end{aligned}$$

6. Maximize $P = x_1 + x_2 + x_3$

subject to $x_1 \geq 0, \quad x_2 \geq 0, \quad x_3 \geq 0$

and
$$\begin{aligned} 2x_1 + 3x_2 + x_3 &= 8 \\ x_1 + 2x_2 + 2x_3 &= 5 \end{aligned}$$

(*Answer:* The optimum value of P is $\frac{13}{3}$.)

7. Maximize $P = 5x_1 + 6x_2 + 7x_3$

subject to $x_1 \geq 0, \quad x_2 \geq 0, \quad x_3 \geq 0$

and
$$\begin{aligned} 3x_1 - x_2 + 2x_3 &\leq 20 \\ 3x_1 - 2x_2 + x_3 &\leq 30 \end{aligned}$$

8. Suppose a linear program possesses a unique optimum point. Is this point necessarily a basic feasible solution?

9. Show that each of the constraints (4), (5), and (6) in Problem 2.14 of this chapter is *equivalent* to any of the other two, in the sense that if a point (x_1, x_2, x_3) satisfies any one of these constraints it will satisfy any one of the others.

10. (a)

$$\text{Maximize} \quad P = x_1 + 3x_2 + 5x_3$$
$$\text{subject to} \quad x_1 \geq 0, \quad x_2 \geq 0, \quad x_3 \geq 0$$
$$\text{and} \quad 4x_1 + 6x_2 + 12x_3 \leq 22$$

(b)

$$\text{Maximize} \quad P = x_1 + 3x_2 + 5x_3$$
$$\text{subject to} \quad x_1 \geq 0, \quad x_2 \geq 0, \quad x_3 \geq 0$$
$$\text{and} \quad 4x_1 + 6x_2 + 12x_3 = 22$$

11.

$$\text{Maximize} \quad P = x_1 + 3x_2 + 5x_3$$
$$\text{subject to} \quad x_1 \geq 0, \quad x_2 \geq 0, \quad x_3 \geq 0$$
$$\begin{aligned} \text{and} \quad 4x_1 + 6x_2 + 12x_3 &= 22 \\ 6x_1 + 9x_2 + 18x_3 &= 33 \\ 2x_1 + 3x_2 + 6x_3 &= 11 \end{aligned}$$

12. Criticize the following statement: Given a system of equations with more unknowns than equations we generate a *basic solution* by giving solution values of 0 to just enough of the unknowns so that once they are thereby disposed of we are left with the same number of equations as unknowns. We then solve for the values of the remaining variables to complete our basic solution.

13. *True or false?*

(a) In a (product-mix) linear program, if two resources are used to produce three products, then there is an optimal solution where at least one product is not produced.

(b) In a diet-type program suppose it is possible to satisfy the minimum requirements for two vitamins by using three foods. Then there is an optimal diet in which at least one of the foods is not used.

14. Consider the points $P_1(1, 2, 3)$ and $P_2(3, 4, 5)$.

(a) Find the coordinates of the point M on the segment joining P_1 and P_2, such that the lengths of the segments P_1M and MP_2 are in the ratio 1:4.

(b) Find the coordinates of the point Q on the segment joining P_1 and P_2, such that the lengths of the segments P_1Q and P_1P_2 are in the ratio 1:10.

(c) If a linear form H assumes the values k_1 and k_2 at the points P_1 and P_2, respectively, find the values assumed by H at the points M and Q.

3
THE LOGIC OF THE SIMPLEX ALGORITHM

In Chapters 1 and 2 we solved some two- and three-dimensional linear-programming problems, using graphical or algebraic techniques. Our major tool was the corner-point theorem, but, as was shown, it cannot always be used easily. The most serious defect in attempting to list all potential corners is that, as the number of variables and constraints grows larger, the number of corners increases much more rapidly. A relatively small problem, containing ten variables and five nontrivial constraints, can have several thousand potential corners and hundreds of feasible corners.

It was especially because of the inefficiency inherent in trying to determine all corners that Dantzig developed the *simplex algorithm* in 1947. The simplex algorithm is a method for moving from one feasible corner to another feasible corner in such a way that the objective function is always improved, or, at the least, never worse. And as we see later, this algorithm not only finds optimal solutions whenever they exist but also automatically indicates the existence of alternative optima, empty regions of feasibility, unbounded solutions, and so on. It is today one of the most efficient and popular methods for solving linear programs.

The simplex algorithm is described in this chapter, using the usual notation for systems of linear equations. In Chapter 5 we describe this algorithm in tableau notation. The use of tableaux is nothing but a systematization and streamlining of the underlying algebra. At first the simplex routine may appear difficult, especially for students weak in algebra. But the algebraic difficulties should disappear with practice, especially after Chapter 4, which is devoted to the study of systems of linear equations.

This chapter is divided into two sections. In Section 3.1 the simplex algorithm is introduced by means of several examples. Although simple, these examples point out the analytic power of the simplex algorithm. In Section 3.2 we consolidate the concepts of Section 3.1 and then continue with further illustrations of the power of the simplex method.

We remind the reader that the terms *potential corner* and *fundamental system* are synonyms.

3.1 INTRODUCTION TO THE SIMPLEX ALGORITHM

Consider the following program:

PROBLEM 3.1. Maximize $P = 3x + 6y$

subject to $x \geq 0, \quad y \geq 0$

and $3x + y \leq 48$

$x + 3y \leq 48$

This program can be solved by finding the corners of the region of feasibility determined by the limiting lines

$$x = 0 \tag{1}$$

$$y = 0 \tag{2}$$

$$3x + y = 48 \tag{3}$$

$$x + 3y = 48 \tag{4}$$

By definition, there is a potential corner corresponding to every system of equations which can be formed from two of these four equations. It turns out that each potential corner is an actual corner, as shown in the following table.

Potential corner	*Coordinates*
1. $[E_1, E_2]$	$x = 0, \quad y = 0$
2. $[E_1, E_3]$	$x = 0, \quad y = 48$
3. $[E_1, E_4]$	$x = 0, \quad y = 16$
4. $[E_2, E_3]$	$x = 16, \quad y = 0$
5. $[E_2, E_4]$	$x = 48, \quad y = 0$
6. $[E_3, E_4]$	$x = 12, \quad y = 12$

The reader can continue the analysis of this problem by the corner-point method. We merely wish here to present a new way of labeling all the potential corners of a linear-programming problem. This labeling will be carried out algebraically by means of a simple but profound ruse. For every nontrivial inequality we will introduce a nonnegative *slack* variable to make up the difference between the left and right side of the inequality and then replace all nontrivial inequalities by equivalent equalities. To illustrate, we can transform the inequality

$$3x + y \leq 48$$

into the equality

$$3x + y + r = 48$$

where r is a nonnegative slack variable. Note that the maximum value of r is 48, and this can only occur if x and y are both 0. Furthermore, the inequality $r > 0$ implies that $3x + y$ is strictly less than 48. But what is perhaps most important, the equation $r = 0$ is a synonym for what we called Eq. (3) in the old system of labeling, that is, $3x + y = 48$.

In a similar fashion, we can replace the inequality

$$x + 3y \leq 48$$

by the equation

$$x + 3y + s = 48$$

where $s \geq 0$. The algebraic equivalent of Eq. (4) is now "$s = 0$."

Note that we have nothing to gain by transforming the trivial inequalities $x \geq 0$, $y \geq 0$ into equalities by means of slack variables. We could have transformed $x \geq 0$ into the equation $x - h = 0$, where h is nonnegative and defined the line associated with $x \geq 0$ by the equation $h = 0$, but we already have a convenient algebraic name for the line associated with $x \geq 0$, namely, $x = 0$.

Using slack variables we can rewrite Problem 3.1 as follows:

$$\begin{aligned}
&\text{Maximize} \quad P = 3x + 6y + 0(r) + 0(s) \\
&\text{subject to} \quad x \geq 0, \quad y \geq 0, \quad r \geq 0, \quad s \geq 0 \\
&\text{and} \quad 3x + y + r = 48 \\
&\qquad\quad x + 3y + s = 48
\end{aligned}$$

Because of the introduction of slack variables we can now algebraically specify the potential corners of our problem by setting any two of the four variables equal to zero. For example, by specifying that $r = s = 0$ we choose the corner which was previously labeled $[E_3, E_4]$. As mentioned, in this problem all potential corners are actual corners, although not necessarily feasible. In other words, each fundamental system does possess a unique solution, which may or may not represent a feasible point.

The region of feasibility is shown in Figure 28. Each corner, whether feasible or not, has been named by setting the appropriate "corner variables" equal to zero. The variables which must be equated to zero in order to define a corner will be called the *corner variables* (at the corner in question). Two *feasible* corners will be said to be *neighboring*, or *adjacent*, corners if the number of corner variables they have in common is one less than the number of variables needed to define a corner. To illustrate, the feasible corners $x = y = 0$ and $s = x = 0$ are neighbors, but $x = y = 0$ and $r = s = 0$ are not neighbors.

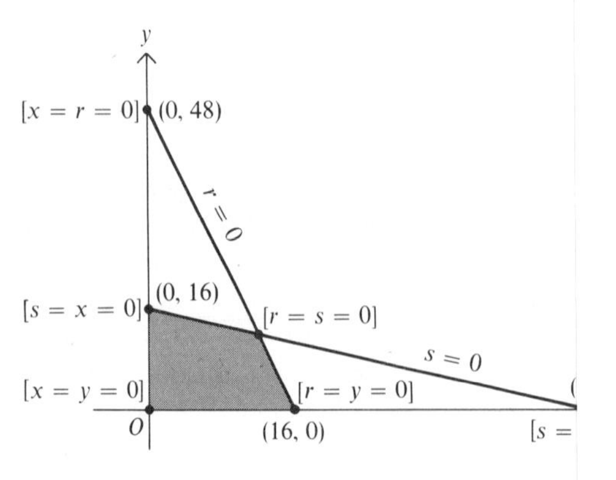

Figure 28

Finally, note that the use of slack variables allows
ascertaining whether "all variables in a program are no
immediately from the definition of slack variables.

We stress that the values of all four variables in Prob
apparent at one feasible corner (without really havi
This corner is the origin O, where $x = y = 0$ and
Furthermore, the profit function is defined in terms
which correspond to the origin, namely, $P = 3x +$
follows.

Starting at the corner where $x = y = 0$, if we in
increase profit, 3 units for every unit increase in x, 6 uni
in y. Suppose we decide to increase *y as much as th*
If there is no limit to how much y can be increased then
and the objective function will both be unbounded. H
and y is increased to the limit, we will automatically
corner.

There are two nontrivial constraints. The first,

$$3x + y + r = 48$$

allows us to increase y to 48, at which point $r = 0$. Bey
y must render r negative and hence be infeasible. The se

$$x + 3y + s = 48$$

allows us to increase y until $y = 16$, at which point s fa
beyond 16, s will become negative.

Since any allowable increase in y must satisfy every
that we can only increase y from 0 to 16. For each unit c
contribution to profit of \$6. Increasing y to 16 autom
corner $[x = s = 0]$. (Linear programming is intrinsically
in making y as large as possible we forced another vari
as possible.)

Equation (5′) tells us to increase x; Eq. (3′) informs us that if x exceeds 48, the variable y will become negative; Eq. (4′) tells us that if we increase x beyond 12, then r will become negative. Hence the maximum allowable increase in x is 12, at which point $r = 0$. This takes us to the next feasible corner, where s is still 0 and r has fallen to 0.

Now what is the profit function in terms of r and s, the corner variables at the newest corner? At the previous corner, the profit function took the form

$$P = 96 + x - 2s$$

The very equation which limited the increase in x, that is, Eq. (4′), can easily be solved for x in terms of r and s:

$$x = 12 + \tfrac{1}{8}s - \tfrac{3}{8}r \qquad (6')$$

Now we substitute this expression for x into the previous profit function, that is, $P = 96 + x - 2s$, and obtain

$$\begin{aligned} P &= 96 - 2s + (12 + \tfrac{1}{8}s - \tfrac{3}{8}r) \\ &= 108 - \tfrac{3}{8}r - \tfrac{15}{8}s \qquad (7') \end{aligned}$$

Equation (7′) is the view on profit from corner $[r = s = 0]$. To move away from this corner we must increase either r or s. But the coefficients of r and s are negative in Eq. (7′), and an increase in either will cause P to decrease. Hence, the corner defined by $r = s = 0$ is optimal. We leave it to the reader to complete the dictionary at this corner.

It is instructive to realize that we pass from one corner to the next by solving a one-dimensional linear program. At corner $[x = s = 0]$ we were instructed to increase x. For $s = 0$, Eqs. (3′) and (4′) read

$$y = 16 - \tfrac{1}{3}x \qquad (3'')$$

$$r = 32 - \tfrac{8}{3}x \qquad (4'')$$

Equations (3″) and (4″) are equivalent to the inequalities

$$\tfrac{1}{3}x \leq 16 \qquad (3''')$$

$$\tfrac{8}{3}x \leq 32 \qquad (4''')$$

Clearly, the one-dimensional problem we solved was really

$$\begin{aligned} &\text{Maximize} \quad x \\ &\text{subject to} \quad x \geq 0 \\ &\text{and} \quad \tfrac{1}{3}x \leq 16 \\ &\qquad \tfrac{8}{3}x \leq 32 \end{aligned}$$

Now we apply the simplex method to a three-dimensional program.

PROBLEM 3.2. Maximize $P = 5x + 6y + 7z$

subject to $x \geq 0, \quad y \geq 0, \quad z \geq 0$

and

$$\begin{aligned} x + y \quad &\leq 10 \\ 2x \quad + z &\leq 18 \end{aligned}$$

We introduce nonnegative slack variables r and s and rewrite the nontrivial constraints as

$$\begin{aligned} x + y \quad + r \quad &= 10 \\ 2x \quad + z \quad + s &= 18 \end{aligned}$$

The origin $[x = y = z = 0]$ in 3-space is clearly a feasible corner. The dictionary associated with this corner can be formed immediately; it reads

Dictionary 1.

$$r = 10 - x - y \tag{1}$$

$$s = 18 - 2x - z \tag{2}$$

$$P = 5x + 6y + 7z \tag{3}$$

Equation (3) shows that P can be increased by increasing either x, y, or z. Since the coefficient of z is largest, we begin by increasing z as much as possible. Note that Eq. (1) puts no limit whatsoever on the allowable increase in z, because the coefficient of z in Eq. (1) is 0. But Eq. (2) informs us that if z is increased beyond 18, s will become negative. We therefore increase z to 18, at which point s becomes 0.

We are now at a new corner, defined by $x = y = s = 0$. To carry out the simplex algorithm in a neat fashion, we should draw up the dictionary appropriate to this corner. The "key to translation" is the equation which limited the increase in z, Eq. (2). It follows easily from this equation that

$$z = 18 - 2x - s \tag{4}$$

The correct expression for r in the new dictionary follows from Eq. (1) and is

$$r = 10 - x - y + 0(s) \tag{5}$$

To find the current form of the profit function, we need only substitute Eq. (4) into Eq. (3). Thus,

$$\begin{aligned} P &= 5x + 6y + 7z \\ &= 5x + 6y + 7(18 - 2x - s) \\ &= -9x + 6y - 7s + 126 \end{aligned} \tag{6}$$

To summarize, the complete dictionary at corner 2 $[x = y = s = 0]$ reads:

$$z = 18 - 2x + 0(y) - s \tag{4}$$

$$r = 10 - x - y + 0(s) \tag{5}$$

$$P = 126 - 9x + 6y - 7s \tag{6}$$

The current form of P motivates us to increase y as much as possible. Equation (4) puts no limit on y, because the coefficient of y is zero. But Eq. (5) limits the increase in y to 10, at which point r is driven to 0.

The new corner variables are x, s, and r. The key to forming the dictionary at corner 3 $[x = s = r = 0]$ is Eq. (5), the equation which limited the increase in y. Equation (5) implies that

$$y = 10 - x + 0(s) - r \tag{7}$$

The current expression for z is also immediate. From Eq. (4) we obtain

$$z = 18 - 2x - s + 0(r) \tag{8}$$

But what is most important, if we substitute Eq. (7) into Eq. (6) we obtain the new form of P. Thus,

$$\begin{aligned} P &= 126 - 9x + 6y - 7s \\ &= 126 - 9x + 6(10 - x - r) - 7s \\ &= 186 - 15x - 6r - 7s \end{aligned} \tag{9}$$

To move away from corner 3 we would have to increase either x, r, or s. Equation (9) shows that doing so can only decrease P. Hence, corner 3 is optimal. From Eqs. (7) and (8) we see that at this corner $y = 10$ and $z = 18$; hence, $P = 186$, obtained either from (9) or from the original profit function.

PROBLEM 3.3.

$$\begin{aligned} &\text{Minimize} \quad C = 3x + 5y \\ &\text{subject to} \quad x \geq 0, \quad y \geq 0 \\ &\text{and} \quad x + 3y \geq 9 \\ &\qquad\quad x + y \geq 5 \\ &\qquad\quad 2x + y \geq 6 \end{aligned}$$

We can convert the first nontrivial constraint into an equality by *subtracting* a positive slack variable r from the left side. Thus

$$x + 3y - r = 9 \tag{1}$$

where $r \geq 0$, is an equation equivalent to

$$x + 3y \geq 9$$

(Because the variable r appears with a negative coefficient it is sometimes called a *surplus* variable.) The other nontrivial inequalities can be converted into equalities by the same procedure:

$$x + y - s = 5 \tag{2}$$

where $s \geq 0$, and

$$2x + y - t = 6 \tag{3}$$

where $t \geq 0$.

We are now ready to solve Problem 3.3 by the simplex technique. Unfortunately, the corner $[x = y = 0]$ is not feasible in Problem 3.3, because if

$x = y = 0$, each of the nontrivial constraints will be violated. One way to start the algorithm is to find a feasible corner by trial and error. Recall that by setting any two of the five variables x, y, r, s, and t equal to 0 we automatically specify a potential corner. There are ten possibilities and we must find one which will define a feasible corner.

Let us try $y = r = 0$. The constraints (1), (2), and (3) give, for $y = r = 0$, the values $x = 9$, $s = 4$, and $t = 12$. Since all five variables are nonnegative, it follows that $y = r = 0$ defines a feasible corner. We must now view the problem from this corner by constructing the appropriate dictionary, that is, by expressing all other variables in terms of y and r.

From Eq. (1) we deduce the key to translation, namely,

$$x = 9 - 3y + r \tag{4}$$

Now substitute Eq. (4) into Eq. (2), obtaining

$$\begin{aligned} s &= x + y - 5 \\ &= (9 - 3y + r) + y - 5 \\ &= 4 - 2y + r \end{aligned} \tag{5}$$

Similarly, by substituting Eq. (4) into Eq. (3) we obtain

$$\begin{aligned} t &= 2x + y - 6 \\ &= 2(9 - 3y + r) + y - 6 \\ &= 12 - 5y + 2r \end{aligned} \tag{6}$$

Finally, to obtain the appropriate form of C, note that

$$\begin{aligned} C &= 3x + 5y \\ &= 3(9 - 3y + r) + 5y \\ &= 27 - 4y + 3r \end{aligned} \tag{7}$$

Equations (4) to (7) constitute the complete dictionary in terms of y and r. At the corner $[y = r = 0]$ these equations show immediately that $x = 9$, $s = 4$, $t = 12$, and $C = 27$. But what is much more important, from Eq. (7) we learn that C can be reduced by increasing y. To find the maximum allowable increase in y, refer to the current dictionary. Equation (4) shows that we cannot increase y beyond 3, Eq. (5) shows that we cannot increase y beyond 2, and Eq. (6) shows that we cannot increase y beyond $\frac{12}{5}$. If we are to satisfy each constraint, we can only increase y to 2, at which point the variable s falls to 0.

The equation which limited the increase in y is Eq. (5). It is the key relation needed to build the next dictionary in terms of the new corner variables r and s. From Eq. (5) we obtain immediately

$$y = 2 + \tfrac{1}{2}r - \tfrac{1}{2}s \tag{8}$$

The new objective function can be found by substituting Eq. (8) into Eq. (7):

$$\begin{aligned} C &= 27 - 4y + 3r \\ &= 27 - 4(2 + \tfrac{1}{2}r - \tfrac{1}{2}s) + 3r \\ &= 19 + r + 2s \end{aligned} \tag{9}$$

Equation (9) shows that the corner $[r = s = 0]$ is optimal since any increase in r or s increases C, and our objective is to minimize C.

The reader can easily verify that the complete dictionary at this corner reads

$$\begin{aligned} y &= \;\; 2 + \tfrac{1}{2}r - \tfrac{1}{2}s \\ x &= \;\; 3 - \tfrac{1}{2}r + \tfrac{3}{2}s \\ t &= \;\; 2 - \tfrac{1}{2}r + \tfrac{5}{2}s \\ C &= 19 + \;\; r + 2s \end{aligned}$$

PROBLEM 3.4. Maximize $P = x + y$

subject to $x \geq 0, \quad y \geq 0$

and $x + y \leq 10$

We have seen by the methods of Chapters 1 and 2 that there is more than one point which yields the optimum value of P. Let us see how the existence of alternate optimum points is indicated by the simplex algorithm.

As an equality, the nontrivial constraint reads

$$x + y + r = 10$$

where $r \geq 0$. We can easily find a first feasible corner, namely, $x = y = 0$; the dictionary at this corner is

$$r = 10 - x - y \tag{1}$$

$$P = \qquad x + y \tag{2}$$

We can increase P by increasing either x or y; suppose we choose x. Equation (1) limits the increase in x to 10, at which point $r = 0$. The second corner is therefore $y = r = 0$. To form the appropriate dictionary at this corner we solve Eq. (1) for x in terms of y and r, obtaining

$$x = 10 - y - r \tag{3}$$

If we substitute Eq. (3) into Eq. (2) we then get

$$\begin{aligned} P &= x + y \\ &= (10 - y - r) + y \\ &= 10 + 0(y) - r \end{aligned} \tag{4}$$

Note that the coefficient of y in the new form of P is 0. This is the indication that we should seek alternative optimum points. Equation (4) indicates that

y can be increased without changing P. Suppose we increase y as much as possible. Equation (3) shows that the maximum allowable increase in y is 10, at which point $x = 0$. Hence $x = r = 0$ is an alternative optimal corner. The dictionary at this corner is easily seen to be

$$y = 10 - x - r \tag{5}$$

$$P = 10 + 0(x) - r \tag{6}$$

If we omit variables with 0 as a coefficient, then Eqs. (4) and (6) both read

$$P = 10 - r$$

Here they are identical, but in unabbreviated form they represent the view on profit from two different optimal corners.

It is left to the reader to show that by following the signal of Eq. (6) and increasing x, we return to the original optimal corner.

The following program has an unbounded region of feasibility.

PROBLEM 3.5. Maximize $P = 5x + 6y + 7z$

subject to $x \geq 0,\ y \geq 0,\ z \geq 0,\ s_1 \geq 0,\ s_2 \geq 0,\ s_3 \geq 0$

and

$$\begin{aligned} x - 4y + z + s_1 \qquad\qquad &= 48 \\ 3x + 3y - z \qquad + s_2 \qquad &= 120 \\ -2x + y - z \qquad\qquad + s_3 &= 50 \end{aligned}$$

It makes little difference whether we think of s_1, s_2, and s_3 as slack variables or not. What does matter is that we can easily express s_1, s_2, and s_3 in terms of x, y, and z and form what must be the dictionary corresponding to the feasible corner $[x = y = z = 0]$. Thus, simply by transposing, we obtain

Dictionary 1.

$$s_1 = 48 - x + 4y - z \tag{1}$$

$$s_2 = 120 - 3x - 3y + z \tag{2}$$

$$s_3 = 50 + 2x - y + z \tag{3}$$

$$P = 5x + 6y + 7z \tag{4}$$

Following the "instructions" of Eq. (4), z is increased as much as possible. Observe that only Eq. (1) puts a limit on z. This limit is 48, at which point $s_1 = 0$. The second corner to examine is therefore $x = y = s_1 = 0$.

To form the new dictionary, we solve Eq. (1) for z and obtain

$$z = 48 - x + 4y - s_1 \tag{5}$$

Now we substitute Eq. (5) into the other equations of Dictionary 1. From Eq. (2),

$$\begin{aligned} s_2 &= 120 - 3x - 3y + z \\ &= 120 - 3x - 3y + (48 - x + 4y - s_1) \\ &= 168 - 4x + y - s_1 \end{aligned} \tag{6}$$

From Eq. (3), we get

$$\begin{aligned} s_3 &= 50 + 2x - y + z \\ &= 50 + 2x - y + (48 - x + 4y - s_1) \\ &= 98 + x + 3y - s_1 \end{aligned} \tag{7}$$

And from Eq. (4),

$$\begin{aligned} P &= 5x + 6y + 7z \\ &= 5x + 6y + 7(48 - x + 4y - s_1) \\ &= 336 - 2x + 34y - 7s_1 \end{aligned} \tag{8}$$

This last form of P indicates that by increasing y the objective can be increased. To find the maximum allowable increase, Eqs. (5), (6), and (7) must be examined. None of these puts a limit on the allowable increase, and, by increasing y, we can make P larger than any preassigned value. The region of feasibility is unbounded and so is P.

To conclude this section, let us return to Problem 2.4. There we solved by graphical methods a balanced transportation problem whose costs and other relevant data were

	D_1	D_2	D_3	
S_1	\$4	\$2	\$1	15
S_2	\$1	\$3	\$5	25
	10	20	10	(40)

This problem was formulated as a linear program by using x to denote the quantity shipped from S_1 to D_1, and y to denote the quantity shipped from S_1 to D_2. With this choice of unknowns the problem took the following form:

$$\begin{aligned} &\text{Minimize} \quad C = 7x + 3y + 60 \\ &\text{subject to} \quad x \geq 0, \quad y \geq 0 \\ &\text{and} \quad x \leq 10 \\ &\qquad y \leq 20 \\ &\qquad x + y \leq 15 \\ &\qquad x + y \geq 5 \end{aligned}$$

Note that the coefficients of x and y in the objective function are both nonnegative. Hence, if the point $x = y = 0$ were feasible, it would be optimal. Unfortunately, this point was infeasible (by the last constraint), and we had to do some calculating to solve the problem.

However, the method used to describe the amounts shipped was arbitrary. For example, we could just as well have let u denote the quantity to be shipped from S_1 to D_1, and v the quantity from S_2 to D_3. Then, the shipping schedule is

	D_1	D_2	D_3	
S_1	u	$5 + v - u$	$10 - v$	15
S_2	$10 - u$	$15 + u - v$	v	25
	10	20	10	(40)

The problem can then be described as follows:

$$\begin{aligned} \text{Minimize} \quad & K = 4u + 3v + 75 \\ \text{subject to} \quad & u \geq 0, \quad v \geq 0 \\ \text{and} \quad & u \leq 10 \\ & v \leq 10 \\ & -u + v \geq -5 \\ & u - v \geq -15 \end{aligned}$$

Now, with the problem expressed in terms of u and v, the optimum solution is self-evident; it is $u = v = 0$. This is because:

(1) The objective function is to be minimized.
(2) The objective function can be expressed solely in terms of u and v; in this form, the coefficients of u and v are both nonnegative.
(3) $u = v = 0$ is feasible.

Of course, we never really know ahead of time how to describe a linear program so as to make its optimum solution appear self-evident. However, the simplex method, by its very lavishness in the use of "nonnatural" slack variables, allows us to modify any given formulation of a program until we do find the "optimal formulation."

Exercises

In most of the following exercises a linear-programming problem is stated and then set up in slack-variable notation. If the initial dictionary produces a feasible corner, apply the simplex method. If the slack-variable dictionary does not provide an immediately obvious feasible corner, find a first feasible corner by trial and error, and then solve. *In all cases the optimal dictionary is given.*

1. (a)

$$\begin{aligned} \text{Maximize} \quad & P = 4x + 3y \\ \text{subject to} \quad & x \geq 0, \quad y \geq 0 \\ \text{and} \quad & 4x + 2y \leq 46 \\ & x + 3y \leq 24 \end{aligned}$$

(b) In slack-variable notation,

$$\begin{aligned} 4x + 2y + r \quad &= 46 \\ x + 3y \quad + s &= 24 \end{aligned}$$

(c) At optimality:

$$\begin{aligned} P &= 51 - \tfrac{9}{10}r - \tfrac{2}{5}s \\ x &= 9 - \tfrac{3}{10}r + \tfrac{1}{5}s \\ y &= 5 + \tfrac{1}{10}r - \tfrac{2}{5}s \end{aligned}$$

2. (a)

$$\begin{aligned} &\text{Maximize} \quad P = 3x + 5y \\ &\text{subject to} \quad x \geq 0, \quad y \geq 0 \\ &\text{and} \quad 2x + 7y \leq 20 \\ &\qquad\quad 6x + 3y \leq 24 \end{aligned}$$

(b) In slack-variable notation,

$$\begin{aligned} 2x + 7y + r \quad &= 20 \\ 6x + 3y \quad + s &= 24 \end{aligned}$$

(c) At optimality:

$$\begin{aligned} P &= 19 - \tfrac{7}{12}r - \tfrac{11}{36}s \\ x &= 3 + \tfrac{1}{12}r - \tfrac{7}{36}s \\ y &= 2 - \tfrac{1}{6}r + \tfrac{1}{18}s \end{aligned}$$

3. (a)

$$\begin{aligned} &\text{Maximize} \quad P = 3x + 2y \\ &\text{subject to} \quad x \geq 0, \quad y \geq 0 \\ &\text{and} \quad x + 4y \leq 10 \\ &\qquad\quad 3x + 5y \leq 15 \end{aligned}$$

(b) In slack-variable notation,

$$\begin{aligned} x + 4y + r \quad &= 10 \\ 3x + 5y \quad + s &= 15 \end{aligned}$$

(c) At optimality:

$$\begin{aligned} P &= 15 - 3y - s \\ x &= 5 - \tfrac{5}{3}y - \tfrac{1}{3}s \\ r &= 5 - \tfrac{7}{3}y + \tfrac{1}{3}s \end{aligned}$$

4. (a)

$$\begin{aligned} &\text{Minimize} \quad C = 48x + 48y \\ &\text{subject to} \quad x \geq 0, \quad y \geq 0 \\ &\text{and} \quad 3x + y \geq 3 \\ &\qquad\quad x + 3y \geq 6 \end{aligned}$$

(b) In slack-variable notation,

$$\begin{aligned} 3x + y - t_1 \quad &= 3 \\ x + 3y \quad - t_2 &= 6 \end{aligned}$$

(c) At optimality:

$$C = 108 + 12t_1 + 12t_2$$
$$x = \tfrac{3}{8} + \tfrac{3}{8}t_1 - \tfrac{1}{8}t_2$$
$$y = \tfrac{15}{8} - \tfrac{1}{8}t_1 + \tfrac{3}{8}t_2$$

5. (a)

$$\text{Minimize } C = x + 3y$$
$$\text{subject to } x \geq 0, \quad y \geq 0$$
$$\text{and } x + 4y \geq 24$$
$$5x + y \geq 25$$

(b) In slack-variable notation,

$$x + 4y - r = 24$$
$$5x + y - s = 25$$

(c) At optimality:

$$C = 19 + \tfrac{14}{19}r + \tfrac{1}{19}s$$
$$x = 4 - \tfrac{1}{19}r + \tfrac{4}{19}s$$
$$y = 5 + \tfrac{5}{19}r - \tfrac{1}{19}s$$

6. (a)

$$\text{Minimize } C = 2x_1 + 2x_2$$
$$\text{subject to } x_1 \geq 0, \quad x_2 \geq 0$$
$$\text{and } 2x_1 + x_2 \geq 10$$
$$x_1 + 2x_2 \geq 10$$

(b) In slack-variable notation,

$$2x_1 + x_2 - t_1 = 10$$
$$x_1 + 2x_2 - t_2 = 10$$

(c) At optimality:

$$C = \tfrac{40}{3} + \tfrac{2}{3}t_1 + \tfrac{2}{3}t_2$$
$$x_1 = \tfrac{10}{3} + \tfrac{2}{3}t_1 - \tfrac{1}{3}t_2$$
$$x_2 = \tfrac{10}{3} - \tfrac{1}{3}t_1 + \tfrac{2}{3}t_2$$

7. (a)

$$\text{Maximize } P = 2x_1 + 5x_2$$
$$\text{subject to } x_1 \geq 0, \quad x_2 \geq 0$$
$$\text{and } x_1 \leq 4$$
$$x_2 \leq 6$$
$$x_1 + x_2 \leq 8$$

(b) Using slack-variable notation,

$$x_1 + s_1 = 4$$
$$x_2 + s_2 = 6$$
$$x_1 + x_2 + s_3 = 8$$

(c) At optimality:

$$\begin{aligned} P &= 34 - 3s_2 - 2s_3 \\ x_1 &= 2 + s_2 - s_3 \\ x_2 &= 6 - s_2 + 0(s_3) \\ s_1 &= 2 - s_2 + s_3 \end{aligned}$$

8. Solve by the simplex algorithm:

$$\begin{aligned} &\text{Maximize} \quad Z = 184 - \tfrac{2}{9}u - \tfrac{13}{9}v \\ &\text{subject to} \quad u \geq 0, \quad v \geq 0 \\ &\text{and} \quad \tfrac{4}{9}u - \tfrac{1}{9}v \leq 8 \\ &\qquad \tfrac{22}{3}u + \tfrac{1}{3}v \leq 24 \end{aligned}$$

9. Relate the following statement to the optimality criterion of linear programming: *Rip Collins holds the record for fewest put-outs by a first baseman in a nine-inning ball game—zero put-outs.*

10. Clarify the relation between the following statement and the simplex technique: Production must be terminated when 5 hours of machine time have been consumed. It then becomes logical to measure profit in terms of machine-hours instead of units produced.

11. Construct a linear program where the slack variables are nonzero at optimality.

3.2 SUMMARY OF THE SIMPLEX ROUTINE

In this section we consolidate and generalize the discussion of Section 3.1. .

The General Problem

For the purposes of this chapter we can define the *general linear-programming problem, in inequality format*, as follows:

$$\begin{aligned} &\text{Maximize} \quad P = c_1x_1 + \cdots + c_nx_n \\ &\text{subject to} \quad x_1 \geq 0, \ldots, x_n \geq 0 \\ &\text{and} \quad a_{11}x_1 + a_{12}x_2 + \cdots + a_{1n}x_n \leq b_1 \\ &\qquad a_{i1}x_1 + a_{i2}x_2 + \cdots + a_{in}x_n \leq b_i \\ &\qquad a_{m1}x_1 + a_{m2}x_2 + \cdots + a_{mn}x_n \leq b_m \end{aligned}$$

Here a_{ij} denotes the coefficient of x_j in the ith nontrivial inequality, or row i. For each row we assume that at least one coefficient $a_{ij} \neq 0$ and that every variable appears in some nontrivial inequality with a nonzero coefficient. Otherwise the a_{ij} may be positive or negative. We make no special assumptions about the b_i; they may be positive, negative, or zero. The variables $x_1, x_2, \ldots, x_n$ are called the *original* or *structural* variables of our problem, the a_{ij} the *structural coefficients*, and the b_i the *upper bounds* or *constants* (b for bound).

We remind the reader that the minimum of an objective function occurs at the same set of values as the maximum of the negative of the given function. Thus, whatever the constraints, the problem of minimizing $C = k_1x_1 + \cdots + k_nx_n$ is equivalent to maximizing $-C = -k_1x_1 - \cdots - k_nx_n$. Also, any inequality can be written in "less than" form, because

$$a_{i1}x_1 + a_{i2}x_2 + \cdots + a_{in}x_n \geq b_i$$

is equivalent to

$$-a_{i1}x_1 - a_{i2}x_2 - \cdots - a_{in}x_n \leq -b_i$$

To illustrate, Problem 3.3 could just as well have been formulated as a maximization problem with upper bounds, as follows:

$$\begin{aligned} &\text{Maximize} \quad -C = -3x - 5y \\ &\text{subject to} \quad x \geq 0, \quad y \geq 0 \\ &\text{and} \quad -x - 3y \leq -9 \\ &\qquad\quad -x - y \leq -5 \\ &\qquad\quad -2x - y \leq -6 \end{aligned}$$

Slack Variables

By introducing slack variables we can pass from the inequality form of the general problem to the equality form:

$$\text{Maximize} \quad P = c_1x_1 + \cdots + c_nx_n + 0x_{n+1} + \cdots + 0x_{n+m}$$

where all variables are nonnegative

$$\begin{aligned} \text{and} \quad a_{11}x_1 + a_{12}x_2 + \cdots + a_{1n}x_n + x_{n+1} \qquad\qquad\qquad &= b_1 \\ a_{i1}x_1 + a_{i2}x_2 + \cdots + a_{in}x_n \qquad + x_{n+i} \qquad &= b_i \\ a_{m1}x_1 + a_{m2}x_2 + \cdots + a_{mn}x_n \qquad\qquad + x_{n+m} &= b_m \end{aligned}$$

The variables $x_{n+1}, \ldots, x_{n+m}$ are the *slack variables*. In solving small illustrative examples in this text it will often be clearer to use the symbols $s_1, \ldots, s_m$, respectively, to denote the above slack variables.

Corners or Basic Solutions

In the slack-variable formulation, the nontrivial constraints are expressed as a system of m equations in $(n + m)$ unknowns. We can always set n of these variables equal to zero and try to solve for the other m variables. Every choice of n different variables to be set equal to zero corresponds to specifying a *potential corner* (that is, a *fundamental system of equations*). A potential corner will be called a *corner*, or *basic solution*, only when the system in question possesses a unique solution. At any corner, the variables whose values are prespecified as zero are called the *corner variables*, more commonly known as the *nonbasic* variables. The other variables are called *basic variables*. Note that there are n nonbasic and m basic variables at a corner.

Basic Feasible Solutions

If at any corner the value of each basic variable is nonnegative, we say that the corner in question is a *feasible corner*. In this case the values of all the variables, both corner and basic, are said to form a *basic feasible solution* (to the system of equations which describes the given constraints).

Corner-Point Theorem

The corner-point theorem states that if a feasible solution exists at which the objective function of a linear-programming problem achieves its maximum value, then there exists a *basic feasible solution at which this maximum is achieved.*

Dictionary

With every corner we can associate a unique dictionary, determined by expressing the objective function and the basic variables in terms of the corner variables. It should be stressed that:

(1) If a system of m equations in $(m + n)$ unknowns has *already been solved* for, say, $x_1, x_2, \ldots, x_m$ in terms of $x_{m+1}, \ldots, x_{m+n}$, then the system which remains when we set $x_{m+1} = \cdots = x_{m+n} = 0$ necessarily has a unique solution—in other words, $x_{m+1}, \ldots, x_{m+n}$ are corner variables for a basic solution.

We repeat: If some m variables, say $x_1, \ldots, x_m$ (the left-hand variables), have been expressed in terms of the remaining n variables (the right-hand variables), then by putting the right-hand variables equal to 0, the values of the left-hand variables are immediately determined. To illustrate, consider the system

$$x_3 = b_0 + \alpha x_1 + \beta x_2$$
$$x_4 = b_1 + \gamma x_1 + \delta x_2$$

If we set $x_1 = x_2 = 0$, then, clearly, $x_3 = b_0$ and $x_4 = b_1$, a unique solution.

We can think of the variables on the left-hand side of the equations as *dependent*, because their values are determined as soon as the values of the right-hand variables have been specified. The variables on the right-hand side are the *independent variables*. The independent variables are precisely the corner, or nonbasic, variables; the dependent variables are the noncorner, or basic, variables.

(2) The preceding definition of "corner" is purely algebraic. In Chapter 10 we demonstrate that such an algebraic corner solution is actually an extreme point of the region of feasibility; that is, it cannot be the average of two other

feasible solutions. (Henceforth, the terms *corner*, *corner point*, *extreme point*, and *basic solution* are used interchangeably.)

Clearly, we cannot speak of the dictionary at a corner until we are sure that the basic, or noncorner, variables can actually be expressed in terms of the nonbasic, or corner, variables. Even if this is possible, the dictionary will not be called *adjusted* until each nonbasic variable, *as well as the profit function*, has been explicitly expressed solely in terms of the appropriate corner variables. To illustrate, in Problem 3.3 we found that $y = r = 0$ determined a feasible corner. Not until P and all the basic variables were expressed as explicit functions of only y and r were we able to see clearly the effects of moving away from this corner. Only then was the dictionary *adjusted.*

The steps in the simplex algorithm are summarized in the following section. In this chapter, it is assumed that the problem has been formulated using slack variables and that a first feasible corner, from which to begin the algorithm, has been found. [The determination of such a corner can be quite difficult and is studied in more detail in Chapter 6. But note that if a program appears as in the subsection on slack variables (page 66), and if each $b_i \geq 0$, then a feasible corner can be immediately obtained by setting $x_1 = x_2 = \cdots = x_n = 0$.]

The Simplex Algorithm Summarized

Step 1. Adjust the dictionary. Once the set of first, or current, feasible corner variables has been determined, make sure that the profit function and all other variables are explicitly expressed only in terms of the corner variables. This is the *adjustment* process.

Step 2. Examine the objective function as expressed in the current feasible dictionary. Suppose that $V_1, \ldots, V_n$ are the corner variables and that the objective function reads

$$P = P_0 + k_1V_1 + \cdots + k_nV_n$$

Then:

(1) If all the k_i's are strictly less than 0, the algorithm is terminated.

(2) If none of the k_i's is strictly positive but some $k_i = 0$, we are at an optimal corner, but there may exist alternative optimal solutions.

(3) If some of the k_i's are positive, then increase any variable V_i whose coefficient k_i is largest (or even any variable whose coefficient is positive).

If there is no limit on how much we can increase V_i, the region of feasibility is unbounded and P can be made arbitrarily large. Otherwise, increase V_i to the maximum allowable extent. In doing so, some other variable will necessarily fall to 0, and a new feasible corner will be determined.

Step 3. Repeat Step 1. The algorithm as previously described will almost always converge. (Later, we discuss *degeneracy* and the possibility of nonconvergence.)

The solutions to the following problems serve to illustrate the simplex rules further. They are somewhat more difficult than those problems previously encountered, and the student will be referred back to them later for a second, more careful reading.

PROBLEM 3.6 (THE CONTINUOUS LOADING PROBLEM).

$$\text{Maximize} \quad P = 3x_1 + 2x_2 + 3x_3 + 4x_4$$

$$\text{subject to} \quad x_1 \geq 0, \quad x_2 \geq 0, \quad x_3 \geq 0, \quad x_4 \geq 0$$

$$\text{and} \quad 3x_1 + x_2 + 4x_3 + 3x_4 = 24 \tag{1}$$

A solution independent of anything previously studied can be found by considering each variable in turn and finding its largest possible total contribution to P. For example, the maximum value of x_1 consistent with constraint (1) is 8. When $x_1 = 8$, all other variables are 0, and $P = (3)(8) = 24$. Proceeding similarly, we obtain the following results:

Maximum value of variable	*Maximum contribution to profit*
$x_1 = 8$	$P = 24$
$x_2 = 24$	$P = 48$
$x_3 = 6$	$P = 18$
$x_4 = 8$	$P = 32$

As demonstrated below, the optimal solution to Problem 3.6 is $P = 48$, occurring at the point $x_1 = 0$, $x_2 = 24$, $x_3 = 0$, $x_4 = 0$. Write any other feasible solution in the form

$$x_1 = \theta_1, \quad x_2 = 24 - \theta_2, \quad x_3 = \theta_3, \quad x_4 = \theta_4$$

where the θ_i, for $i = 1, 2, 3, 4$, are not all 0. These θ_i are related to each other by Eq. (1), which implies that

$$3\theta_1 + (24 - \theta_2) + 4\theta_3 + 3\theta_4 = 24$$

and hence

$$\theta_2 = 3\theta_1 + 4\theta_3 + 3\theta_4 \tag{2}$$

The value of P at this feasible solution can be calculated easily:

$$\begin{aligned} P &= 3x_1 + 2x_2 + 3x_3 + 4x_4 \\ &= 3\theta_1 + 2(24 - \theta_2) + 3\theta_3 + 4\theta_4 \\ &= 48 + 3\theta_1 + 3\theta_3 + 4\theta_4 - 2\theta_2 \end{aligned} \tag{3}$$

Now substitute Eq. (2) into Eq. (3), obtaining

$$P = 48 - 3\theta_1 - 5\theta_3 - 2\theta_4 \tag{4}$$

Since θ_1, θ_3, and θ_4 are not all 0, the value of P at any feasible solution other than $(x_1 = 0, x_2 = 24, x_3 = 0, x_4 = 0)$ is strictly less than 48. This proves our contention.

Let us now study Problem 3.6 by the simplex algorithm. We can solve for any one variable in terms of the other three and hopefully construct a feasible dictionary. Suppose we choose to examine the dictionary expressed in terms of x_1, x_3, and x_4. From Eq. (1),

$$x_2 = 24 - 3x_1 - 4x_3 - 3x_4$$

and

$$\begin{aligned} P &= 3x_1 + 2x_2 + 3x_3 + 4x_4 \\ &= 3x_1 + 2(24 - 3x_1 - 4x_3 - 3x_4) + 3x_3 + 4x_4 \\ &= 48 - 3x_1 - 5x_3 - 2x_4 \end{aligned}$$

This form of P immediately indicates that the optimum value of profit is 48 and that it occurs at the corner $[x_1 = x_3 = x_4 = 0]$. The profit function above is identical to Eq. (4), except for notation. [Note that if Eq. (1) is replaced by the inequality $3x_1 + x_2 + 4x_3 + 3x_4 \leq 24$ we still obtain the same solution as above.]

In Chapter 9 we explain why this program is called the *continuous loading problem*. Can you guess the reason?

PROBLEM 3.7. Maximize $P = 3x_1 + 2x_2 + 3x_3 + 4x_4$

where all variables are nonnegative

and

$$3x_1 + x_2 + 4x_3 + 3x_4 = 24 \quad (1)$$

$$2x_1 + 3x_2 + 5x_3 + 2x_4 = 30 \quad (2)$$

Observe that the objective function in this problem is the same as that in Problem 3.6, and that here every feasible solution must be a feasible solution of Problem 3.6. Hence the optimal value here can not exceed 48, which is the optimal value of Problem 3.6.

Usually, when there are two equations in four unknowns, any two unknowns can be expressed in terms of the other two. If this is possible we can construct a complete dictionary associated with a well-defined corner, although this corner may not be feasible. But we cannot always express *any* two of the variables in terms of the other two. For example, it is impossible to solve Eqs. (1) and (2) for x_1 and x_4 in terms of x_2 and x_3. To see why, suppose it were actually possible; then set $x_2 = x_3 = 0$. This would fix values of x_1 and x_4 which would have to satisfy the system

$$3x_1 + 3x_4 = 24 \quad (1')$$

$$2x_1 + 2x_4 = 30 \quad (2')$$

But Eq. (1′) implies that $x_1 + x_4 = 8$, and Eq. (2′) implies that $x_1 + x_4 = 15$, which is clearly inconsistent.

To solve Problem 3.7 by the simplex algorithm, we must find a feasible corner from which to begin. (As seen below, $x_3 = x_4 = 0$ defines such a corner.)

First, *eliminate* x_2 by adding -3 times Eq. (1) to Eq. (2), obtaining

$$-7x_1 - 7x_3 - 7x_4 = -42$$

or

$$x_1 = 6 - x_3 - x_4 \tag{3}$$

Similarly, to eliminate x_1, add 2 times Eq. (1) to -3 times Eq. (2) to obtain

$$-7x_2 - 7x_3 = -42$$

or

$$x_2 = 6 - x_3 + 0x_4 \tag{4}$$

Equations (3) and (4) show that $x_3 = x_4 = 0$ defines a feasible corner, because, at this point, $x_1 \geq 0$ and $x_2 \geq 0$. To complete the dictionary, substitute Eqs. (3) and (4) into the original objective function:

$$\begin{aligned} P &= 3x_1 + 2x_2 + 3x_3 + 4x_4 \\ &= 3(6 - x_3 - x_4) + 2(6 - x_3) + 3x_3 + 4x_4 \\ &= 30 - 2x_3 + x_4 \end{aligned} \tag{5}$$

Equation (5) shows that the current corner is not optimal. We can increase P by increasing x_4. The maximum allowable increase in x_4 is 6, as determined by Eq. (3). When $x_4 = 6$, the variable x_1 falls to 0. Hence the new corner variables are x_3 and x_1. We leave it to the reader to verify that the dictionary at this corner reads:

$$x_4 = 6 - x_1 - x_3 \tag{6}$$

$$x_2 = 6 - x_3 \tag{7}$$

$$P = 36 - x_1 - 3x_3 \tag{8}$$

This dictionary shows that the optimal solution to Problem 3.7 is $P = 36$, achieved at the point $x_1 = x_3 = 0$; $x_2 = 6$, $x_4 = 6$.

PROBLEM 3.8.

$$\begin{aligned} &\text{Maximize} \quad P = 5x_1 + 6x_2 \\ &\text{subject to} \quad x_1 \geq 0, \quad x_2 \geq 0 \\ &\text{and} \quad 3x_1 + x_2 \leq 48 \\ &\qquad\quad 3x_1 + 4x_2 \leq 192 \end{aligned}$$

We convert the nontrivial constraints into equalities by introducing slack variables s_1 and s_2, as follows:

$$3x_1 + x_2 + s_1 = 48 \tag{1}$$

$$3x_1 + 4x_2 + s_2 = 192 \tag{2}$$

Equations (1) and (2) show that $x_1 = x_2 = 0$ defines a first feasible corner. The profit function indicates that P can be increased by increasing x_2. The maximum allowable increase in x_2 is 48, at which point both s_1 and s_2 are 0. *We are thus faced with an ambiguity, since either s_1 or s_2 can be the new corner variable.* The logic for choosing a new corner has *degenerated* in the sense that a definite set of instructions is no longer available.

Suppose we decide to make s_1 the new corner variable. Equation (1) is then the key to forming the new dictionary. Solving this equation for x_2 in terms of x_1 and s_1, we obtain

$$x_2 = 48 - s_1 - 3x_1 \tag{3}$$

From Eqs. (2) and (3) we see that

$$\begin{aligned} s_2 &= 192 - 3x_1 - 4x_2 \\ &= 192 - 3x_1 - 4(-s_1 - 3x_1 + 48) \\ &= 0 + 9x_1 + 4s_1 \end{aligned} \tag{4}$$

Observe that the constant in Eq. (4) has fallen to 0; this is because we decided to take s_1 as the new corner variable forcing s_2 to be a basic variable with value of 0 at the next corner. Substituting Eq. (3) into the old profit function gives

$$\begin{aligned} P &= 5x_1 + 6x_2 \\ &= 5x_1 + 6(-s_1 - 3x_1 + 48) \\ &= 288 - 13x_1 - 6s_1 \end{aligned} \tag{5}$$

Equation (5) shows that we reach optimality at the corner defined by $x_1 = s_1 = 0$. Here $P = 288$, and the values of the basic variables are $x_2 = 48$ and $s_2 = 0$. See Figure 29 for a sketch of the region of feasibility.

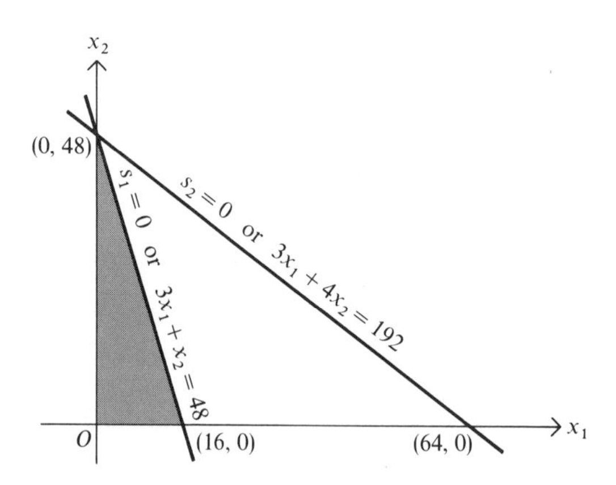

Figure 29

Whenever a basic variable turns 0, we say that the solution at hand is *degenerate*. A degenerate basic feasible solution may or may not be optimal. In this problem the optimal solution is degenerate. We stress that a degenerate solution is no different from an ordinary one. The phrase "degenerate" is a reminder that an ambiguity arose in the application of the simplex algorithm—in the case at hand, whether s_1 or s_2 should have been increased in moving away from the origin. We subsequently point out the possible dangers in an ambiguous algorithm.

PROBLEM 3.9. Maximize $P = 5x_1 + 6x_2 + 0(s_1) + 0(s_2) + 0(s_3)$

where all variables are nonnegative

and
$$3x_1 + x_2 + s_1 = 48 \quad (1)$$
$$3x_1 + 4x_2 + s_2 = 120 \quad (2)$$
$$3x_1 + 5x_2 + s_3 = 150 \quad (3)$$

Evidently $x_1 = x_2 = 0$ determines a feasible corner. The profit function indicates that P can be increased by increasing x_2. Equation (1) limits the increase in x_2 to 48, but both Eqs. (2) and (3) limit the increase in x_2 to 30. *Hence there is ambiguity in the choice of the next corner variable. Whether s_2 or s_3 is chosen as the next corner variable, the next basic feasible solution will be degenerate.*

Suppose we choose s_3 to be the next corner variable. The key equation for building the new dictionary is then Eq. (3), so we start the new dictionary with

$$x_2 = 30 - \tfrac{3}{5}x_1 - \tfrac{1}{5}s_3 \quad (4)$$

The reader can easily verify that the other equations in the dictionary are

$$s_1 = 18 - \tfrac{12}{5}x_1 + \tfrac{1}{5}s_3 \quad (5)$$
$$s_2 = 0 - \tfrac{3}{5}x_1 + \tfrac{4}{5}s_3 \quad (6)$$
$$P = 180 + \tfrac{7}{5}x_1 - \tfrac{6}{5}s_3 \quad (7)$$

Equation (6) shows that we are in the presence of a degenerate solution, because if we set $x_1 = s_3 = 0$, then the basic variable s_2 turns 0. Equation (7) motivates us to increase x_1 as much as possible, because the coefficient of x_1 in Eq. (7) is positive. *But Eq. (6) limits the increase in x_1 to 0.*

We may or may not be at optimality. To see clearly, let us continue the simplex routine as indicated by Eq. (7), that is, increase x_1 until s_2 falls to 0. The new corner variables will then be s_2 and s_3. The dictionary at this new corner can be obtained as usual. We leave it to the reader to verify that it reads:

$$x_1 = 0 - \tfrac{5}{3}s_2 + \tfrac{4}{3}s_3 \quad (8)$$
$$x_2 = 30 + s_2 - s_3 \quad (9)$$
$$s_1 = 18 + 4s_2 - 3s_3 \quad (10)$$
$$P = 180 - \tfrac{7}{3}s_2 + \tfrac{2}{3}s_3 \quad (11)$$

Note that both Eqs. (7) and (11) show profit to be 180 but in terms of different corner variables. Equation (11) indicates that we should increase s_3 as much as possible. Equation (8) puts no limit on the allowable increase in s_3. Equations (9) and (10), however, do limit s_3. The limiting equation is Eq. (10), which shows that s_3 can only increase to 6, at which point $s_1 = 0$.

The next group of corner variables is s_1 and s_2. The dictionary at this corner reads:

$$s_3 = 6 - \tfrac{1}{3}s_1 + \tfrac{4}{3}s_2 \tag{12}$$

$$x_1 = 8 - \tfrac{4}{9}s_1 + \tfrac{1}{9}s_2 \tag{13}$$

$$x_2 = 24 + \tfrac{1}{3}s_1 - \tfrac{1}{3}s_2 \tag{14}$$

$$P = 184 - \tfrac{2}{9}s_1 - \tfrac{13}{9}s_2 \tag{15}$$

Hence the optimal solution occurs at the corner defined by $s_1 = s_2 = 0$. At this corner, $P = 184$, $s_3 = 6$, $x_1 = 8$, $x_2 = 24$. It is *not* a degenerate solution.

See Figure 30 for a sketch of the region of feasibility.

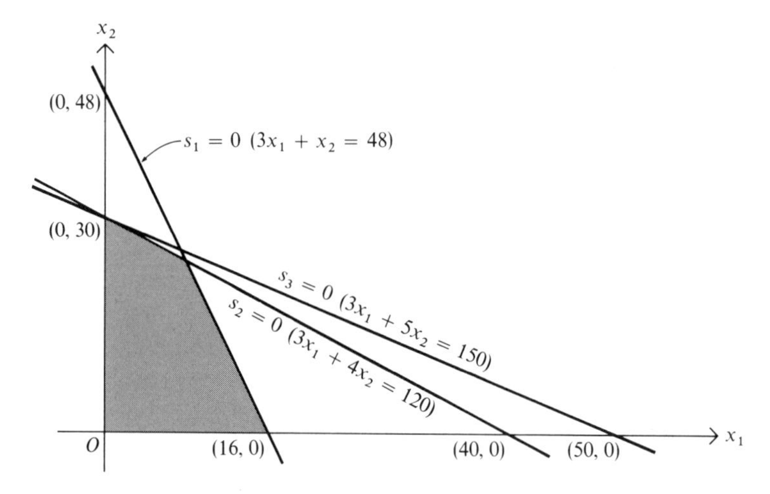

Figure 30

Note that the value of P remained unchanged as we passed from the second to the third dictionary in Problem 3.9. Could the simplex algorithm cycle endlessly, repeating alternative representations of the same nonoptimal feasible corner, so that P would never increase and the algorithm never terminate?

Examples have been constructed to show that the simplex algorithm, as we have formulated it, can actually cycle endlessly around a loop of nonoptimal corners. However, it can be amended slightly so as to prevent cycling. Since cycling is so rare in actual practice, the theoretical refinements needed to prevent cycling are not even written into the computer codes used in solving non-textbook-type problems, and we do not discuss them in this introductory text.

Here we emphasize the following aspects of cycling and termination:

(1) Cycling is no problem except when the calculations are carried out on an electronic computer.

(2) We cannot eliminate the possibility of cycling by simply giving the machine some specific set of rules to avoid ambiguity of choice. What is needed

is a set of rules *designed* to avoid cycling. Cycling occurs precisely because of the existence of the wrong systematic rules. In fact, if we instruct the computer to work in a random manner whenever it has a choice between two new corner variables, then cycling will almost always be avoided.

(3) Many standard computer programs do omit the rules that guarantee termination of the simplex algorithm. Such rules are necessary, however, if we are to use the simplex algorithm as a theoretical tool.

(4) Although cycling is rare, degeneracy is fairly common. For example, if a nontrivial constraint specifies a definite relationship between variables which does not involve a constant, for instance,

$$2x \leq y + z$$

then the initial dictionary will have some $b_i = 0$, arising from the slack-variable equivalent of the above inequality, namely, $s_i \geq 0$ and

$$2x - y - z + s_i = 0$$

(5) The simplex algorithm proceeds by moving from one feasible corner to a *neighboring* feasible corner where the objective function is never worse than previously. In the absence of degeneracy and cycling, the objective function will *strictly improve* until the optimum is attained, or until it is discovered that the objective function is unbounded—this must be so, because there are only a finite number of corners and, in the nondegenerate case, we can never revisit a corner.

What matters is that the objective function is always improving. This will occur under either of the following rules for choosing the next corner variable to be increased:

(a) Increase any current corner variable whose coefficient in the objective function is positive.

(b) Increase the current corner variable whose coefficient in the objective function is positive and as large as any other positive objective coefficient (we will usually follow this rule).

(6) The effectiveness of the algorithm is not so evident from our textbook problems. If a problem contains 30 variables and 10 nontrivial constraints, there are several million potential corners, but the simplex solution will usually involve only 20 or 30 dictionaries.

We stress that the relative efficiency of the simplex algorithm is an empirical fact and not a deduction from some underlying theory. Experience suggests that the number of dictionaries needed is about 3 times as great as the number of nontrivial constraints.

(7) In the present chapter we have assumed that all variables are nonnegative and that a first feasible corner (initial basic feasible solution) is apparent. In Chapter 6 we will see how to handle programs in which these assumptions are not satisfied.

Exercises

Solve the following linear programs first by the simplex algorithm, then by the corner-point method.

1.

$$\text{Maximize} \quad P = 5x_1 + 6x_2 + 7x_3$$
$$\text{where} \quad x_1 \geq 0, \quad x_2 \geq 0, \quad x_3 \geq 0$$
$$\text{and} \quad 3x_1 + 4x_2 + 50x_3 \leq 60$$

2.

$$\text{Maximize} \quad P = 5x_1 + 6x_2 + 7x_3$$
$$\text{where} \quad x_1 \geq 0, \quad x_2 \geq 0, \quad x_3 \geq 0$$
$$\text{and} \quad 3x_1 + 4x_2 + 50x_3 = 60$$

3. Compare the two preceding problems and their answers. Are the solutions degenerate?

4.

$$\text{Minimize} \quad C = x_1 + x_2 + x_3 + x_4$$
$$\text{where} \quad x_1 \geq 0, \quad x_2 \geq 0, \quad x_3 \geq 0, \quad x_4 \geq 0$$
$$\begin{aligned} \text{and} \quad x_1 + 2x_2 \qquad\quad + x_4 &\leq 14 \\ x_2 + x_3 + 3x_4 &\leq 6 \end{aligned}$$

5.

$$\text{Minimize} \quad C = x_1 + x_2 + x_3 + x_4$$
$$\text{where} \quad x_1 \geq 0, \quad x_2 \geq 0, \quad x_3 \geq 0, \quad x_4 \geq 0$$
$$\begin{aligned} \text{and} \quad x_1 + 2x_2 \qquad\quad + x_4 &= 14 \\ x_2 + x_3 + 3x_4 &= 6 \end{aligned}$$

(*Answer:* Optimum $C = 8$.)

6. Compare the two preceding problems and their answers.

7. Solve the following one-dimensional program by the simplex algorithm:

$$\text{Maximize} \quad P = 7x$$
$$\text{where} \quad x \geq 0$$
$$\begin{aligned} \text{and} \quad 3x &\leq 12 \\ 2x &\leq 6 \end{aligned}$$

8.

$$\text{Minimize} \quad Z = 18x_1 + 3x_2 + 22x_3 + 18x_4 + 7x_5$$

where all variables are nonnegative

$$\begin{aligned} \text{and} \quad 3x_1 + 2x_2 + x_3 \qquad\qquad\quad &= 8 \\ x_2 \qquad + 2x_4 + x_5 &= 5 \\ x_3 \qquad\quad + x_5 &= 10 \end{aligned}$$

9. Consider the following program:

$$\text{Maximize} \quad P = 2x_1 + x_2$$
$$\text{where} \quad x_1 \geq 0, \quad x_2 \geq 0$$
$$\text{and} \quad x_1 + 2x_2 \leq 10$$
$$x_1 + x_2 \leq 6$$
$$x_1 - x_2 \leq 2$$
$$x_1 - 2x_2 \leq 1$$
$$2x_1 - 3x_2 \leq 3$$

(a) Something peculiar happens at $x_1 = 3$, $x_2 = 1$. What is it?
(b) What does this indicate?
(c) Is there a feasible solution?
(d) Solve the given program by the simplex algorithm.

10. Solve Problem 2.14 by the simplex algorithm.

11. A company manufactures hexnuts, screws, and bolts. Daily production is governed by the following table of requirements and availabilities.

	Products			
Resources	*100 hexnuts*	*100 screws*	*100 bolts*	*Maximum availability*
Man-hours	4	2	2	24
Lathe-hours	1	0	1	4
Grinder-hours	0	1	3	8

Profits are \$6 per 100 hexnuts, \$4 per 100 screws, \$5 per 100 bolts. Use the simplex algorithm to determine the product mix which maximizes total profit.

12. Two types of summer homes are produced at a factory, types X_1 and X_2. Net gain per type X_1 is \$4000, per type X_2 is \$3000. The following facts limit production: (1) The cost of producing a home of type X_1 is \$8000; the cost for a home of type X_2 is \$10,000. (2) The maximum capital available for production during the period in question is \$36 million. (3) Market research indicates that the firm should not produce homes having a total selling price of more than \$48 million. (4) A home of type X_1 occupies 420 square yards of the floor space available for storage; a home of type X_2 occupies 330 square yards. (5) Plant storage space is limited to 100,000 square yards of floor space.

The objective is to determine the product mix which maximizes total profit. To this end, let x_1 denote the number of homes of type X_1 to be manufactured, and let x_2 denote the number of homes of type X_2. Accordingly show that:

(a) Production cost $= 8000x_1 + 10{,}000x_2$ dollars.

(b) $8x_1 + 10x_2 \leq 36{,}000$.

(c) $12x_1 + 13x_2 \leq 48{,}000$.

(d) $42x_1 + 33x_2 \leq 10{,}000$.

(e) Profit $= 4000x_1 + 3000x_2$ dollars.

Solve this program by the simplex algorithm.

13. Solve as a one-dimensional linear program: A moving company charges \$0.40 per pound for moving furniture from New York to Los Angeles and \$0.25 per pound for moving crates on the same trip across country. If at least one-third of each load must be furniture, and at least one-fifth of each load consists of crates, find the minimum and maximum cost per pound for a load.

14. *True or false?*

(a) The function $f(x)$ achieves its maximum at the same set of points where $-f(x)$ achieves its minimum.

(b) $\min f(x) = \max\,[-f(x)]$.

(c) $\max f(x) = -\min\,[-f(x)]$.

15. Clarify the following statement: A basic variable is not a corner variable, but a basic feasible solution is a corner solution.

4

SYSTEMS OF LINEAR EQUATIONS

4.1 INTRODUCTION

We interrupt our examination of the structure of linear programming and turn to the study of systems of linear equations. This will help to clarify the calculations performed in Chapter 3 and, at the same time, serve as preparation for the tableau version of the simplex algorithm to be discussed in Chapter 5.

Consider the following system of two equations in two unknowns:

$$3x + y = 7 \tag{1}$$

$$2x - y = 3 \tag{2}$$

To solve this system of equations means to find *all* pairs of numbers x, y which satisfy *both* Eqs. (1) and (2).

Adding Eqs. (1) and (2) allows us to *eliminate* y and thereby arrive at a relation involving only the one unknown x. Thus

$$\begin{array}{r} 3x + y = 7 \\ +2x - y = 3 \\ \hline 5x \phantom{{}-y} = 10 \end{array}$$

and so

$$x = 2$$

Now, if we *substitute* the value $x = 2$ into either Eq. (1) or (2) above, we find that $y = 1$.

We could equally as well have arrived at the value $y = 1$ by multiplying Eq. (1) by the number 2 and Eq. (2) by the number 3, obtaining the new system

$$6x + 2y = 14 \tag{1'}$$

$$6x - 3y = 9 \tag{2'}$$

If we subtract Eq. (2') from Eq. (1'), x will be eliminated; we obtain $5y = 5$, or $y = 1$, as before. Substitution of this value in either equation gives $x = 2$.

Hence if any numbers satisfy Eqs. (1) and (2), they must be $x = 2$ and $y = 1$. Furthermore, on substituting $x = 2$, $y = 1$ into the original system we get the identities

$$(3)(2) + 1 = 7$$

$$(2)(2) - 1 = 3$$

which actually proves that $x = 2$, $y = 1$ is a solution.

We solved the above system by the helter-skelter methods of high-school algebra, without asking ourselves whether our methods are mathematically legitimate. The following questions arise naturally:

(1) What operations on equations should be considered legitimate? In fact, what exactly is meant by the concept of *legitimate operation*?

(2) If by some process we do effectively get a solution to a system of linear equations, for example, the solution $x = 2, y = 1$ to the system above, when can we be sure that there are no other possible solutions, obtainable perhaps by some other method?

4.2 EQUIVALENCE OF EQUATIONS

Two equations are said to be *equivalent* if every solution of one is a solution of the other. For example, the equations

$$x^2 - x - 2 = 0 \qquad \text{and} \qquad 7(x - 2)(x + 1) = 0$$

are equivalent. So are the equations

$$2x - 3y + 1 = 0 \qquad \text{and} \qquad y = \tfrac{2}{3}x + \tfrac{1}{3}$$

Two equivalent equations are often considered to be the "same" equation. In particular, if every term of an equation is multiplied by the same *nonzero number*, the new equation is equivalent to the original.

Note that $x^2 = 25$ and $x = 5$ are *not* equivalent equations. If $x = 5$, it follows that $x^2 = 25$, but not conversely. This example suggests the following definition:

An equation E′ is a *consequence* of equation E if every solution of E is also a solution of E′. Thus $x^2 = 25$ is a consequence of $x = 5$, but not conversely. We can now say that two equations are equivalent if and only if each is a consequence of the other.

4.3 EQUIVALENT SYSTEMS OF LINEAR EQUATIONS

The general system of two linear equations in two unknowns can be denoted as follows:

$$a_1x + b_1y + c_1 = 0$$
$$a_2x + b_2y + c_2 = 0$$

DEFINITION 1. Any ordered couple (x_0, y_0), such that

$$a_1x_0 + b_1y_0 + c_1 = 0$$

and

$$a_2x_0 + b_2y_0 + c_2 = 0$$

is said to be *a solution* to the given system.

To illustrate, consider the system

$$x + 2y + 1 = 0$$
$$2x + 4y + 2 = 0$$

The couple $x = 1$, $y = -1$ is a solution, as is the couple $x = 3$, $y = -2$. In fact, the set of all solutions to this system is infinite, as we shall see later.

For another example, consider the system

$$3x + y - 7 = 0$$
$$2x - y - 3 = 0$$

This system has one and only one solution, namely, $x = 2$, $y = 1$.

The definition of a solution generalizes immediately to the case of m equations in n unknowns.

DEFINITION 2. Two systems of equations are *equivalent* if they have the same set of solutions, that is, if every solution of one system is a solution of the other.

For example, the following three systems are all equivalent:

$$\text{(a)} \begin{cases} 3x + y = 7 \\ 2x - y = 3 \end{cases} \qquad \text{(b)} \begin{cases} 5x \phantom{{}-y} = 10 \\ 2x - y = 3 \end{cases} \qquad \text{(c)} \begin{cases} x = 2 \\ y = 1 \end{cases}$$

The process of solving a linear system consists in replacing the given system successively by equivalent systems until a particularly simple system is reached. Any operation by which one may pass from a given system to an equivalent system is said to be *legitimate*. There are three legitimate operations, the so-called *elementary operations*, which are always sufficient to "solve" a given system of linear equations. (We use the word "solve" in the broad sense—in other words, if the solution to a system is not unique, or if the system has no solution, then the use of elementary operations will always tell us so.)

The elementary operations are of the following three types:

Type 1. The interchange of two equations.

Type 2. The multiplication of each term of an equation by a nonzero number.

Type 3. The addition of a multiple of one equation to another equation of the system.

Proof. An operation of type 1 certainly does not alter the solutions of the system, while an operation of type 2 merely replaces an equation by another equation equivalent to it. In considering operations of type 3, we must show that the systems

$$\begin{aligned} a_1x + b_1y + c_1 &= 0 \\ a_2x + b_2y + c_2 &= 0 \end{aligned} \qquad \text{and} \qquad \begin{aligned} a_1x + b_1y + c_1 &= 0 \\ a_2x + b_2y + c_2 + k(a_1x + b_1y + c_1) &= 0 \end{aligned}$$

are equivalent for every number k.

Let (x_1, y_1) denote a solution of the first system. Then, by definition,

$$\begin{aligned} a_1x_1 + b_1y_1 + c_1 &= 0 \\ a_2x_1 + b_2y_1 + c_2 &= 0 \end{aligned}$$

Hence,

$$a_2x_1 + b_2y_1 + c_2 + k(a_1x_1 + b_1y_1 + c_1) = 0 + k(0) = 0$$

for every k, and therefore (x_1, y_1) is a solution of the second system.

Conversely, let (x_2, y_2) be a solution of the second system. From the equations

$$\begin{aligned} a_1x_2 + b_1y_2 + c_1 &= 0 \\ (a_2x_2 + b_2y_2 + c_2) + k(a_1x_2 + b_1y_2 + c_1) &= 0 \end{aligned}$$

it follows immediately that (x_2, y_2) is also a solution of the first system. This proves that the two systems are equivalent.

To illustrate, we now apply the theory of equivalent systems to the first system encountered in this chapter:

$$\text{(I)} \qquad \begin{cases} 3x + y = 7 & (1) \\ 2x - y = 3 & (2) \end{cases}$$

This system is equivalent to the following system, obtained by adding Eq. (2) to Eq. (1), a type-3 operation:

$$\text{(I}'\text{)} \qquad \begin{cases} 5x \phantom{{}- y} = 10 \\ 2x - y = 3 \end{cases}$$

Using a type-2 operation, we transform system (I′) into the equivalent system

$$\text{(I}''\text{)} \qquad \begin{cases} x \phantom{{}- y} = 2 \\ 2x - y = 3 \end{cases}$$

Finally, we need only "substitute" $x = 2$ into the second equation of (I″) to get $4 - y = 3$, or $y = 1$. This "substitution" is really equivalent to carrying out the following elementary operations: Add -2 times the first equation in (I″) to the second equation to obtain

$$(\mathrm{I}''') \quad \begin{cases} x = 2 \\ -y = -1 \end{cases} \quad \text{or} \quad \begin{cases} x = 2 \\ y = 1 \end{cases}$$

We stress that since systems (I) and (I‴) are equivalent, they must have the same set of solutions. But (I‴) has the obvious and unique solution $x = 2$ *and* $y = 1$. *This tells us that* $x = 2$, $y = 1$ *gives all the possible solutions to (I).*

When dealing with systems of linear equations, we need not carry back our answers to the original system, provided, of course, that we have used only legitimate operations to isolate our answers. *But consider the following situation, where checking the solution is imperative:* Given the equation

$$+\sqrt{x^2 - 16} = \tfrac{3}{5}x$$

solve for x. Squaring both sides, we see that $x^2 - 16 = \frac{9}{25}x^2$, or $x^2 = 25$, whence $x = \pm 5$.

DEFINITION 3. A system of linear equations is said to be *diagonalized* if each equation contains a variable which appears in that equation with a coefficient of 1 and this variable appears in no other equation.

Thus, systems (A) and (B) below are diagonalized:

$$(\mathrm{A}) \quad \begin{cases} y = b \\ x = a \\ z = c \end{cases} \quad \text{and} \quad (\mathrm{B}) \quad \begin{cases} x_1 + 4x_4 = 1 \\ x_2 - 2x_4 = 2 \\ x_3 + 7x_4 = 3 \end{cases}$$

Note that in writing these systems we have omitted all terms such as $0(x)$, etc. Without these omissions, system (A) would read:

$$\begin{cases} 0(x) + 1(y) + 0(z) = b \\ 1(x) + 0(y) + 0(z) = a \\ 0(x) + 0(y) + 1(z) = c \end{cases}$$

A *diagonalized system* is both a system of equations and the "solution" to itself; therein lies its power as a prototype. (The general solution to system (B) is $x_1 = 1 - 4x_4$, $x_2 = 2 + 2x_4$, $x_3 = 3 - 7x_4$, where x_4 is arbitrary.)

Now consider the following systems:

$$(\mathrm{C}) \quad \begin{cases} 2x = 2 \\ x + 2y = 1 \end{cases} \quad \text{and} \quad (\mathrm{D}) \quad \begin{cases} x = 1 \\ 2x + 3y = 8 \\ 3x - y + 2z = 7 \end{cases}$$

Systems like (C) and (D) are said to be in *echelon* or *triangular form* because of the visual pattern exhibited by the unknowns. Such systems can be thought of as *almost solved.* They can be converted to the completely diagonalized form either by using elementary operations or by substituting back from an appropriate corner of the triangle. For example, the first equation of system (C) implies that $x = 1$, and, if $x = 1$ is substituted into the second equation, it can be seen immediately that $y = 0$. As an exercise the reader should now diagonalize system (D).

One more step-by-step example will help to fix these ideas: Consider the system

$$\text{(II)} \qquad \begin{cases} x - 2y + \ \ 5 = 0 \\ 3x + \ \ y - 13 = 0 \end{cases}$$

If we add twice the second equation to the first equation, we obtain the equivalent system

$$\text{(II}'\text{)} \qquad \begin{cases} 7x \qquad - 21 = 0 \\ 3x + y - 13 = 0 \end{cases}$$

Note that (II′) is in echelon form. To diagonalize completely, divide the first equation of (II′) by 7:

$$\text{(II}''\text{)} \qquad \begin{cases} x \qquad - \ \ 3 = 0 \\ 3x + y - 13 = 0 \end{cases}$$

This system is still in echelon form, but if we multiply the first equation by -3 and add the product to the second, we obtain the equivalent diagonalized system

$$\text{(II}'''\text{)} \qquad \begin{cases} x \qquad - 3 = 0 \\ \qquad y - 4 = 0 \end{cases}$$

4.4 TYPICAL PROBLEMS

The following examples are worked out in partial detail. The reader should try to solve them before studying the solutions.

Example (*a*). Consider the system

$$\text{(III)} \qquad \begin{cases} 2x + \ \ y = 2 \\ 6x + 3y = 7 \end{cases}$$

Multiply the upper equation by $\frac{1}{2}$ to obtain the equivalent system

$$\text{(III}'\text{)} \qquad \begin{cases} x + \frac{1}{2}y = 1 \\ 6x + 3y = 7 \end{cases}$$

If we add -6 times the first equation of (III′) to the second equation, we get

$$\text{(III}''\text{)} \qquad \begin{cases} x + \ \ \frac{1}{2}y = 1 \\ 0(x) + 0(y) = 1 \end{cases}$$

Systems (III) and (III″) are equivalent. But (III″) has *no solutions*, since it is impossible that $0(x) + 0(y) = 1$. Hence (III) has no solutions.

Example (*b*). Consider the system

$$\text{(IV)} \quad \begin{cases} 2x + y = 2 & (1) \\ 6x + 3y = 6 & (2) \end{cases}$$

By following the same sequence of elementary operations as in Example (a), it follows that (IV) is equivalent to

$$\text{(IV}'\text{)} \quad \begin{cases} x + \tfrac{1}{2}y = 1 & (1') \\ 0(x) + 0(y) = 0 & (2') \end{cases}$$

Equation (2′) of (IV′) is *not* impossible. In fact, it holds true for any x and any y. This means that the only conditions imposed on solutions to (IV′) are embodied in Eq. (1′). But this equation has an infinite number of solutions, for example, $x = 0$, $y = 2$; $x = 1$, $y = 0$; and so on. All we need do to find a solution is to give x a value, say m, and then solve for y, obtaining $y = 2 - 2m$. The "auxiliary letter" or *parameter* m is merely a linguistic device. To say that the solutions of (IV′) are all x and y satisfying $x + \frac{1}{2}y = 1$ is synonymous with saying that all our solutions can be given by the *parametric representation*

$$\begin{aligned} x &= m \\ y &= 2 - 2m \end{aligned}$$

Systems (IV) and (IV′) and the above parametric representation are all descriptions of the same straight line, every point of which is a solution to the system under study.

Example (*c*). Consider the system

$$\text{(V)} \quad \begin{cases} 3x - 2y + 4z = 9 & (1) \\ 2x + 5y - 3z = 13 & (2) \\ 7x + 8y - z = 36 & (3) \end{cases}$$

Multiply the first equation by $\frac{1}{3}$, then subtract 2 times the new first equation from the second, then subtract 7 times the new first equation from the third, obtaining

$$\begin{aligned} x \qquad - \tfrac{2}{3}y \qquad\qquad + \tfrac{4}{3}z &= 3 \\ (5 + \tfrac{4}{3})y + (-3 - \tfrac{8}{3})z &= 13 - 6 = 7 \\ (8 + \tfrac{14}{3})y + (-1 - \tfrac{28}{3})z &= 36 - 21 = 15 \end{aligned}$$

or, after simplification,

$$\text{(V}'\text{)} \quad \begin{cases} x - \tfrac{2}{3}y + \tfrac{4}{3}z = 3 & (1') \\ \qquad \tfrac{19}{3}y - \tfrac{17}{3}z = 7 & (2') \\ \qquad \tfrac{38}{3}y - \tfrac{31}{3}z = 15 & (3') \end{cases}$$

The variable x has been eliminated from all except Eq. (1′). Let us now eliminate y from Eq. (3′). To do this, we subtract twice Eq. (2′) from Eq. (3′), obtaining

$$(\mathrm{V}'')\quad \begin{cases} x - \frac{2}{3}y + \frac{4}{3}z = 3 & (1'') \\ \frac{19}{3}y - \frac{17}{3}z = 7 & (2'') \\ z = 1 & (3'') \end{cases}$$

System (V″) is in echelon form. To diagonalize, either continue to use elementary operations, or else "substitute back" from $z = 1$. By either method we obtain

$$(\mathrm{V}''')\quad \begin{cases} x = 3 \\ y = 2 \\ z = 1 \end{cases}$$

Example (*d*). Consider the system

$$\begin{aligned} x_1 \qquad\qquad + 4x_4 + 2x_5 &= 1 \\ x_2 \qquad - 2x_4 \qquad\quad &= 2 \\ x_3 + 7x_4 + 3x_5 &= 3 \end{aligned}$$

This system is already diagonalized. The variables x_4 and x_5 can be taken as arbitrary parameters. Once we fix the values of these parameters, the values of x_1, x_2, and x_3 are immediately apparent. (Note that two of the variables became parameters because we had two more variables than equations. A system of m equations in n variables, where $n > m$, will usually lead to $n - m$ parameters.)

Exercises

1. Show that the equations $x - y = 0$ and $x^2 - y^2 = 0$ are *not* equivalent. Which is a consequence of the other?

2. Are the equations $x - y = 0$ and $x^3 - y^3 = 0$ equivalent?

3. Which of the following systems of equations are equivalent:

(a) $\begin{aligned} x - 3y &= -7 \\ 2x + 2y &= 14 \end{aligned}$ (b) $\begin{aligned} x - 3y &= -7 \\ 8y &= 28 \end{aligned}$ (c) $\begin{aligned} x &= \tfrac{7}{2} \\ y &= \tfrac{7}{2} \end{aligned}$

4. Solve the following systems of equations by transforming them into equivalent diagonalized systems:

(a) $\begin{aligned} 6x - 2y + 10 &= 0 \\ 4x + 3y - 2 &= 0 \end{aligned}$ (b) $\begin{aligned} x + 2y &= 0 \\ -2x + 3y &= 0 \end{aligned}$

(c) $\begin{aligned} x - 3y &= -7 \\ 2x - 6y &= -14 \end{aligned}$ (d) $\begin{aligned} x - 3y &= -7 \\ 2x - 6y &= -15 \end{aligned}$

5. Show that the following systems are equivalent:

(a) $\begin{aligned} 5x + 2y + z &= 36 \\ x + 7y + 3z &= 63 \\ 2x + 3y + 8z &= 81 \end{aligned}$ (b) $\begin{aligned} x &= 3.6 \\ y &= 5.4 \\ z &= 7.2 \end{aligned}$

6. Describe all solutions to the following system:

$$\begin{aligned} x_1 \qquad\qquad + x_4 + 5x_5 &= 10 \\ x_3 + 2x_4 + 6x_5 &= 11 \\ x_2 \qquad + 3x_4 \qquad &= 12 \end{aligned}$$

7. *True or false?* The equation $0(x) + 0(y) + 0(z) = 0$ is a consequence of $x + 2y + 3z = 6$.

8. Clarify the following statement: The use of elementary row operations does not affect the informational content of a system of linear equations.

4.5 RECTANGULAR SYSTEMS

Most of the previous illustrations involve "square" systems, that is, systems with the same number of equations and unknowns. Whenever the number of equations is *less than* the number of unknowns, we call the system *rectangular*, or, more precisely, *flat rectangular*.

Consider the system

$$\text{(VI)} \qquad \begin{cases} 3x + y + r = 48 \\ x + 3y + 2r = 48 \end{cases}$$

containing two equations in three unknowns. Such flat rectangular systems are often said to be *underdetermined*, because, in order to determine unique values of x, y, and r, we ordinarily should have three equations, whereas we only have two. One way out of this difficulty is to assign values arbitrarily to one of the three unknowns, say r, and then attempt to solve for the remaining unknowns. Thus, if we let $r = 0$ in system (VI), then this system reduces to

$$\begin{aligned} 3x + y &= 48 \\ x + 3y &= 48 \end{aligned}$$

which implies $x = 12$, $y = 12$. *This means that* $x = 12$, $y = 12$, $r = 0$ *is a solution of system* (*VI*). If we let $r = 8$, then system (VI) reduces to

$$\begin{aligned} 3x + y &= 40 \\ x + 3y &= 32 \end{aligned}$$

which implies $x = 11$, $y = 7$; *hence* $x = 11$, $y = 7$, $r = 8$ *is another solution of* (*VI*).

By proceeding as above, we can find an infinite number of solutions to (VI), one solution for every choice of r. Such a procedure is, however, quite inane; a much wiser procedure is to think of r not as an unknown, but as a constant, or parameter, which can be specified later. Then we can immediately solve for the unknowns x and y in terms of r. Thus, transposing r to the right-hand side of (VI), we obtain

$$\text{(VI}'\text{)} \qquad \begin{cases} 3x + y = 48 - r \\ x + 3y = 48 - 2r \end{cases}$$

Subtracting the second equation from 3 times the first gives

$$(\text{VI}'')\qquad \begin{cases} 8x = 96 - r \\ x + 3y = 48 - 2r \end{cases}$$

and hence

$$(\text{VI}''')\qquad \begin{cases} x = 12 - \frac{1}{8}r \\ y = 12 - \frac{5}{8}r \\ r \text{ arbitrary} \end{cases}$$

System (VI‴) describes the totality of solutions to the original system (VI). For any given choice of r, it gives the corresponding values of x and y.

It is important to realize that we could just as well have solved the given system for x and r in terms of y, or for y and r in terms of x. For example, to solve for x and r in terms of y, we start from

$$3x + r = 48 - y$$
$$x + 2r = 48 - 3y$$

and finally obtain the solution

$$(\text{VI}^*)\qquad \begin{cases} x = \dfrac{48 + y}{5} \\ r = \dfrac{96 - 8y}{5} \\ y \text{ arbitrary} \end{cases}$$

Solutions (VI‴) and (VI*) are different representations of the same set of solutions. They only differ in the parameter by means of which they are expressed. In solution (VI‴) we get the particular solution $x = 12$, $y = 12$, $r = 0$ by stipulating that $r = 0$. This solution follows from (VI*) when we stipulate that $y = 12$.

To conclude this section, we point out the power and generality of the theory of equivalent systems: It can be employed on any number of linear equations, involving any number of unknowns; it has the capacity to answer questions concerning the nature of a solution, and at the same time it yields all solutions (if any exist).

Exercises

1. Solve the following system in terms of x:

$$x + y + z = 4$$
$$2x - y + 3z = 8$$

2. Consider the system of equations

$$x + 3y - 2z + t = 10$$
$$x - 2y + z - 2t = 5$$

As z and t take on the values in the following table, find the corresponding values of x and y.

z	0	1	2	3	4	5
t	1	2	13	4	14	15

(*Hint:* If you decide to solve six sets of equations, you have missed the point.)

3. Given the system

$$x - y + z = 0$$
$$2x + y - 2z = 0$$

(a) Show that $x = 1, y = 4, z = 3$ is a solution.

(b) Find a *complete* solution by solving for x and y in terms of z.

(*Solution:* $x = \frac{1}{3}z,\ y = \frac{4}{3}z$.)

4. Find the complete solution to the system in Exercise 3 in terms of x as parameter.

5. Find all solutions to the following system (see Exercise 11, page 49):

$$4x_1 + 6x_2 + 12x_3 = 22$$
$$6x_1 + 9x_2 + 18x_3 = 33$$
$$2x_1 + 3x_2 + 6x_3 = 11$$

4.6 RECTANGULAR SYSTEMS AND LINEAR PROGRAMMING

As we saw in Chapter 3, the simplex method for solving linear programs requires different dictionaries, that is, different ways of expressing the general solution of a flat rectangular system of constraints. Recall the following system

$$\text{(VII)} \quad \begin{cases} 3x + y + r = 48 & (1) \\ x + 3y + s = 48 & (2) \end{cases}$$

This system occurred in Chapter 3, where r and s were introduced to take up slack in the inequalities $3x + y \leq 48$, $x + 3y \leq 48$. *Note that it is diagonalized in the variables r and s. Each of these variables appears in one of the equations with the coefficient* 1 *and in the other equation with the coefficient* 0. It is this fact which enables us to solve (VII) immediately for r and s in terms of x and y. Merely by transposition, we have

$$r = 48 - 3x - y$$
$$s = 48 - x - 3y$$

We have already seen how to solve (VII) for x and y in terms of r and s—any of several methods and formats will do the job. In linear programming it is usually the standard practice to keep all variables on the left-hand side of the equation. As an illustration, we shall solve (VII) for x and y in terms of r and s, adhering to that convention. We must, by elementary operations, create a diagonal in x and y, that is, make x appear in one and only one equation and make y appear only in the other equation. Note that x already appears in Eq. (2) with the coefficient 1. Thus we can begin diagonalizing by subtracting 3 times Eq. (2) from Eq. (1), obtaining

$$\text{(VII}')\qquad \begin{cases} \quad\; - 8y + r - 3s = -96 & (1') \\ x + 3y \qquad\; + \; s = 48 & (2') \end{cases}$$

Dividing Eq. (1′) by -8 gives

$$\text{(VII}'')\qquad \begin{cases} \qquad\; y - \frac{1}{8}r + \frac{3}{8}s = 12 & (1'') \\ x + 3y \qquad\quad + \; s = 48 & (2'') \end{cases}$$

To complete our task we need only "annihilate" the $3y$ in Eq. (2″). To this end, subtract 3 times Eq. (1″) from Eq. (2″), obtaining the diagonalized system

$$\text{(VII*)}\qquad \begin{cases} \quad\; y - \frac{1}{8}r + \frac{3}{8}s = 12 \\ x \qquad + \frac{3}{8}r - \frac{1}{8}s = 12 \end{cases}$$

4.7 TABLEAU NOTATION

Much of the repetition involved in solving systems of equations can be avoided by use of *tableaux*, wherein column headings, or *positions*, play the role usually played by the symbols representing unknowns in equations.

Consider the system

$$\text{(VIII)}\qquad \begin{cases} 3x + y = 7 \\ 2x - y = 3 \end{cases}$$

We can write this system *positionally* as

x	y	‖
3	1	7
2	−1	3

(VIII)

The double vertical bar in the tableau is synonymous with the equals sign. Henceforth, we shall call it the *equality bar*. Row 1 of the tableau is, of course, synonymous with $3x + y = 7$, and so on. This tableau, or positional, notation is sometimes referred to as the method of "detached coefficients," for reasons which are self-evident.

All the *elementary operations* we performed on our old systems can just as well be carried out on the *rows* of tableaux, namely,

(1) Interchanging two rows
(2) Multiplying a row by a constant
(3) Adding a multiple of one row to another

Thus tableau (VIII) is *equivalent* to

x	y	
5	0	10
2	−1	3

(VIII′)

We passed from tableau (VIII) to tableau (VIII′) by adding the bottom row of tableau (VIII) to the top row. Now, if we divide the top row of tableau (VIII′) by 5, we get

x	y	
1	0	2
2	−1	3

(VIII″)

Tableau (VIII″) is in echelon form. To push to complete diagonalization, subtract twice the top row of tableau (VIII″) from the bottom row, obtaining

x	y	
1	0	2
0	−1	−1

(VIII*)

Tableau (VIII*), spelled out, gives $x = 2$, $-y = -1$, that is, $x = 2$, $y = 1$. *We urge the reader to practice working with tableaux by re-solving in tableau notation several of the systems previously solved.*

4.8 CONTROL SUMS

Consider tableau (VIII) of the previous section, namely,

x	y	
3	1	7
2	−1	3

We can control, or check, all elementary operations on such tableaux by adjoining a dummy column (labeled n for *nullify*), the entries of which are chosen so as to make all row sums equal to 0. In this tableau, the first row sum is 11, the second row sum is 4. Hence, we would enter −11 and −4 in the n column:

x	y		n
3	1	7	−11
2	−1	3	−4

The advantage of using a nullified tableau lies in the fact that such a tableau must remain nullified after any elementary operation performed on it; for if we multiply a nullified row by a constant k it remains nullified, while the sum of two nullified rows must again be nullified. *We therefore have a check on the work involved in applying elementary operations to a nullified tableau. If ever it ceases to be nullified, an error has been made somewhere.*

We illustrate by studying tableau (VIII) using the method of nullification. We begin with the preceding nullified tableau. Adding the bottom row to the top row, we obtain

x	y		n
5	0	10	−15
2	−1	3	−4

Dividing the top row by 5 gives us

x	y		n
1	0	2	−3
2	−1	3	−4

Subtracting twice the top row from the bottom row, we obtain the final tableau, which is still nullified, showing that we probably made no mistakes:

x	y		n
1	0	2	−3
0	−1	−1	+2

Exercises

1. Solve all the systems studied in Section 4.4 using tableaux, with or without a nullification column as check.

2. Rework all the exercises following Section 4.5 using tableau notation.

4.9 RECORDING OPERATIONS PERFORMED

In manipulating a system of linear equations by elementary row operations it is often useful to record at the side of a given tableau the history of the manipulations that led from the previous tableau to the current tableau. To illustrate, consider the passage from Tableau 1 below to Tableau 2:

Tableau 1

x	y	s_1	s_2		
3	1	1	0	8	r_1
4	2	0	1	4	r_2

Tableau 2

x	y	s_1	s_2		
1	$\frac{1}{3}$	$\frac{1}{3}$	0	$\frac{8}{3}$	$R_1 = \frac{1}{3}r_1$
0	$\frac{2}{3}$	$-\frac{4}{3}$	1	$-\frac{20}{3}$	$R_2 = r_2 - 4R_1$

The notation $R_1 = \frac{1}{3}r_1$ tells us that to form the first row of Tableau 2 we divided the first row of Tableau 1 by 3. $R_2 = r_2 - 4R_1$ tells us that to form the new second row we subtracted 4 times the new first row from the old second row.

4.10 UNIT COLUMNS

Consider Tableau 1 in Section 4.9. Columns like s_1 and s_2 are said to be *unit columns.* In general, a column containing the number 1 in a certain row and zeros everywhere else is said to be a *unit column.* Two unit columns are called *distinct* if the entry 1 appears in different rows of these columns. Thus, the tableau

x_1	x_2	x_3	x_4	
1	0	0	a_1	b_1
0	0	1	a_2	b_2
0	1	0	a_3	b_3

contains three distinct unit columns.

Clearly, if a tableau contains n rows, to say that the corresponding system is diagonalized is synonymous with saying that it contains n distinct unit columns.

4.11 THE PIVOT RULES

We have, until now, reduced the study of systems of linear equations to the row manipulation of tableaux (with or without control sums as a check, with or without a system for recording the sequence of operations performed). The manipulation of tableaux can be streamlined even further by applying the *pivot rule*. Consider the passage from Tableau 1 to Tableau 2:

Tableau 1

x	y	z		
Ⓐ a (circled)	b	c	f	r_1
A	B	C	F	r_2

Tableau 2

x	y	z		
1	$\frac{b}{a}$	$\frac{c}{a}$	$\frac{f}{a}$	$R_1 = \frac{1}{a} r_1$
0	$B - \frac{bA}{a}$	$C - \frac{cA}{a}$	$F - \frac{fA}{a}$	$R_2 = r_2 - AR_1 = r_2 - \frac{A}{a} r_1$

We formed row 1 of Tableau 2 by dividing row 1 of Tableau 1 by the number a. And we formed row 2 of Tableau 2 by subtracting A/a times row 1 of Tableau 1 from row 2 of Tableau 1. This is the way to convert the x column into a unit column, with the coefficient 1 appearing in row 1 of the x column. And, of course, creating (distinct) unit columns by elementary operations is equivalent to diagonalizing, or solving for certain variables in terms of others.

The creation of a unit column under the x heading can be described by the following *pivot rules:*

(1) *Circle the entry in the x column which is to become* 1. This entry is called the *pivot*. *Clearly, the pivot must be a nonzero entry*. The row and column which contain the pivot are called the pivot row and column. Any entry which lies neither in the pivot row nor pivot column is called an *off-pivot entry*. The new x column will be a unit column, with 1 in the place of the pivot.

(2) The pivot row is transformed by dividing each of its entries by the pivot.

(3) To find the new entry in any off-pivot position we must subtract from the old entry an "adjustment" term which is best held in mind by invoking the following picture:

This picture is a rectangle, with the pivot entry a at one vertex and the off-pivot entry C at the diagonally opposite vertex. The other entries in the rectangle, namely, A and c, are said to be the *complementary*, or *adjacent*, entries.

In the new tableau the entry in the same position as C will be

$$C - \frac{cA}{a}$$

The reader should memorize this result. In words:

New off-pivot entry = (old off-pivot entry) − (adjustment term)
Adjustment term = (inverse of pivot) × (product of complementary entries)

To illustrate the pivot rules, we return to a system previously studied by other methods:

$$\begin{aligned} 3x + y + r \quad &= 48 \\ x + 3y \quad + s &= 48 \end{aligned}$$

We diagonalize this system in the variables x and y, using tableau notation and pivot methods. Accordingly, consider

Tableau 1

x	y	r	s	
③	1	1	0	48
1	3	0	1	48

We begin by converting the x column into a unit column, with the number 1 in row 1. Hence the pivot must be the entry circled. Column 1 and row 1 of the new tableau can be formed quite easily, using pivot rules (1) and (2). The off-pivot entries are, in order,

$$3, \quad 0, \quad 1, \quad 48$$

The new, adjusted off-pivot entries are

$$3 - \frac{(1)(1)}{3}, \quad 0 - \frac{(1)(1)}{3}, \quad 1 - \frac{(0)(1)}{3}, \quad 48 - \frac{(48)(1)}{3}$$

that is, $\frac{8}{3}, \quad -\frac{1}{3}, \quad 1, \quad 32$

This gives us Tableau 2:

Tableau 2

x	y	r	s	
1	$\frac{1}{3}$	$\frac{1}{3}$	0	16
0	($\frac{8}{3}$)	$-\frac{1}{3}$	1	32

To complete the diagonalization in x and y we must pivot in row 2 of the y column. The new row 2 is found by multiplying each entry in row 2 of Tableau 2 by $\frac{3}{8}$. The off-pivot entries in Tableau 2 are

$$1, \quad \frac{1}{3}, \quad 0, \quad 16$$

The adjusted off-pivot entries must be

$$1 - \frac{0(\frac{1}{3})}{\frac{8}{3}}, \quad \frac{1}{3} - \frac{\frac{1}{3}(-\frac{1}{3})}{\frac{8}{3}}, \quad 0 - \frac{1(\frac{1}{3})}{\frac{8}{3}}, \quad 16 - \frac{\frac{1}{3}(32)}{\frac{8}{3}}$$

or

$$1, \quad \tfrac{3}{8}, \quad -\tfrac{1}{8}, \quad 12$$

Thus we obtain the following diagonalized system.

Tableau 3

x	y	r	s	
1	0	$\frac{3}{8}$	$-\frac{1}{8}$	12
0	1	$-\frac{1}{8}$	$\frac{3}{8}$	12

Exercises

1. Do all the exercises following Sections 4.4 and 4.5 using the pivot method.

2. Solve the following systems by any method:

$$\text{(I)} \quad \begin{cases} a + 2.000b = 3 \\ a + 2.001b = 1 \end{cases} \qquad \text{(II)} \quad \begin{cases} a + 2.000b = 3 \\ a + 1.999b = 1 \end{cases}$$

(a) Are systems (I) and (II) "alike"?
(b) Are the solutions to systems (I) and (II) alike?
(c) If we solve either system (I) or (II) by pivoting, can we say anything about the relative sizes of the pivots?

3. Obtain the solutions of

$$\begin{aligned} x \qquad + 1.52y &= 1 \\ 2x + (3.05 + \delta)y &= 1 \end{aligned}$$

for the following values of δ,

(a) -0.2 (b) -0.1 (c) 0 (d) 0.1 (e) 0.2

(*Hint:* Solve for x and y in terms of δ as a parameter, then substitute in the relevant numerical values of δ.)

4. Solve by pivoting:

(a)
$$\begin{aligned} x + y + z &= 1 \\ 2x + 2y + 2z &= 3 \end{aligned}$$

(b)
$$\begin{aligned} x + y + z &= 1 \\ 2x + 2y + 2z &= 3 \\ 5x - y + z &= 7 \end{aligned}$$

(c)
$$\begin{aligned} x + y + z &= 1 \\ 2x + 2y + 2z &= 2 \\ 3x + 3y + 3z &= 3 \end{aligned}$$

5. We pivot on the nonzero element a_{ij} in row i, column j of a tableau. Show that:

(a) If column j contains a zero in row r, then all of row r is unchanged by the pivot operation.
(b) If row i contains a zero in column k, all of column k is unchanged.

5

THE SIMPLEX TABLEAU

In this chapter we begin to write all programs in tableau format. By reconsidering some problems previously studied in Chapter 3, we transform the "simplex rules for revising a dictionary" into an equivalent set of instructions for revising a tableau. *Throughout this chapter we only consider programs for which a first feasible corner is immediately apparent.*

Let us reconsider a problem previously solved using the simplex algorithm in dictionary form (see Problem 3.1).

PROBLEM 5.1. Maximize $P = 3x + 6y$

subject to $x \geq 0, \quad y \geq 0$

and

$$3x + y \leq 48$$

$$x + 3y \leq 48$$

By introducing the nonnegative slack variables r and s we can transform inequalities (1) and (2) into the equalities

$$3x + y + r = 48 \tag{1'}$$

$$x + 3y + s = 48 \tag{2'}$$

This program can be expressed in tableau form, as follows:

Tableau 1

x	y	r	s	
3	1	1	0	48
1	3	0	1	48
3	6	0	0	P

Row 1 of this tableau corresponds to Eq. (1′), row 2 to Eq. (2′), and row 3 to the objective function. (It is implicitly understood that all variables must be nonnegative and that the objective is to maximize P.)

We refer to such a tableau as a *simplex tableau* and, in referring to its parts, we employ the following terms:

(1) The *right-hand column*, or *stub*, is the column of entries to the right of the equality bar.

(2) The last row is called the *objective row*. All rows other than the objective row will be called *basic rows*.

Except for the fact that all variables appear to the left of the equality bar, Tableau 1 is the exact analog of Dictionary 1, as defined in Chapter 3. Note that, in Tableau 1, the *noncorner*, or *basic*, variables appear as the headings of the *distinct unit columns*. They are r and s. The other column headings correspond to the nonbasic, or corner, variables.

It is very convenient to use the left margin of a tableau to record the current basic variable associated with each row. Thus, Tableau 1 can be rewritten as

	x	y	r	s	
r	3	1	1	0	48
s	1	3	0	1	48
	3	6	0	0	P

Since P is expressed solely in terms of the nonbasic variables, we are now ready to consider ways of passing to a more profitable solution. The procedure described below is nothing other than the "simplex algorithm for revising a dictionary," reexpressed in a language adapted to tableaux.

In order to pass from one feasible corner (basic feasible solution) to another, we proceed as follows:

Step 0. Make sure that the objective function is expressed solely in terms of the corner variables, that is, *adjust* the objective row if necessary.

Step 1. Examine the objective row for a corner variable to increase. If there is a positive entry in the adjusted objective row then revision is worthwhile. This is the case in Tableau 1. We place a vertical arrow ↑ under the largest positive coefficient in the objective row to show that we have decided to increase y. We call the y column the *pivot column*. In the next tableau y will be a basic variable. For this reason y is called the *incoming variable*.

Step 2. Determine the maximum possible increase in the incoming variable. Recall from Chapter 3 that when a nonbasic variable is increased, the current basic variables usually decrease. The limit beyond which we cannot increase the incoming variable is determined by the first current basic variable which falls to 0.

In Chapter 3, to determine the maximum possible increase in y, we examined the following equations (each of which represents a basic row of Tableau 1, with the nonincoming corner variable x set equal to 0):

$$y + r \qquad = 48 \tag{1''}$$

$$3y \qquad + s = 48 \tag{2''}$$

Equation (1″) permits y to reach 48, at which point r falls to 0. Equation (2″) permits y to reach $\frac{48}{3} = 16$, at which point s falls to 0. Since s falls to 0 first, this variable becomes nonbasic in the revised tableau. We say that s is the *outgoing variable* (outgoing in the sense that it will not be listed in the left margin of the next tableau).

The outgoing variable can be determined by direct examination of Tableau 1. Consider the entries in the basic rows of the pivot column. Divide each such entry into the right-hand constant in the same row and record the quotients obtained, in parentheses, to the right of the stub. The row determined by the *smallest positive quotient* is marked with a horizontal arrow. We call it the *pivot row*—it is the row of the outgoing basic variable. We call the entry which lies in both the pivot column and pivot row the *pivot* (from the current tableau to the following tableau). We always circle the pivot, as in Tableau 1 below:

Tableau 1

	x	y	r	s		
r	3	1	1	0	48	(48)
s	1	③	0	1	48	←(16)
	3	6 ↑	0	0	P	

Later we examine in greater detail the process of choosing the pivot. At this point we stress that if we *pivot on the circled pivot entry* of Tableau 1, the resulting tableau will represent the dictionary at the new corner. Thus we obtain:

Tableau 2

	x	y	r	s	
r	$\frac{8}{3}$	0	1	$-\frac{1}{3}$	32
y	$\frac{1}{3}$	1	0	$\frac{1}{3}$	16
	1	0	0	-2	$P - 96$

Pivoting is nothing more than a neat way of performing elementary row operations. Passage from Tableau 1 to Tableau 2 via pivoting is simply another way of performing the algebraic calculations needed to transform Dictionary 1 into Dictionary 2. And the heart of this transformation is choosing the new basic (incoming) variable and the new corner (outgoing) variable.

In general, if the reader is ever in doubt as to what information is contained in a tableau, he should write out the rows in dictionary form. Thus the profit row of Tableau 2, when written out, reads

$$1(x) + 0(y) + 0(r) - 2s = P - 96$$

Hence profit can be further increased by increasing x.

The arrows and quotients below indicate how to determine the pivot entry of Tableau 2.

Tableau 2

	x	y	r	s		
r	$\frac{8}{3}$ (circled)	0	1	$-\frac{1}{3}$	32	←(12)
y	$\frac{1}{3}$	1	0	$\frac{1}{3}$	16	(48)
	1 ↑	0	0	−2	$P - 96$	

After pivoting on Tableau 2, we obtain

Tableau 3

	x	y	r	s	
x	1	0	$\frac{3}{8}$	$-\frac{1}{8}$	12
y	0	1	$-\frac{1}{8}$	$\frac{3}{8}$	12
	0	0	$-\frac{3}{8}$	$-\frac{15}{8}$	$P - 108$

The coefficients in the objective row of this tableau are all nonpositive, which shows that we are at optimality. The stub indicates the optimum values of P and of the basic variables, namely, $P = 108$, $x = 12$, $y = 12$.

The reader should study carefully the tableaux of Problem 5.1 and compare the process of revising a tableau with the equivalent way of passing to a better dictionary. He should understand clearly that these two procedures differ only in form. In substance they are one and the same, the "simplex algorithm" itself.

In order to show more clearly the parallelism between the dictionary form and the tableau form of the simplex algorithm, we solve the following problem in "parallel columns," dictionary form to the left and tableau form to the right.

PROBLEM 5.2. Maximize $E = 4x + 3y + 7z$

subject to $x \geq 0, \quad y \geq 0, \quad z \geq 0$

and
$$x + 3y + 2z \leq 120 \tag{1}$$
$$2x + y + 3z \leq 120 \tag{2}$$

See Problem 2.11 for the corner-point solution of this problem. Here the program is solved by the simplex algorithm, using s_1 and s_2 to denote the slack in Eqs. (1) and (2), respectively.

CORNER 1: $x = y = z = 0$

Dictionary 1:

$$s_1 = 120 - x - 3y - 2z \quad (1)$$
$$s_2 = 120 - 2x - y - 3z \quad (2)$$
$$E = 4x + 3y + 7z \quad (3)$$

The algorithm instructs us to increase z as much as possible, and yet not violate the nonnegativity constraint on each variable. Equation (1) shows that we can increase z to 60, at which point s_1 will have fallen to 0. Equation (2) shows that if we increase z beyond 40, then s_2 will become negative.

If both s_1 and s_2 are to remain non-negative, we can only increase z to 40. Equation (2) is the limiting constraint, and s_2 will be the new corner variable. Hence the next dictionary will be expressed in terms of x, y, and s_2.

The key to building Dictionary 2 is the limiting constraint of Dictionary 1, namely, Eq. (2). It is easy to solve this equation for z in terms of x, y, and s_2. Furthermore, the new expression for z allows us to complete the passage from Dictionary 1 to Dictionary 2. If we substitute the new form of z into the old expressions for s_1 and E, we obtain the new expressions for s_1 and E.

The details of substitution can be found in Chapter 3. The end result is the dictionary which corresponds to corner 2, as shown on the next page.

Tableau 1:

	x	y	z	s_1	s_2		
s_1	1	3	2	1	0	120	(60)
s_2	2	1	③	0	1	120	←(40)
	4	3	7 ↑	0	0	E	

The objective row tells us to increase z as much as possible; that is, the pivot column is the z column.

The smallest "quotient in parentheses" is 40 and hence the pivot row is row 2. The pivot of Tableau 1 is the circled entry which lies in both the pivot column and pivot row. If we pivot on this entry, the unit columns of Tableau 2 will be the z and s_1 columns; that is, z and s_1 will be the new basic variables, and thus x, y, and s_2 will be the new corner variables.

After pivoting as indicated, we obtain Tableau 2, as shown on the next page.

CORNER 2: $x = y = s_2 = 0$

Dictionary 2:

$$s_1 = 40 + \tfrac{1}{3}x - \tfrac{7}{3}y + \tfrac{2}{3}s_2 \quad (1)$$
$$z = 40 - \tfrac{2}{3}x - \tfrac{1}{3}y - \tfrac{1}{3}s_2 \quad (2)$$

$$E = 280 - \tfrac{2}{3}x + \tfrac{2}{3}y - \tfrac{7}{3}s_2 \quad (3)$$

We can increase E by increasing y. Equation (2) of the current dictionary shows that if we increase y beyond 120, z will fall below 0. Equation (1) shows that if we increase y beyond $\frac{120}{7}$, then s_1 will become negative. Since the simplex algorithm never permits a variable to become negative, we can only increase y to $\frac{120}{7}$, at which point $s_1 = 0$.

Hence the next set of corner variables is x, s_1, and s_2. The key equation for constructing the next dictionary is the limiting constraint of Dictionary 2, namely, Eq. (2). Details can be found in Chapter 3.

Tableau 2:

	x	y	z	s_1	s_2		
s_1	$-\frac{1}{3}$	($\frac{7}{3}$)	0	1	$-\frac{2}{3}$	40	←($\frac{120}{7}$)
z	$\frac{2}{3}$	$\frac{1}{3}$	1	0	$\frac{1}{3}$	40	(120)
	$-\frac{2}{3}$	$\frac{2}{3}$ ↑	0	0	$-\frac{7}{3}$	$E - 280$	

The objective row of Tableau 2 instructs us to pivot in the y column. To determine the pivot row we divide the coefficients in the first two rows of the y column into the constants, obtaining the limiting quotients shown.

The smallest of these quotients shows that row 1 is the pivot row. The pivot of Tableau 2 is the circled entry in column y, row 1. We obtain Tableau 3 by pivoting on this entry.

CORNER 3: $x = s_1 = s_2 = 0$

Dictionary 3:

$$y = \tfrac{120}{7} + \tfrac{1}{7}x - \tfrac{3}{7}s_1 + \tfrac{2}{7}s_2 \quad (1)$$
$$z = \tfrac{240}{7} - \tfrac{5}{7}x + \tfrac{1}{7}s_1 - \tfrac{3}{7}s_2 \quad (2)$$

$$E = 291\tfrac{3}{7} - \tfrac{4}{7}x - \tfrac{2}{7}s_1 - \tfrac{15}{7}s_2 \quad (3)$$

We are now at a maximum, because any increase in x, s_1, or s_2 can only reduce E.

Tableau 3:

	x	y	z	s_1	s_2	
y	$-\frac{1}{7}$	1	0	$\frac{3}{7}$	$-\frac{2}{7}$	$\frac{120}{7}$
z	$\frac{5}{7}$	0	1	$-\frac{1}{7}$	$\frac{3}{7}$	$\frac{240}{7}$
	$-\frac{4}{7}$	0	0	$-\frac{2}{7}$	$-\frac{15}{7}$	$E - 291\frac{3}{7}$

The signs of the entries in the objective row of Tableau 3 show that this tableau is optimal.

The following problem should help the reader better understand why the simplex algorithm can only be applied to tableaux which are adjusted.

PROBLEM 5.3. In order to manufacture three products, X, Y, and Z, assume that we must utilize three resources, namely, time on machines I and II, and specialized labor. Within a production period, we have available at most 48 units of time on machine I, at most 120 units of time on machine II, and exactly 80 units of special man-time.

The production of 1 unit of product X requires 3 units of time on machine I, 3 units of time on machine II, and 2 units of special man-time; the production of a unit of Y requires 1 unit of time on machine I, 4 units of time on machine II, and 4 units of special man-time; the production of 1 unit of Z requires 2 units of time on machine I, 3 units of time on machine II, and 1 unit of special man-time.

Let x denote units of product X, y denote units of product Y, and z denote units of product Z. Then the constraints on production are

$$3x + y + 2z + s_1 = 48 \quad (1)$$

$$3x + 4y + 3z + s_2 = 120 \quad (2)$$

$$2x + 4y + 1z + s_3 = 80 \quad (3)$$

where all variables are nonnegative; s_1 and s_2 denote slack on machines I and II, respectively, and s_3 denotes unutilized man-time.

As for contributions to profit, product X contributes 5 monetary units; product Y, 6 monetary units; and product Z, 14 monetary units—but, by contractual agreement, for every unit of unutilized man-time, the company must pay 4 monetary units to the guild of specialists. Clearly, the objective is to maximize

$$P = 5x + 6y + 14z - 4s_3$$

The above constraints and objective function can be expressed in tableau form:

Tableau 0

	x	y	z	s_1	s_2	s_3	
s_1	3	1	2	1	0	0	48
s_2	3	4	3	0	1	0	120
s_3	2	4	1	0	0	1	80
	5	6	14	0	0	−4	P

We call this "tableau zero" because it is nonadjusted—the objective function is not expressed solely as a function of the current corner variables. In order to adjust we must eliminate s_3 from P, as follows:

$$\begin{aligned} P &= 5x + 6y + 14z - 4s_3 \\ &= 5x + 6y + 14z - 4(80 - 2x - 4y - z) \\ &= 13x + 22y + 18z - 320 \end{aligned}$$

A tableau-type rule for determining whether or not P is adjusted is to look down the columns of the basic variables and check to see if all the objective coefficients in these columns are zero. If these coefficients are not all zero, adjustment must be made before we can begin to apply the simplex algorithm. We can adjust Tableau 0 by adding 4 times row 3 to the objective row—or equivalently *by pivoting on the unit entry in the s_3 row and s_3 column.* Thus we obtain the adjusted Tableau 1:

Tableau 1

	x	y	z	s_1	s_2	s_3		
s_1	3	1	2	1	0	0	48	(48)
s_2	3	4	3	0	1	0	120	(30)
s_3	2	④	1	0	0	1	80	←(20)
	13	22 ↑	18	0	0	0	$P + 320$	

Now we can apply the simplex algorithm. We obtain the following sequence of adjusted tableaux:

Tableau 2

	x	y	z	s_1	s_2	s_3		
s_1	$\frac{5}{2}$	0	$\left(\frac{7}{4}\right)$	1	0	$-\frac{1}{4}$	28	←(16)
s_2	1	0	2	0	1	-1	40	(20)
y	$\frac{1}{2}$	1	$\frac{1}{4}$	0	0	$\frac{1}{4}$	20	(80)
	2	0	$\frac{25}{2}$ ↑	0	0	$-\frac{11}{2}$	$P - 120$	

Tableau 3

	x	y	z	s_1	s_2	s_3	
z	$\frac{10}{7}$	0	1	$\frac{4}{7}$	0	$-\frac{1}{7}$	16
s_2	$-\frac{13}{7}$	0	0	$-\frac{8}{7}$	1	$-\frac{5}{7}$	8
y	$\frac{1}{7}$	1	0	$-\frac{1}{7}$	0	$\frac{2}{7}$	16
	$-\frac{111}{7}$	0	0	$-\frac{50}{7}$	0	$-\frac{26}{7}$	$P - 320$

Tableau 3 is terminal.

Now suppose we had failed to observe that Tableau 0 is nonadjusted. Blind application of the pivot rules would have led to the following sequence:

Tableau 0

	x	y	z	s_1	s_2	s_3		
s_1	3	1	(2)	1	0	0	48	←(24)
s_2	3	4	3	0	1	0	120	(40)
s_3	2	4	1	0	0	1	80	(80)
	5	6	14 ↑	0	0	-4	P	

Tableau 1

	x	y	z	s_1	s_2	s_3	
z	$\frac{3}{2}$	$\frac{1}{2}$	1	$\frac{1}{2}$	0	0	24
s_2	$-\frac{3}{2}$	$\frac{5}{2}$	0	$-\frac{3}{2}$	1	0	48
s_3	$\frac{1}{2}$	$\frac{7}{2}$	0	$-\frac{1}{2}$	0	1	56
	-16	-1	0	-7	0	-4	$P - 336$

An uncritical examination of Tableau 1 suggests that this tableau is terminal. This is not true, however, because Tableau 1 is nonadjusted. (As an exercise, the reader should adjust Tableau 1 and, if revision is then indicated, continue with the simplex algorithm.)

Before turning to a last numerical example, let us sum up. For clarity of exposition, think of the following tableau as being representative of the general case. Our interest is in moving from Tableau r to Tableau (ı + 1).

Tableau r

	x_1	x_2	x_3	x_4	x_5	x_6		
x_2	a_{11}	1	(a_{13})	0	a_{15}	0	b_1	←(θ_1*)
x_6	a_{21}	0	a_{23}	0	a_{25}	1	b_2	(θ_2)
x_4	a_{31}	0	a_{33}	1	a_{35}	0	b_3	(θ_3)
	c_1	0	c_3* ↑	0	c_5	0	P	

We assume that $x_i \geq 0$ for $i = 1, \ldots, 6$, and furthermore, that $b_i \geq 0$, $i = 1, 2, 3$. Then a first feasible corner is immediately apparent, namely, $x_1 = x_3 = x_5 = 0$; $x_2 = b_1$, $x_6 = b_2$, $x_4 = b_3$. Tableau r is adjusted and we can therefore apply the simplex algorithm. But first some remarks:

(1) *Choosing the pivot column.* The usual rule is to take the column with the largest positive objective coefficient. This can be thought of as a miniature optimization problem, namely, to find the maximum positive c_j. But, as pointed out in Chapter 3, the simplex algorithm works if we pivot in any column whose

coefficient is strictly positive. This means that when a tie occurs among the largest positive c_j's, we can choose to pivot in any of the associated columns without worrying about the consequences.

(2) *Choosing the pivot row.* Referring to Tableau r, recall that, by definition, $\theta_1 = b_1/a_{13}$, $\theta_2 = b_2/a_{23}$, $\theta_3 = b_3/a_{33}$. The student should remember that the θ_i value represents the maximum number of units of the variable "coming into solution" that may be introduced without violating the ith constraint and that, as a consequence, the basic variable in the row with the *minimum positive* θ_i must be selected as the variable leaving solution to prevent violating the constraints. In theory, ties among the minimum θ_i values must be resolved in a special way if cycling is to be prevented. We leave such theoretical considerations to a second course in linear programming and repeat again that the possibility of cycling can be ignored in practice.

Since we have agreed to work with tableaux whose right-hand constants are nonnegative, we can improve upon the previously formulated rule for choosing the pivot row, as follows: In forming the list of θ_i, only divide strictly *positive* entries in the pivot column into the corresponding b_i; ignore 0 or negative divisors. (To see why this is true, consider a constraint like

$$x + 3y - 5z = 20$$

Because the coefficient of z is negative, we can make it as large as we desire, say $z = 1000$, provided we compensate by taking $x = 5000$.)

Now after pivoting on the circled entry of Tableau r, we again obtain an adjusted tableau whose right-hand basic constants are nonnegative.

Tableau (r + 1)

	x_1	x_2	x_3	x_4	x_5	x_6	
x_3	a'_{11}	a'_{12}	1	0	a'_{15}	0	b'_1
x_6	a'_{21}	a'_{22}	0	1	a'_{25}	1	b'_2
x_4	a'_{31}	a'_{32}	0	0	a'_{35}	0	b'_3
	c'_1	c'_2	0	0	c'_5	0	$P - \theta_1^* c_3^*$

At the corner described by this tableau, $P = \theta_1^* c_3^*$; this is clearly an improvement if neither θ_1^* nor c_3^* is 0.

The following solution illustrates which "limiting quotients θ_i" may be disregarded in choosing the pivot row.

PROBLEM 5.4. Maximize $P = 4x_1 + 12x_2 + 3x_3$
subject to the usual nonnegativity constraints

$$\text{and}\quad x_1 \leq 100 \qquad (1)$$

$$x_2 \leq 50 \qquad (2)$$

$$x_3 \leq 150 \qquad (3)$$

$$3x_1 + 6x_2 + 2x_3 \leq 675 \qquad (4)$$

Tableau 1

	x_1	x_2	x_3	s_1	s_2	s_3	s_4		
s_1	1	0	0	1	0	0	0	100	
s_2	0	①	0	0	1	0	0	50	←($\frac{50}{1}$)
s_3	0	0	1	0	0	1	0	150	
s_4	3	6	2	0	0	0	1	675	($\frac{675}{6}$)
	4	12	3	0	0	0	0	P	
		↑							

(The x_2 column is the column of the incoming variable. Note that neither row 1 nor row 3 puts a limit on how much we can increase x_2, essentially because these rows do not contain x_2 and, hence, say nothing about this variable.)

Tableau 2

	x_1	x_2	x_3	s_1	s_2	s_3	s_4		
s_1	①	0	0	1	0	0	0	100	←(100)
x_2	0	1	0	0	1	0	0	50	
s_3	0	0	1	0	0	1	0	150	
s_4	3	0	2	0	−6	0	1	375	(125)
	4	0	3	0	−12	0	0	$P - 600$	
	↑								

Tableau 3

	x_1	x_2	x_3	s_1	s_2	s_3	s_4		
x_1	1	0	0	1	0	0	0	100	
x_2	0	1	0	0	1	0	0	50	
s_3	0	0	1	0	0	1	0	150	(150)
s_4	0	0	②	−3	−6	0	1	75	←($\frac{75}{2}$)
	0	0	3	−4	−12	0	0	$P - 1000$	
			↑						

Tableau 4

	x_1	x_2	x_3	s_1	s_2	s_3	s_4		
x_1	1	0	0	1	0	0	0	100	(100)
x_2	0	1	0	0	1	0	0	50	
s_3	0	0	0	($\frac{3}{2}$)	3	1	$-\frac{1}{2}$	$\frac{225}{2}$	←(75)
x_3	0	0	1	$-\frac{3}{2}$	−3	0	$\frac{1}{2}$	$\frac{75}{2}$	
	0	0	0	$\frac{1}{2}$	−3	0	$-\frac{3}{2}$	$P - \frac{2225}{2}$	
				↑					

(Note how we determined the pivot needed for revising Tableau 4. The incoming variable is s_1. In dividing the entries in the s_1 column into the stub, we can ignore the entries in rows 2 and 4. It should be stressed that the reason we can ignore the $-\frac{3}{2}$ term in row 4 is because the fourth row of Tableau 4 affirms that

$$x_3 - \tfrac{3}{2}s_1 - 3s_2 + \tfrac{1}{2}s_4 = \tfrac{75}{2}$$

an equation which puts absolutely no limitation on the maximum feasible value of s_1.)

Tableau 5

	x_1	x_2	x_3	s_1	s_2	s_3	s_4	
x_1	1	0	0	0	-2	$-\frac{2}{3}$	$\frac{1}{3}$	25
x_2	0	1	0	0	1	0	0	50
s_1	0	0	0	1	2	$\frac{2}{3}$	$-\frac{1}{3}$	75
x_3	0	0	1	0	0	1	0	150
	0	0	0	0	-4	$-\frac{1}{3}$	$-\frac{4}{3}$	$P - 1150$

Tableau 5 is terminal.

At this point the reader should turn back to the illustrative problems described in Section 3.2. As an exercise he should rewrite the solutions in tableau format.

Exercises

1. Do all the exercises proposed in Chapters 2 and 3 using simplex tableaux.

2. Prove the following rule:

(a) If there is a 0 in the pivot row, the column in which it lies is unchanged after pivoting.

(b) If there is a 0 in the pivot column, the row in which it lies is unchanged after pivoting.

3. *True or false?*

(a) In the simplex method, a variable that is *outgoing* in one tableau cannot be *incoming* in the next tableau.

(b) A variable that is *incoming* in one tableau cannot be *outgoing* in the next tableau.

4. What can we say about the nature of the basic feasible solutions described in the following tableaux?

(a)

	x_1	x_2	x_3	x_4	
x_3	0	$\frac{5}{2}$	1	$-\frac{1}{2}$	18
x_1	1	$\frac{3}{4}$	0	$\frac{1}{4}$	15
	0	$-\frac{1}{2}$	0	0	$P - 10$

(b)

	x_1	x_2	x_3	x_4	
x_3	0	-1	1	$-\frac{1}{2}$	18
x_1	1	-2	0	$\frac{1}{4}$	15
	0	3	0	$-\frac{3}{2}$	$P - 10$

(c)

	x_1	x_2	x_3	x_4	
x_3	0	-1	1	$-\frac{1}{2}$	18
x_1	1	-2	0	$\frac{1}{4}$	0
	0	-1	0	$-\frac{3}{2}$	$P - 10$

(d)

	x_1	x_2	x_3	x_4	
x_3	0	2	1	$-\frac{1}{3}$	18
x_1	1	3	0	$\frac{1}{4}$	0
	0	0	0	$-\frac{3}{2}$	$P - 10$

(e)

	x_1	x_2	x_3	x_4	
x_3	0	2	1	$-\frac{1}{3}$	18
x_1	1	3	0	$\frac{1}{4}$	0
	0	7	0	$-\frac{3}{2}$	$P - 10$

5. Clarify the following statements:

(a) Degeneracy can occur in the initial tableau; but if it occurs in Tableau n its presence has already been signalled in Tableau (n − 1).

(b) An unbounded solution will be detected by the simplex algorithm if, at some iteration, *any* of the candidates for the incoming variable can be increased without limit.

(c) Whenever alternative optimal solutions exist, there is a nonbasic optimal solution.

(d) In a degenerate situation, we should continue the simplex algorithm until the coefficients of the simplex algorithm indicate optimality.

6

ARTIFICIAL VARIABLES AND FREE VARIABLES

Chapter 5 was based upon two important assumptions, namely, that a first feasible corner solution was immediately evident and that all variables were constrained to be nonnegative. In Section 6.1 we treat problems in which a feasible corner is not immediately evident; in Section 6.2 we show how to work with variables which are free to take on both positive and negative values. In Section 6.3 the main ideas of Chapters 3 to 6 are summarized.

6.1 ARTIFICIAL VARIABLES

We begin by considering the following program:

PROBLEM 6.1. Minimize $C = 2x + 5y$

where $x \geq 0, \quad y \geq 0$

and $3x + 3y \geq 5$

$x + 4y \geq 6$

We can transform the nontrivial inequalities into equalities by using the nonnegative *surplus* variables s_1 and s_2:

$$3x + 3y - s_1 \qquad = 5$$
$$x + 4y \qquad - s_2 = 6$$

But the corner defined by $x = y = 0$ is not feasible, because if x and y are 0, then $s_1 = -5$ and $s_2 = -6$. To overcome this difficulty, we invent an *auxiliary problem* for which an initial feasible corner can easily be found and which is so formulated that its solution furnishes the solution to the original problem. To this end, consider

PROBLEM 6.1A (AUXILIARY).

$$\text{Minimize} \quad C = 2x + 5y + MA_1 + MA_2$$

where all variables are nonnegative

$$\begin{aligned} \text{and} \quad 3x + 3y - s_1 \qquad + A_1 \qquad &= 5 \\ x + 4y \qquad - s_2 \qquad + A_2 &= 6 \end{aligned}$$

Note that Problem 6.1A has two more variables than Problem 6.1, namely, A_1 and A_2. But because this auxiliary problem has more variables, it possesses an obvious feasible corner, that is, an initial basic feasible solution, namely, $A_1 = 5$, $A_2 = 6$; $x = y = s_1 = s_2 = 0$. Note further that Problem 6.1A reduces to Problem 6.1 if and only if $A_1 = A_2 = 0$. Now, because A_1 and A_2 have large positive coefficients M in the objective function of the auxiliary problem, the very logic of the simplex algorithm dictates that A_1 and A_2 will be driven to 0 at optimality (minimality), provided, of course, that the original problem does actually possess a feasible corner. We say that the variables A_1 and A_2 are *artificial variables* (as opposed to the *original*, or *natural*, *variables*).

In tableau form, Problem 6.1A reads

Tableau 0

	x	y	s_1	s_2	A_1	A_2	
A_1	3	3	−1	0	①	0	5
A_2	1	4	0	−1	0	①	6
	2	5	0	0	M	M	C

This tableau is not adjusted because A_1 and A_2 represent the first set of basic variables, but the objective row contains nonzero coefficients in the A_1 and A_2 columns. After adjusting (for example, by pivoting as indicated in Tableau 0) we obtain the following tableau:

Tableau 1

	x	y	s_1	s_2	A_1	A_2	
A_1	3	3	−1	0	1	0	5
A_2	1	4	0	−1	0	1	6
	$2 - 4M$	$5 - 7M$	M	M	0	0	$C - 11M$

Usually in dealing with objective rows involving M, we shall divide such a row into two rows, one for ordinary numbers and one for "prohibitively large numbers," as shown below:

Tableau 1'

	x	y	s_1	s_2	A_1	A_2		
A_1	3	3	-1	0	1	0	5	$(\frac{5}{3})$
A_2	1	④	0	-1	0	1	6	$\leftarrow(\frac{3}{2})$
	2	5	0	0	0	0	C	
	$-4M$	$-7M$	M	M	0	0	$-11M$	
		↑						

Now we are ready to apply the simplex algorithm. To choose the pivot column we need only examine the "basement row" of Tableau 1'. Since the objective is to minimize cost, we are instructed to increase y as much as possible. By the usual pivot mechanics we obtain

Tableau 2

	x	y	s_1	s_2	A_1	A_2		
A_1	$\frac{9}{4}$ (circled)	0	-1	$\frac{3}{4}$	1	$-\frac{3}{4}$	$\frac{1}{2}$	$\leftarrow(\frac{2}{9})$
y	$\frac{1}{4}$	1	0	$-\frac{1}{4}$	0	$\frac{1}{4}$	$\frac{3}{2}$	(6)
	$\frac{3}{4}$	0	0	$\frac{5}{4}$	0	$-\frac{5}{4}$	$C-\frac{15}{2}$	
	$-\frac{9}{4}M$	0	M	$-\frac{3}{4}M$	0	$\frac{7}{4}M$	$-\frac{1}{2}M$	
	↑							

Tableau 2 is not terminal. Cost can be further decreased by increasing x. The maximum possible increase in x is $\frac{2}{9}$, and hence the pivot is the entry $\frac{9}{4}$ in row 1 of the x column. After pivoting we obtain

Tableau 3

	x	y	s_1	s_2	A_1	A_2	
x	1	0	$-\frac{4}{9}$	$\frac{1}{3}$	$\frac{4}{9}$	$-\frac{1}{3}$	$\frac{2}{9}$
y	0	1	$\frac{1}{9}$	$-\frac{1}{3}$	$-\frac{1}{9}$	$\frac{1}{3}$	$\frac{13}{9}$
	0	0	$\frac{1}{3}$	1	$-\frac{1}{3}$	-1	$C-\frac{23}{3}$
	0	0	0	0	M	M	0

This tableau is terminal. The optimal solution to Problem 6.1A is $x = \frac{2}{9}$, $y = \frac{13}{9}$, $s_1 = s_2 = A_1 = A_2 = 0$, and $C = \frac{23}{3}$; hence, the optimal solution to Problem 6.1 is $x = \frac{2}{9}$, $y = \frac{13}{9}$, $s_1 = s_2 = 0$, $C = \frac{23}{3}$.

In computer programs based on this approach a value has to be assigned to M. One rule of thumb formerly popular was to take M equal to 1000 times the highest coefficient belonging to a legitimate variable in the original objective function. Another rule was to take $M = 9{,}999{,}999{,}999$. However, these large numerical values of M can create round-off errors. In order to avoid manipulating large values of M, the so-called "two-phase method" was devised, as follows:

(1) Introduce artificial variables into the given constraints, so as to create an auxiliary region of feasibility having a feasible corner immediately available. For example, in Problem 6.1A we introduced the artificial variables A_1 and A_2 to form the auxiliary region describable by

$$x \geq 0, \quad y \geq 0, \quad s_1 \geq 0, \quad s_2 \geq 0, \quad A_1 \geq 0, \quad A_2 \geq 0$$

and

$$\begin{aligned} 3x + 3y - s_1 \qquad + A_1 \qquad &= 5 \\ x + 4y \qquad - s_2 \qquad + A_2 &= 6 \end{aligned}$$

Clearly, at the corner $[x = y = s_1 = s_2 = 0]$, the variables A_1 and A_2 are nonnegative; that is, this corner is feasible.

(2) Now consider the special objective function

$$S = A_1 + \cdots + A_n$$

where S equals the sum of all the artificial variables needed to describe the auxiliary region of feasibility. For example, in Problem 6.1A, $S = A_1 + A_2$. *Phase I consists in finding the minimum of S over the auxiliary region of feasibility*. Furthermore, it is an established fact that if the minimum of S is strictly greater than 0, then the region of feasibility of the *original* problem is empty; that is, the set of original constraints is contradictory. We shall show later why this is true.

(3) Suppose that the minimum of S is 0. Then clearly both A_1 and A_2 must be 0. In addition, suppose that the optimal tableau is not degenerate; that is, both A_1 and A_2 are corner variables. If we ignore the A_1 and A_2 columns of this tableau, what remains is an adjusted tableau which corresponds to a feasible corner of the original problem. The purpose of Phase I is to determine systematically a feasible corner (as well as to form the tableau corresponding to this corner).

(4) Phase II begins by using the feasible corner obtained at the end of Phase I as a starting point for solving the original problem by the simplex algorithm.

The original objective function is almost never adjusted to the nonbasic variables of the tableau obtained at the termination of Phase I. One way to avoid the need to adjust the original objective function is to carry it along during Phase I, in its own separate row, and permit it to be adjusted while we minimize the sum of the artificialities. An illustration of this procedure is given in Problem 6.2.

The M method and the two-phase method can only differ in minor details. Both methods solve the original problem by progressing over essentially the same sequence of auxiliary and original corners. That is, (a) both methods start from the same auxiliary region of feasibility, (b) although the objective functions used are different, the coefficients of these functions are proportionately related to each other—the M's in the M objective function belong to the variables whose coefficients are 1 in the expression for S, and the ordinary coefficients in the M objective function belong to the variables whose coefficients are 0 in the expression for S. *And, of course, a pivot column is chosen based only on relative size.*

To show the similarities between these methods, we give the two-phase solution to Problem 6.1 below.

Tableau 0 (Phase I)

	x	y	s_1	s_2	A_1	A_2	
A_1	3	3	−1	0	①	0	5
A_2	1	4	0	−1	0	①	6
	0	0	0	0	1	1	S

Tableau 1

	x	y	s_1	s_2	A_1	A_2		
A_1	3	3	−1	0	1	0	5	$(\frac{5}{3})$
A_2	1	④	0	−1	0	1	6	$\leftarrow(\frac{3}{2})$
	−4	−7 ↑	1	1	0	0	$S - 11$	

Tableau 2

	x	y	s_1	s_2	A_1	A_2		
A_1	$\frac{9}{4}$ (circled)	0	−1	$\frac{3}{4}$	1	$-\frac{3}{4}$	$\frac{2}{4}$	$\leftarrow(\frac{2}{9})$
y	$\frac{1}{4}$	1	0	$-\frac{1}{4}$	0	$\frac{1}{4}$	$\frac{3}{2}$	(6)
	$-\frac{9}{4}$ ↑	0	1	$-\frac{3}{4}$	0	$\frac{7}{4}$	$S - \frac{1}{2}$	

Tableau 3

	x	y	s_1	s_2	A_1	A_2	
x	1	0	$-\frac{4}{9}$	$\frac{1}{3}$	$\frac{4}{9}$	$-\frac{1}{3}$	$\frac{2}{9}$
y	0	1	$\frac{1}{9}$	$-\frac{1}{3}$	$-\frac{1}{9}$	$\frac{1}{3}$	$\frac{13}{9}$
	0	0	0	0	1	1	$S - 0$

Tableau 3 marks the end of Phase I. Both A_1 and A_2 have become corner variables and $S = 0$. To initiate Phase II we delete the artificial columns and the objective row of Tableau 3 and enter the original objective function:

Tableau 0 (Phase II)

	x	y	s_1	s_2	
x	①	0	$-\frac{4}{9}$	$\frac{1}{3}$	$\frac{2}{9}$
y	0	①	$\frac{1}{9}$	$-\frac{1}{3}$	$\frac{13}{9}$
	2	5	0	0	C

Note that this tableau is not adjusted. After adjusting, we obtain

Tableau 1 (Phase II)

	x	y	s_1	s_2	
x	1	0	$-\frac{4}{9}$	$\frac{1}{3}$	$\frac{2}{9}$
y	0	1	$\frac{1}{9}$	$-\frac{1}{3}$	$\frac{13}{9}$
	0	0	$\frac{1}{3}$	1	$C - \frac{23}{3}$

Since the objective is to minimize C, this tableau is terminal.

Problem 6.1 is so simple that it fails to illustrate another similarity between the M method and two-phase method. This similarity results from the fact that under the M method, once all the artificial variables have been driven to zero, then no artificial variable can subsequently become basic and nonzero. To see why this is so, suppose the objective of the auxiliary problem is to maximize

$$P = c_1x_1 + c_2x_2 + c_3x_3 - MA_1 - MA_2$$

As we perform the simplex algorithm, P must never decrease. Suppose that, at some tableau, $A_1 = A_2 = 0$. Then the value of P at this tableau is an ordinary number, not involving M. If, at a subsequent tableau, an artificial variable ceases to be 0, then clearly P will be smaller than at the tableau where $A_1 = A_2 = 0$. This is clearly impossible.

PROBLEM 6.2. Maximize $P = 4x + 3y + 7z$

where $x \geq 0, \quad y \geq 0, \quad z \geq 0$

and $2x + y + 3z \leq 120 \qquad (1)$

$x + 3y + 2z = 120 \qquad (2)$

(See Problem 2.12 for a solution of this program via a listing of all feasible corner points.) We shall solve this problem by considering the auxiliary problem.

PROBLEM 6.2A. Maximize $P = 4x + 3y + 7z - MA$
where all variables are nonnegative
and
$$2x + y + 3z + s = 120$$
$$x + 3y + 2z + A = 120$$

Clearly, s is a slack variable and A an artificial variable. The sequence of tableaux which solve this problem is the following:

Tableau 0

	x	y	z	s	A	
s	2	1	3	1	0	120
A	1	3	2	0	①	120
	4	3	7	0	0	P
	0	0	0	0	$-M$	

Tableau 0 is not adjusted. After adjustment, we obtain

Tableau 1

	x	y	z	s	A		
s	2	1	3	1	0	120	(120)
A	1	③	2	0	1	120	←(40)
	4	3	7	0	0	P	
	M	$3M$	$2M$	0	0	$120M$	
		↑					

Tableau 2

	x	y	z	s	A		
s	$\frac{5}{3}$	0	($\frac{7}{3}$)	1	$-\frac{1}{3}$	80	←($\frac{240}{7}$)
y	$\frac{1}{3}$	1	$\frac{2}{3}$	0	$\frac{1}{3}$	40	(60)
	3	0	5	0	-1	$P - 120$	
	0	0	0	0	$-M$	0	
			↑				

Tableau 2 does not give the optimal solution of Problem 6.2A but we can think of it as signifying the end of "Phase I" of Problem 6.2. (In Tableau 2, A is non-basic, and if we ignore the A column and M row we have the correct tableau for beginning "Phase II.") After pivoting as indicated above, we obtain

Tableau 3

	x	y	z	s	A	
z	$\frac{5}{7}$	0	1	$\frac{3}{7}$	...	$\frac{240}{7}$
y	$-\frac{1}{7}$	1	0	$-\frac{2}{7}$	...	$\frac{120}{7}$
	$-\frac{4}{7}$	0	0	$-\frac{15}{7}$	...	$P - \frac{2040}{7}$

Tableau 3 exhibits the solution to the original problem.

PROBLEM 6.3. Maximize $P = 3x_1 + 4x_2 + 5x_3 + 6x_4$

where all variables are nonnegative

and $2x_1 + 4x_2 + 3x_3 - 4x_4 = 20$ (1)

$3x_1 + 2x_2 + 2x_3 + 5x_4 = 30$ (2)

As seen in Chapter 3, Eqs. (1) and (2) can usually be solved for any two variables in terms of any other two. Suppose we can solve for x_i and x_k in terms of the other two variables. We can then substitute these expressions for x_i and x_k into P, and thereby reduce P to a function of two "corner" variables. There are at most six potential corners and six possible views on profit.

But we do not want to find a first feasible corner in such an unsystematic fashion. An initial feasible corner can be found by solving the *Phase I problem:*

Minimize $S = A_1 + A_2$

where all variables are nonnegative

and $2x_1 + 4x_2 + 3x_3 - 4x_4 + A_1 \qquad = 20$

$3x_1 + 2x_2 + 2x_3 + 5x_4 \qquad + A_2 = 30$

In what follows we carry along the original objective function as we maximize $-S$. This will give us an already adjusted tableau with which to begin Phase II.

Tableau 0

	x_1	x_2	x_3	x_4	A_1	A_2	
A_1	2	4	3	−4	1	0	20
A_2	3	2	2	5	0	1	30
	3	4	5	6	0	0	P
	0	0	0	0	−1	−1	$-S$

Tableau 1

	x_1	x_2	x_3	x_4	A_1	A_2		
A_1	2	④	3	−4	1	0	20	←(5)
A_2	3	2	2	5	0	1	30	(15)
	3	4	5	6	0	0	P	
	5	6	5	1	0	0	$-S+50$	
		↑						

Tableau 2

	x_1	x_2	x_3	x_4	A_1	A_2		
x_2	$\frac{1}{2}$	1	$\frac{3}{4}$	−1	$\frac{1}{4}$	0	5	
A_2	2	0	$\frac{1}{2}$	⑦	$-\frac{1}{2}$	1	20	←($\frac{20}{7}$)
	1	0	2	10	−1	0	$P-20$	
	2	0	$\frac{1}{2}$	7	$-\frac{3}{2}$	0	$-S+20$	
				↑				

Tableau 3

	x_1	x_2	x_3	x_4	A_1	A_2	
x_2	$\frac{11}{14}$	1	$\frac{23}{28}$	0	. . .	. . .	$\frac{55}{7}$
x_4	$\frac{2}{7}$	0	$\frac{1}{14}$	1	$-\frac{1}{14}$	$\frac{1}{7}$	$\frac{20}{7}$
	$-\frac{13}{7}$	0	$\frac{9}{7}$	0			$P-\frac{340}{7}$
	0	0	0	0	. . .	. . .	$-S+0$

Tableau 3 completes Phase I of our problem, because A_1 and A_2 have become corner variables. We did not even bother to fill in all the entries in the A_1 and A_2 columns since these columns are not needed during Phase II. We begin Phase II as shown below.

Tableau 1 (Phase II)

	x_1	x_2	x_3	x_4		
x_2	$\frac{11}{14}$	1	$\frac{23}{28}$ (circled)	0	$\frac{55}{7}$	←($\frac{220}{23}$)
x_4	$\frac{2}{7}$	0	$\frac{1}{14}$	1	$\frac{20}{7}$	(40)
	$-\frac{13}{7}$	0	$\frac{9}{7}$	0	$P-\frac{340}{7}$	

Tableau 2

	x_1	x_2	x_3	x_4	
x_3	$\frac{22}{23}$	$\frac{28}{23}$	1	0	$\frac{220}{23}$
x_4	$\frac{5}{23}$	$-\frac{2}{23}$	0	1	$\frac{50}{23}$
	$-\frac{71}{23}$	$-\frac{36}{23}$	0	0	$P - \frac{1400}{23}$

With this tableau, Phase II is ended. The optimal solution to the original problem is $x_1 = 0$, $x_2 = 0$, $x_3 = \frac{220}{23}$, $x_4 = \frac{50}{23}$, and $P = \frac{1400}{23}$.

The following problem illustrates the behavior of the auxiliary problem when the original problem is contradictory.

PROBLEM 6.4.

$$\text{Minimize} \quad I = 3x + 2y$$
$$\text{where} \quad x \geq 0, \quad y \geq 0$$
$$\text{and} \quad x + y \leq 10$$
$$2x + y \geq 30$$

We studied this program in Problem 2.7 by graphical methods. We solve it below by the M method, with slack and artificial variables as indicated.

Tableau 0

	x	y	s_1	s_2	A_2	
s_2	1	1	1	0	0	10
A_2	2	1	0	−1	①	30
	3	2	0	0	M	I

Tableau 1

	x	y	s_1	s_2	A_2		
s_1	①	1	1	0	0	10	←(10)
A_2	2	1	0	−1	1	30	(15)
	3	2	0	0	0	$I - 0$	
	$-2M$	$-M$	0	M	0	$-30M$	
	↑						

Tableau 2

	x	y	s_1	s_2	A_2	
x	1	1	1	0	0	10
A_2	0	−1	−2	−1	1	10
	0	−1	−3	0	0	$I - 30$
	0	M	$2M$	M	0	$-10M$

This tableau shows that the optimal solution to the auxiliary problem is $I = 30 + 10M$, occurring at the corner $[y = s_1 = s_2 = 0]$, where $x = 10$ and $A_2 = 10$.

For purposes of comparison we now solve Problem 6.4 by the two-phase method.

Tableau 0

	x	y	s_1	s_2	A_2	
s_1	1	1	1	0	0	10
A_2	2	1	0	−1	①	30
	0	0	0	0	1	S

The objective is to minimize $S = A_2$. *Tableau 0 is defective*—we want A_2 to be a basic variable, but A_2 appears in the objective row with a nonzero coefficient. The way to adjust Tableau 0 is to subtract row 2 from row 3, or equivalently, to pivot on the circled entry in Tableau 0, obtaining

Tableau 1

	x	y	s_1	s_2	A_2		
s_1	①	1	1	0	0	10	←(10)
A_2	2	1	0	−1	1	30	(15)
	−2 ↑	−1	0	1	0	$S - 30$	

(Note carefully that when we minimize S the coefficients in the objective row are exactly the same as the coefficients of M which appear in the M row when we use the M method.)

Tableau 2

	x	y	s_1	s_2	A_2	
x	1	1	1	0	0	10
A_2	0	−1	−2	−1	1	10
	0	1	2	1	0	$S - 10$

Tableau 2 is terminal. Since the minimum value of S is 10 and not 0, we know that the original problem does not possess a feasible corner.

It is a demonstrated fact that, whenever a linear program possesses any feasible solution whatsoever, it must also possess a feasible corner solution. From this theorem it follows that *whenever* $S \neq 0$, *the region of feasibility of* *the given problem is empty. In the M method, the presence of M in the optimum value of the objective function is the analog of $S \neq 0$ in the two-phase method; this serves to indicate that the original problem is contradictory.*

The following problem points up some interesting differences between the M method and the two-phase method.

PROBLEM 6.5. Maximize $P = 3x_1 + 4x_2 + 5x_3 + 6x_4$

subject to the nonnegativity of all variables

$$\text{and} \quad 2x_1 + 4x_2 + 3x_3 + 4x_4 = 20$$
$$3x_1 + 2x_2 + 2x_3 + 5x_4 = 30$$

Note that Problem 6.5 is almost identical to Problem 6.3. These problems differ only in the sign of the coefficient of x_4 in the first nontrivial constraint We solve Problem 6.5 by the two-phase method.

Tableau 0 (Phase I)

	x_1	x_2	x_3	x_4	A_1	A_2	
A_1	2	4	3	4	①	0	20
A_2	3	2	2	5	0	①	30
	0	0	0	0	1	1	S

Tableau 1

	x_1	x_2	x_3	x_4	A_1	A_2		
A_1	2	4	3	④	1	0	20	←(5)
A_2	3	2	2	5	0	1	30	(6)
	−5	−6	−5	−9	0	0	$S - 50$	
				↑				

Tableau 2

	x_1	x_2	x_3	x_4	A_1	A_2		
x_4	$\textcircled{\frac{1}{2}}$	1	$\frac{3}{4}$	1	$\frac{1}{4}$	0	5	$\overset{?}{\longleftarrow}$(10)
A_2	$\frac{1}{2}$	−3	$-\frac{7}{4}$	0	$-\frac{5}{4}$	1	5	$\overset{?}{\longleftarrow}$(10)
	$-\frac{1}{2}$	3	$\frac{7}{4}$	0	$\frac{9}{4}$	0	$S - 5$	
	↑							

The limiting numbers in Tableau 2 are equal. This creates ambiguity in the choice of pivot. Suppose we choose to pivot on the circled entry in row 1. Then we obtain:

Tableau 3

	x_1	x_2	x_3	x_4	A_1	A_2	
x_1	1	2	$\frac{3}{2}$	2	$\frac{1}{2}$	0	10
A_2	0	−4	$-\frac{5}{2}$	−1	$-\frac{3}{2}$	1	0
	0	4	$\frac{5}{2}$	1	$\frac{5}{2}$	0	$S - 0$

In Tableau 3, $S = 0$. Phase I has been completed. Although both A_1 and A_2 equal 0, they are not both corner variables. The variable A_2 is basic, and Tableau 3 is degenerate. If we delete the A_1 and A_2 columns of Tableau 3 and place the original objective function $P = 3x_1 + 4x_2 + 5x_3 + 6x_4$ in the objective row of Tableau 3, we obtain Tableau 0 of Phase II:

Tableau 0 (Phase II)

	x_1	x_2	x_3	x_4	
x_1	①	2	$\frac{3}{2}$	2	10
?	0	−4	$-\frac{5}{2}$	−1	0
	3	4	5	6	P

How shall we adjust the objective row of Tableau 0? Because Tableau 3 of Phase I is degenerate, only one of the four columns in Tableau 0 above corresponds to a basic variable in Tableau 3 of Phase I, namely, the x_1 column. (This problem never arises if we use the M method and never delete artificial variables.)

Using the two-phase method, one way out of the above dilemma is to carry the degenerate basic column into the "Phase II calculations." Then, in order to make sure that A_2 remains 0, we can add the new constraint

$$x_0 + A_2 = 0$$

where x_0 is a new nonnegative variable whose objective coefficient can be taken as 0. Accordingly, we begin Phase II as follows:

Tableau 0 (Phase II)

	x_0	x_1	x_2	x_3	x_4	A_2	
x_0	1	0	0	0	0	1	0
x_1	0	①	2	$\frac{3}{2}$	2	0	10
A_2	0	0	−4	$-\frac{5}{2}$	−1	①	0
	0	3	4	5	6	0	P

Tableau 0 is not adjusted. After adjusting, we obtain the following sequence:

Tableau 1

	x_0	x_1	x_2	x_3	x_4	A_2		
x_0	1	0	4	$\textcircled{\frac{5}{2}}$	1	0	0	←(0)
x_1	0	1	2	$\frac{3}{2}$	2	0	10	
A_2	0	0	−4	$-\frac{5}{2}$	−1	1	0	
	0	0	−2	$\frac{1}{2}$ ↑	0	0	$P - 30$	

Tableau 2

	x_0	x_1	x_2	x_3	x_4	A_2	
x_3	$\frac{2}{5}$	0	$\frac{8}{5}$	1	$\frac{2}{5}$	0	0
x_1	$-\frac{3}{5}$	1	$-\frac{2}{5}$	0	$\frac{7}{5}$	0	10
A_2	1	0	0	0	0	1	0
	$-\frac{1}{5}$	0	$-\frac{14}{5}$	0	$-\frac{1}{5}$	0	$P - 30$

Tableau 2 is terminal. (Delete the A_2 row and both the x_0 and A_2 columns.)

We pause to remark that the best features of both the M method and two-phase method have been combined in today's computer codes. In theory and in computation, M has become purely conceptual—as if the ordinary part of $(a + bM)$ were stored in one place and the M part in another place.

For another illustration of the usefulness of artificial variables, consider the following problem.

PROBLEM 6.6. Maximize $P = x_1 + 2x_2 + x_3 + 3x_4$

where $x_i \geq 0, \quad i = 1, 2, 3, 4$

and
$$x_1 + x_2 + x_3 + x_4 = 2 \tag{1}$$
$$x_1 + 2x_2 + 3x_3 + 4x_4 = 6 \tag{2}$$
$$x_2 + 2x_3 + 3x_4 = 4 \tag{3}$$

Observe that the nontrivial constraints are equalities. It would seem that we could solve for any three variables in terms of a fourth, but note that Eq. (2) is the sum of Eqs. (1) and (3). If Eq. (2) were to be discarded, we would then have two nontrivial equations in four unknowns, suggesting that we could solve for two variables in terms of the other two.

Below is the complete sequence of tableaux needed to solve Problem 6.6.

We use the two-phase method, maximizing $-S = -A_1 - A_2 - A_3$ and, at the same time, adjusting $P = x_1 + 2x_2 + x_3 + 3x_4$.

Tableau 0

	x_1	x_2	x_3	x_4	A_1	A_2	A_3	
A_1	1	1	1	1	①	0	0	2
A_2	1	2	3	4	0	①	0	6
A_3	0	1	2	3	0	0	①	4
	1	2	1	3	0	0	0	P
	0	0	0	0	-1	-1	-1	$-S$

Tableau 1

	x_1	x_2	x_3	x_4	A_1	A_2	A_3		
A_1	1	1	1	1	1	0	0	2	(2)
A_2	1	2	3	4	0	1	0	6	$(\frac{3}{2})$
A_2	0	1	2	③	0	0	1	4	$\leftarrow(\frac{4}{3})$
	1	2	1	3	0	0	0	P	
	2	4	6	8 ↑	0	0	0	$-S + 12$	

Tableau 2

	x_1	x_2	x_3	x_4	A_1	A_2	A_3		
A_1	1	$\frac{2}{3}$	$\frac{1}{3}$	0	1	0	$-\frac{1}{3}$	$\frac{2}{3}$	$\overset{?}{\longleftarrow}(\frac{2}{3})$
A_2	①	$\frac{2}{3}$	$\frac{1}{3}$	0	0	1	$-\frac{4}{3}$	$\frac{2}{3}$	$\overset{?}{\longleftarrow}(\frac{2}{3})$
x_4	0	$\frac{1}{3}$	$\frac{2}{3}$	1	0	0	$\frac{1}{3}$	$\frac{4}{3}$	
	1	1	-1	0	0	0	-1	$P - 4$	
	2 ↑	$\frac{4}{3}$	$\frac{2}{3}$	0	0	0	$-\frac{8}{3}$	$-S + \frac{4}{3}$	

Tableau 3

	x_1	x_2	x_3	x_4	A_1	A_2	A_3	
A_1	0	0	0	0	1	-1	1	0
x_1	1	$\frac{2}{3}$	$\frac{1}{3}$	0	0	1	$-\frac{4}{3}$	$\frac{2}{3}$
x_4	0	$\frac{1}{3}$	$\frac{2}{3}$	1	0	0	$\frac{1}{3}$	$\frac{4}{3}$
	0	$\frac{1}{3}$	$-\frac{4}{3}$	0	0	-1	$\frac{1}{3}$	$P - \frac{14}{3}$
	0	0	0	0	0	-2	0	$-S + 0$

Observe that all artificial variables have been driven to 0. Although A_1 appears basic, row 1 is really nonexistent. Hence we can begin Phase II simply by crossing out all trace of artificial variables. Thus, we obtain the following nonredundant system:

Tableau 1 (Phase II)

	x_1	x_2	x_3	x_4		
x_1	1	$\textcircled{\frac{2}{3}}$	$\frac{1}{3}$	0	$\frac{2}{3}$	←(1)
x_4	0	$\frac{1}{3}$	$\frac{2}{3}$	1	$\frac{4}{3}$	(4)
	0	$\frac{1}{3}$ ↑	$-\frac{4}{3}$	0	$P - \frac{14}{3}$	

Pivoting as indicated, we obtain

Tableau 2 (Phase II)

	x_1	x_2	x_3	x_4	
x_2	$\frac{3}{2}$	1	$\frac{1}{2}$	0	1
x_4	$-\frac{1}{2}$	0	$\frac{1}{2}$	1	1
	$-\frac{1}{2}$	0	$-\frac{3}{2}$	0	$P - 5$

Tableau 2 is optimal.

Exercises

Use the M method to solve Exercises 1 to 8 below. Given for each exercise are: (1) the statement of the problem, (2) the restatement including our choice of slack and artificial variables, and (3) the complete optimal tableau (artificial variables included in Exercises 1 to 7).

1. (a) Minimize $I = x + 3y$

where $x, y \geq 0$

and $3x + 4y \geq 30$

$x + 2y \geq 12$

(b) $3x + 4y - s_1 + A_1 = 30$

$x + 2y - s_2 + A_2 = 12$

(c) At optimality:

	x	y	s_1	s_2	A_1	A_2	
x	1	2	0	−1	0	1	12
s_1	0	2	1	−3	−1	3	6
	0	1	0	1	0	−1	$I - 12$
	0	0	0	0	M	M	0

2. (a) Minimize $C = 48x + 48y$
where $x, y \geq 0$
and $3x + y \geq 3$
$x + 3y \geq 6$

(b) $3x + y - s_1 + A_1 = 3$
$x + 3y - s_2 + A_2 = 6$

(c) At optimality:

	x	y	s_1	s_2	A_1	A_2	
x	1	0	$-\frac{3}{8}$	$\frac{1}{8}$	$\frac{3}{8}$	$-\frac{1}{8}$	$\frac{3}{8}$
y	0	1	$\frac{1}{8}$	$-\frac{3}{8}$	$-\frac{1}{8}$	$\frac{3}{8}$	$\frac{15}{8}$
	0	0	12	12	-12	-12	$C - 108$
	0	0	0	0	M	M	0

3. (a) Minimize $K = 2x_1 + 5x_2$
where $x_1, x_2 \geq 0$
and $3x_1 + 3x_2 \geq 5$
$x_1 + 4x_2 \geq 6$

(b) $3x_1 + 3x_2 - x_3 + A_1 = 5$
$x_1 + 4x_2 - x_4 + A_2 = 6$

(c) At optimality:

	x_1	x_2	x_3	x_4	A_1	A_2	
x_1	1	0	$-\frac{4}{9}$	$\frac{1}{3}$	$\frac{4}{9}$	$-\frac{1}{3}$	$\frac{2}{9}$
x_2	0	1	$\frac{1}{9}$	$-\frac{1}{3}$	$-\frac{1}{9}$	$\frac{1}{3}$	$\frac{13}{9}$
	0	0	$\frac{1}{3}$	1	$-\frac{1}{3}$	-1	$K - \frac{23}{3}$
	0	0	0	0	M	M	0

4. (a) Minimize $C = x_1 + 3x_2$
where $x_1, x_2 \geq 0$
and $x_1 + 4x_2 \geq 24$
$5x_1 + x_2 \geq 25$

(b) $x_1 + 4x_2 - x_3 + A_1 = 24$
$5x_1 + x_2 - x_4 + A_2 = 25$

(c) At optimality:

	x_1	x_2	x_3	x_4	A_1	A_2	
x_2	0	1	$-\frac{5}{19}$	$\frac{1}{19}$	$\frac{5}{19}$	$-\frac{1}{19}$	5
x_1	1	0	$\frac{1}{19}$	$-\frac{4}{19}$	$-\frac{1}{19}$	$\frac{4}{19}$	4
	0	0	$\frac{14}{19}$	$\frac{1}{19}$	$-\frac{14}{19}$	$-\frac{1}{19}$	$C - 19$
	0	0	0	0	M	M	0

5. (a) Minimize $C = 4x_1 + 6x_2 + x_3$

where $x_1, x_2, x_3 \geq 0$

and
$$x_1 + 2x_2 \geq 10$$
$$x_2 + 4x_3 \geq 20$$
$$3x_1 + x_3 \geq 40$$

(b)
$$x_1 + 2x_2 - x_4 + A_1 = 10$$
$$x_2 + 4x_3 - x_5 + A_2 = 20$$
$$3x_1 + x_3 - x_6 + A_3 = 40$$

(c) At optimality:

	x_1	x_2	x_3	x_4	x_5	x_6	A_1	A_2	A_3	
x_1	1	2	0	−1	0	0	1	0	0	10
x_3	0	−6	1	3	0	−1	−3	0	1	10
x_5	0	−25	0	12	1	−4	−12	−1	4	20
	0	4	0	1	0	1	−1	0	−1	$C - 50$
	0	0	0	0	0	0	M	M	M	0

6. (a) Minimize $C = 3x_1 + 2x_2$

where $x_1, x_2 \geq 0$

and
$$x_1 + x_2 \leq 8$$
$$6x_1 + 4x_2 \geq 12$$
$$5x_1 + 8x_2 = 20$$

(b)
$$x_1 + x_2 + x_3 = 8$$
$$6x_1 + 4x_2 - x_4 + A_1 = 12$$
$$5x_1 + 8x_2 + A_2 = 20$$

(c) At optimality:

	x_1	x_2	x_3	x_4	A_1	A_2	
x_3	0	0	1	$\frac{3}{28}$	$-\frac{3}{28}$	$-\frac{1}{14}$	$\frac{37}{7}$
x_1	1	0	0	$-\frac{2}{7}$	$\frac{2}{7}$	$-\frac{1}{7}$	$\frac{4}{7}$
x_2	0	1	0	$\frac{5}{28}$	$-\frac{5}{28}$	$\frac{3}{14}$	$\frac{15}{7}$
	0	0	0	$\frac{1}{2}$	$-\frac{1}{2}$	0	$C - 6$
	0	0	0	0	M	M	0

7. (a) Maximize $P = 3x_1 + 4x_2 + 5x_3 + 6x_4$

where all variables are nonnegative

and
$$2x_1 + 4x_2 + 3x_3 + 4x_4 = 20$$
$$3x_1 + 2x_2 + 2x_3 + 5x_4 = 40$$

(b)
$$2x_1 + 4x_2 + 3x_3 + 4x_4 + A_1 = 20$$
$$3x_1 + 2x_2 + 2x_3 + 5x_4 + A_2 = 40$$

(c) Optimal tableau to auxiliary problem:

	x_1	x_2	x_3	x_4	A_1	A_2	
x_1	1	2	$\frac{3}{2}$	2	$\frac{1}{2}$	0	10
A_2	0	-4	$-\frac{5}{2}$	-1	$-\frac{3}{2}$	1	10
	0	-2	$\frac{1}{2}$	0	$-\frac{3}{2}$	0	$P - 30$
	0	$-4M$	$-\frac{5}{2}M$	$-M$	$-\frac{5}{2}M$	0	$10M$

8. (a) Maximize $P = 4x_1 + 3x_2 - x_3 + 2x_4 + 6x_5$

where all variables are nonnegative

and
$$\begin{aligned} 3x_1 \quad + x_3 \quad - x_5 &= 5 \\ x_1 + x_2 \quad - 3x_4 \quad &= -12 \\ x_2 + x_3 \quad + x_5 &= 4 \end{aligned}$$

(b)
$$\begin{aligned} 3x_1 \quad + x_3 \quad - x_5 + A_1 &= 5 \\ -x_1 - x_2 \quad + 3x_4 \quad + A_2 &= 12 \\ x_2 + x_3 \quad + x_5 + A_3 &= 4 \end{aligned}$$

(c) Optimal tableau:

	x_1	x_2	x_3	x_4	x_5	
x_1	1	$\frac{1}{3}$	$\frac{2}{3}$	0	0	3
x_4	0	$-\frac{2}{9}$	$\frac{2}{9}$	1	0	5
x_5	0	1	1	0	1	4
	0	$-\frac{35}{9}$	$-\frac{91}{9}$	0	0	$P - 46$

9. The objective is to maximize P. Is the following tableau optimal?

	x_1	x_2	x_3	x_4	A_1	A_2	
	1	1	-1	0	1	0	10
	3	2	0	-1	0	1	35
	-6	-5	0	0	$-M$	$-M$	P

10. Form Tableau 0 and Tableau 1 for the following problem, to be solved by the M method:

Minimize $Z = 3x_1 + 4x_2$

where $x_1 \geq 0, \quad x_2 \geq 0$

and
$$\begin{aligned} 3x_1 + 4x_2 &\geq 4 \\ 7x_1 - 2x_2 &= 10 \end{aligned}$$

11. What conclusions can be drawn from the following optimal tableau?

	x_1	x_2	x_3	x_4	x_5	
x_2	$\frac{8}{3}$	1	1	0	0	10
x_5	-7	0	-1	-1	1	30
	$-\frac{2}{3}$	0	-1	0	0	$C - 10$
	M	0	M	M	0	$-30M$

12. Can you obtain an initial basic feasible solution to the following program without using artificial variables?

$$\text{Maximize} \quad P = x_1 + 2x_2 + 3x_3$$

where all variables are nonnegative

$$\begin{aligned} \text{and} \quad 3x_1 + 2x_2 \qquad &= 9 \\ 3x_2 + 4x_3 &= 16 \end{aligned}$$

13. The National Factory produces an alloy which must meet the following specifications: The specific gravity of this alloy must not exceed 1.00; it must contain more than 20 percent magic metal; its melting point must exceed 1000°F. The alloy is a blend of three raw materials, X_1, X_2, and X_3, which have the following properties:

	X_1	X_2	X_3
Specific gravity	0.88	0.94	1.06
Percentage of magic metal	16%	24%	28%
Melting point	1200°F	1000°F	900°F

The costs of the raw materials are in the ratio of cost (unit X_1) to cost (unit X_2) to cost (unit X_3) equal to 2:5:1. The objective is to find the optimal proportion in which to blend the raw materials to obtain an alloy meeting all specifications at least cost. To this end, let x_1 denote the percentage of X_1 in the alloy, x_2 the percentage of X_2, x_3 the percentage of X_3. Show that these variables must satisfy the following nontrivial constraints:

$$\begin{aligned} 0.88x_1 + 0.94x_2 + 1.06x_3 &\le 1.00 \\ 0.16x_1 + 0.24x_2 + 0.28x_3 &\ge 0.20 \\ 12\ x_1 + 10\ x_2 + 9\ x_3 &\ge 10 \\ x_1 + x_2 + x_3 &= 1 \end{aligned}$$

Use the simplex algorithm to find the optimal values of x_1, x_2, x_3.

14. The marketing department must decide how to reach, at least cost, 20 million potential customers, 75 percent of whom should have a disposable income of at least $12,000 per annum. Research has furnished the following data:

Medium	*Cost per unit of medium*	*Potential audience (millions)*	*Audience with desired income (millions)*
Magazine X	90	3	2
Television program Y	600	15	5
Radio program Z	50	2	3

Use the simplex algorithm to determine the best way of allocating the advertising budget.

15. Find a nonnegative solution to the system of equations

$$3x_1 + 4x_2 + 5x_3 + 2x_4 = 9$$

$$x_1 + x_2 + 4x_3 + x_4 = 5$$

$$2x_1 + 3x_2 + x_3 + 5x_4 = 22$$

16. Clarify the following remarks: In small programs we can scan the original set of constraints and insert just enough artificial variables to get a first feasible tableau for the auxiliary problem. In practice, however, it is generally more expensive to have the computer scan the constraints than to instruct it to insert an artificial variable for every nontrivial constraint, whether or not these constraints already contain slack variables.

17. A 60-minute television program consists of a male guitarist, a female vocalist, and commercials sponsored by a political candidate. The guitar player insists on performing at least twice as long as the vocalist; the sponsor requires at least 5 minutes of commercials; the station requires the singing performance to last at least as long as the commercial. The guitar player, the vocalist, and the commercial cost $200, $150, and $100 per minute, respectively.

(a) Which of the constraints causes the initial tableau to be degenerate?

(b) Find the minimum cost of programming the show by applying the simplex algorithm.

18. Criticize the following exercise: Discuss the similarities and differences between a maximization problem and a minimization problem using the simplex algorithm.

19. *True or false?*

(a) We can see immediately the dimension of a linear program, that is, the number of basic and nonbasic variables at optimality.

(b) The number of basic variables in Phase I is the same as in Phase II.

20. Clarify the following statements:

(a) Each addition of an artificial variable causes violation of the original constraint.

(b) In the nondegenerate case, the number of iterations in Phase I is never greater than the maximum number of basic feasible solutions in the auxiliary problem.

6.2 FREE VARIABLES

The numerical difference between an artificial variable and the corresponding slack variable is unconstrained in sign and can be considered a *free variable.* More precisely, if $u_1 \geq 0$ and $u_2 \geq 0$, then the variable $u = u_1 - u_2$ is unconstrained in sign. For example, if $u_1 = 0$, then $u = -u_2$, which is nonpositive; but if $u_2 = 0$, then $u = u_1$, which is nonnegative.

Below, we solve a linear program in which some of the variables are free. *It must be stressed again that the simplex algorithm presupposes that all variables are constrained to be nonnegative.* The rule for choosing the pivot row depends upon two assumptions:

(1) All stub values b_i are nonnegative
(2) All variables x_i are constrained to be nonnegative

To illustrate, consider the constraint

$$3x - y + s_1 = b_1 \tag{1}$$

Then the statement that $\frac{1}{3}b_1$ is the maximum value of x consistent with $y = 0$ and $s_1 \geq 0$ depends on both the sign of b_1 and the nonnegativity of x. Furthermore, suppose $b_1 > 0$. Then the statement that Eq. (1) puts no restriction on the size of y still depends on the assumption that $y \geq 0$.

Now we turn to the following program:

PROBLEM 6.7.

$$\begin{aligned} \text{Maximize} \quad & P = 5u + 6v \\ \text{where} \quad & 3u + v \leq 1 \\ & 3u + 4v \leq 0 \end{aligned}$$

Here u and v are unconstrained.

One way to solve this problem is to let $u = u_1 - u_2$, $v = v_1 - v_2$, where $u_1 \geq 0$, $u_2 \geq 0$, $v_1 \geq 0$, $v_2 \geq 0$. Then Problem 6.7 is transformed into the following auxiliary problem.

PROBLEM 6.7A.

$$\begin{aligned} \text{Maximize} \quad & P = 5u_1 - 5u_2 + 6v_1 - 6v_2 \\ & \text{where all variables are nonnegative} \\ \text{and} \quad & 3u_1 - 3u_2 + v_1 - v_2 \leq 1 && (1) \\ & 3u_1 - 3u_2 + 4v_1 - 4v_2 \leq 0 && (2) \end{aligned}$$

Problem 6.7A will be solved by the simplex algorithm; then we shall discuss why Problem 6.7A is equivalent to Problem 6.7. In the following tableaux, s_1 and s_2 denote slack in Eqs. (1) and (2), respectively.

Tableau 1

	u_1	u_2	v_1	v_2	s_1	s_2		
s_1	3	−3	1	−1	1	0	1	(1)
s_2	3	−3	④	−4	0	1	0	←(0)
	5	−5	6	−6	0	0	P	
			↑					

Tableau 2

	u_1	u_2	v_1	v_2	s_1	s_2		
s_1	$\frac{9}{4}$	$-\frac{9}{4}$	0	0	1	$-\frac{1}{4}$	1	$(\frac{4}{9})$
v_1	$\textcircled{\frac{3}{4}}$	$-\frac{3}{4}$	1	-1	0	$\frac{1}{4}$	0	$\leftarrow(0)$
	$\frac{2}{4}$ ↑	$-\frac{2}{4}$	0	0	0	$-\frac{3}{2}$	P	

Tableau 3

	u_1	u_2	v_1	v_2	s_1	s_2		
s_1	0	0	-3	$\textcircled{3}$	1	-1	1	$\leftarrow(\frac{1}{3})$
u_1	1	-1	$\frac{4}{3}$	$-\frac{4}{3}$	0	$\frac{1}{3}$	0	
	0	0	$-\frac{2}{3}$	$\frac{2}{3}$ ↑	0	$-\frac{5}{3}$	P	

Tableau 4

	u_1	u_2	v_1	v_2	s_1	s_2	
v_2	0	0	-1	1	$\frac{1}{3}$	$-\frac{1}{3}$	$\frac{1}{3}$
u_1	1	-1	0	0	$\frac{4}{9}$	$-\frac{1}{9}$	$\frac{4}{9}$
	0	0	0	0	$-\frac{2}{9}$	$-\frac{13}{9}$	$P-\frac{2}{9}$

Tableau 4 is terminal. The optimum solution to Problem 6.7A is $u_1 = \frac{4}{9}$, $u_2 = 0$, $v_1 = 0$, $v_2 = \frac{1}{3}$, and $P = \frac{2}{9}$. Hence the optimum solution to Problem 6.7 is $u = u_1 - u_2 = \frac{4}{9}$, $v = v_1 - v_2 = 0 - \frac{1}{3} = -\frac{1}{3}$, and $P = \frac{2}{9}$.

The optimum of the auxiliary problem always yields the optimum of the original problem because:

(1) With any solution to the original problem we can always associate one or more solutions to the auxiliary problem in such a way that the objective functions in both problems take on the same value.

To illustrate, note that $u = 1$ and $v = -2$ is a feasible point for the original problem. Clearly, to find a solution to the auxiliary problem we need only consider the possibilities defined by

$$1 = u = u_1 - u_2$$
$$-2 = v = v_1 - v_2$$

for $u_1 \geq 0$, $u_2 \geq 0$, $v_1 \geq 0$, $v_2 \geq 0$. This implies that

$$u_1 = 1 + u_2, \qquad u_2 \geq 0$$
$$v_1 = -2 + v_2, \qquad v_2 \geq 2$$

For example, if $u_2 = 1$ and $v_2 = 5$, then $u_1 = 2$ and $v_1 = 3$; if $u_2 = 0$ and $v_2 = 2$, then $u_1 = 1$ and $v_1 = 0$; and so on. By their very definition, the objective functions in both problems must take on the same value at corresponding points.

(2) Now suppose that (u_1, u_2, v_1, v_2) denotes a feasible solution to the auxiliary problem. With this solution we can associate a unique feasible solution to the original problem by means of the equations $u = u_1 - u_2$ and $v = v_1 - v_2$. Obviously, for each of these solutions the auxiliary and original objective functions are equal.

(3) Let $(u_1^*, u_2^*, v_1^*, v_2^*)$ denote an optimal solution to the auxiliary problem. Then $u^* = u_1^* - u_2^*$ and $v^* = v_1^* - v_2^*$ is an optimal solution to the original problem. To prove this point, suppose that u^* and v^* *do not* constitute an optimal solution. Then there is a better solution to the original problem, with which, by Eq. (1) above, we can associate a solution to the auxiliary problem better than $(u_1^*, u_2^*, v_1^*, v_2^*)$. This, however, is contrary to the original hypothesis.

Using reasoning similar to that employed above, we can conclude that if the auxiliary region of feasibility is empty, then the original region is also empty, and if the auxiliary objective function is unbounded then the original objective function is also unbounded.

The reader should not think that free variables are academic or unnatural. In general, whenever a variable x represents a *change* we must treat it as unconstrained in sign. For example, x might be change in inventory, which could certainly be either an increase or a decrease. When we put $x = x_1 - x_2$, where $x_1 \geq 0$ and $x_2 \geq 0$, we are implicitly saying that we might have originally formulated our problem somewhat differently—for example, in terms of x_1 equals quantity manufactured and x_2 equals quantity sold.

As an illustration, consider Problems 1.1, 1.2, and 6.7. The optimum solution of Problem 1.1 is $x = 8$, $y = 24$; this occurs at the corner where no time remains on either machine I or machine II. If, afterward, one more unit of time does become available on machine I, we might ask ourselves how best to *change* the prior optimum values of x and y. The answer to this question is what we seek in Problem 6.7, where u represents the change in x and v the change in y. Since the solution to Problem 6.7 is $u = \frac{4}{9}$, $v = -\frac{1}{3}$, it is clear that, if one more unit of time is available on machine I, to maximize P we should increase x from 8 to $8\frac{4}{9}$ and decrease y from 24 to $23\frac{2}{3}$. This agrees with the solution to Problem 1.2.

6.3 SUMMARY OF THE BASIC FEATURES OF THE SIMPLEX ALGORITHM

The procedure for solving a linear program is the reverse of the order of exposition employed in Chapters 3 to 6. To analyze a linear program by the simplex method, we proceed as follows:

(1) Replace each free variable by the difference of two nonnegative variables.

(2) If any b_i is negative, multiply constraint (i) by -1.

(3) Convert all inequalities to equalities by the use of nonnegative slack or surplus variables, as the case may be. (Keep all $b_i \geq 0$.)

(4) Form an auxiliary problem, using artificial variables as needed. (We can use an artificial variable for every equality, whether needed or not—or else, we can add only enough artificial variables to complete the set of distinct unit columns already available.)

(5) Use either the M method or the sum-of-the-artificialities method to drive all artificial variables to 0. Then: (a) If all artificial variables cannot be driven to 0, the original problem has no feasible solution; (b) if all artificial variables can be driven to 0, the original problem has a feasible point and also an extreme point.

(6) Delete all artificial variables and associated rows and columns, as soon as this necessarily reduces the auxiliary tableau to an *adjusted tableau* of the original problem. This terminates Phase I.

(7) Begin Phase II with the above reduced tableau. From now on, the simplex algorithm either converges to an optimum solution, or else it informs us that the original problem possesses unbounded solutions. (If we prefer to think of the M method as a one-phase method; that is, *if we never delete artificial variables*, then the auxiliary problem either converges to a solution whose objective function does or does not involve M. In the first case the solution to the original problem is found by disregarding the artificial variables; in the second case, the original problem has no feasible solution.)

Exercises

1.

$$\text{Maximize} \quad Z = 3x_1 + 4x_2$$
$$\text{where} \quad x_1 \geq 0, \text{ but } x_2 \text{ is free}$$
$$\text{and} \quad 4x_1 + 2x_2 \leq 3$$
$$2x_1 + x_2 \leq 2$$

2. See Problem 1.1.

(a) Find the effect on optimal profit of relaxing constraint (2) so that it reads "$3x + 4y \leq 121$," keeping all other constraints fixed.

(b) Find the effect on optimal profit of relaxing constraint (1) so that it reads "$3x + y \leq 148$," keeping all other constraints fixed.

3.

$$\text{Minimize} \quad Z = 4x_1 + 6x_2 + 2x_3$$
$$\text{where} \quad x_1 \geq 0, \quad x_2 \geq 0, \quad x_3 \text{ is unrestricted}$$
$$\text{and} \quad 2x_1 - 3x_2 + x_3 = 4$$
$$7x_1 - 10x_2 + x_3 \geq 9$$

4. In Exercise 3, allow both x_2 and x_3 to be free. How does this affect the minimum of Z?

5. Show how to use a computer code for the simplex algorithm to:

(a) Find a nonnegative solution, if one exists, to the system:

$$\begin{aligned} 5x_1 + 8x_2 - 8x_3 - 2x_4 &= 0 \\ 2x_1 - 6x_2 + 5x_3 - 8x_4 &= 0 \\ x_1 + x_2 + x_3 + x_4 &= 1 \end{aligned}$$

(b) Find any solution to the system in (a).

6. Consider the system of equations:

$$\begin{aligned} a_{11}x_1 + a_{12}x_2 + a_{13}x_3 + a_{14}x_4 &= b_1 \\ a_{21}x_1 + a_{22}x_2 + a_{23}x_3 + a_{24}x_4 &= b_2 \\ a_{31}x_1 + a_{32}x_2 + a_{33}x_3 + a_{34}x_4 &= b_3 \end{aligned}$$

Show that this system is equivalent to the following system of inequalities:

$$\begin{aligned} a_{11}x_1 + a_{12}x_2 + a_{13}x_3 + a_{14}x_4 &\leq b_1 \\ a_{21}x_1 + a_{22}x_2 + a_{23}x_3 + a_{24}x_4 &\leq b_2 \\ a_{31}x_1 + a_{32}x_2 + a_{33}x_3 + a_{34}x_4 &\leq b_3 \\ c_1x_1 + c_2x_2 + c_3x_3 + c_4x_4 &\geq b_1 + b_2 + b_3 \end{aligned}$$

where $c_1 = a_{11} + a_{21} + a_{31}$, $c_2 = a_{12} + a_{22} + a_{32}$, $c_3 = a_{13} + a_{23} + a_{33}$, $c_4 = a_{14} + a_{24} + a_{34}$.

7. Show that any set of m linear equations is equivalent to a set of $(m + 1)$ linear inequalities.

7
MULTIPLIER THEORY

In this chapter we describe in greater detail how pivoting in the simplex algorithm transforms the rows of one tableau into those of another. Our objective in emphasizing the *multiplier theorems* is to show that the terminal tableau actually contains the *record*, or *imprint*, of how it evolved from the initial tableau. In Chapter 8 this record of past calculations is applied to the study of implicit prices and postoptimality analysis. The material of the present chapter is also needed for the *multiplier method for transportation problems* in Chapter 14.

The reader must be patient; the importance of the theory of multipliers cannot be fully appreciated until after it has been put into action.

7.1 VECTORS

A vector is an ordered list of numbers, usually enclosed in parentheses. For example, (2, 3, 4) and $\begin{pmatrix}1\\2\end{pmatrix}$ are vectors. The numbers contained in a vector are called its *components*. When the components are listed in a row they are often separated by commas; such a vector is called a *row vector*. When the components are listed in a column it is called a *column vector*. The number of components in a vector is its *dimension*. Thus (5, 6, 7, 8910) is a four-dimensional (row) vector, and the real number 7 can be thought of as a one-dimensional (row, or column) vector. Real numbers will sometimes be called *scalars*.

Two row vectors are *equal*, or *identical*, if they have the same dimension and their components are equal, position by position. Thus (0, 1) and (0, 1, 0) are not equal. Furthermore, if $(x, 1) = (3, 1)$ then $x = 3$; and if $(x + y, 2) = (3, x)$, then $x = 2$ and $y = 1$. A similar definition holds for the equality of column vectors.

Vectors are sometimes designated by arrows, for example $\vec{x}$, or imprinted material by bold-face type, for example, $\mathbf{x}$, especially when they are referred to without listing their components.

Often, to save space, rather than display a column vector vertically we will write it in a row with some distinguishing mark to remind us that it has been *transposed*. Thus, in place of

$$\mathbf{x} = \begin{pmatrix} x_1 \\ x_2 \\ \cdot \\ \cdot \\ \cdot \\ x_n \end{pmatrix}$$

we will write $(\mathbf{x})' = (x_1, x_2, \ldots, x_n)'$. The prime superscript denotes transposition and we call $(\mathbf{x})'$ the transpose of $\mathbf{x}$.

If $\mathbf{a}$ and $\mathbf{b}$ are two (row or column) vectors of the same dimensions, the vector whose components are found by adding corresponding components of $\mathbf{a}$ and $\mathbf{b}$ is called the *sum* of $\mathbf{a}$ and $\mathbf{b}$ and denoted by $\mathbf{a} + \mathbf{b}$. Accordingly, if $\mathbf{a} = (a_1, a_2, \ldots, a_n)$ and $\mathbf{b} = (b_1, b_2, \ldots, b_n)$, then

$$\mathbf{a} + \mathbf{b} = (a_1 + b_1, a_2 + b_2, \ldots, a_n + b_n)$$

The sum of two vectors is defined only if both are row vectors or both column vectors, and then only if they have the same dimension. Thus, if $\mathbf{a} = (1, 2)$, $\mathbf{b} = (1, 2, 3)$, $\mathbf{c} = (4, 5, 6)$, and $\mathbf{d} = \begin{pmatrix} 7 \\ 8 \\ 9 \end{pmatrix}$, only $\mathbf{b} + \mathbf{c}$ is defined, and

$$\mathbf{b} + \mathbf{c} = (5, 7, 9)$$

By definition, "k times the vector $\mathbf{b}$" is the vector obtained by multiplying each component of $\mathbf{b}$ by k. This multiple is written $k\mathbf{b}$ and called a *scalar multiple* of $\mathbf{b}$. To illustrate, if $\mathbf{b} = (1, 2)$, then $2\mathbf{b} = (2, 4) = \mathbf{b} + \mathbf{b}$ and $\sqrt{3}\,\mathbf{b} = (\sqrt{3}, 2\sqrt{3})$.

The vector $(-1)\mathbf{b}$ is called the *negative* of $\mathbf{b}$. It may be written $-\mathbf{b}$. Subtraction of vectors can be defined in terms of addition, in the usual way. Thus $\mathbf{a} - \mathbf{b} = \mathbf{a} + (-1)\mathbf{b}$, by definition.

The vector all of whose components are 0 is known as the "zero vector" and designated by the symbol $\mathbf{0}$. It is easy to see that $(0)\mathbf{b} = \mathbf{0}$ and $\mathbf{b} - \mathbf{b} = \mathbf{0}$.

The three simple operations on vectors defined above, namely, addition, subtraction, and scalar multiplication, allow us to represent a system of linear equations as a *combination of column vectors.* To illustrate, consider the system

$$\begin{aligned} a_{11}x_1 + a_{12}x_2 + a_{13}x_3 + a_{14}x_4 &= b_1 \\ a_{21}x_1 + a_{22}x_2 + a_{23}x_3 + a_{24}x_4 &= b_2 \\ a_{31}x_1 + a_{32}x_2 + a_{33}x_3 + a_{34}x_4 &= b_3 \end{aligned}$$

We can represent this system as

$$x_1\begin{pmatrix} a_{11} \\ a_{21} \\ a_{31} \end{pmatrix} + x_2\begin{pmatrix} a_{12} \\ a_{22} \\ a_{32} \end{pmatrix} + x_3\begin{pmatrix} a_{13} \\ a_{23} \\ a_{33} \end{pmatrix} + x_4\begin{pmatrix} a_{14} \\ a_{24} \\ a_{34} \end{pmatrix} = \begin{pmatrix} b_1 \\ b_2 \\ b_3 \end{pmatrix}$$

The reader who does not see this should perform the indicated multiplications and additions, treating the x_i as scalars and remembering that two vectors are equal only if all pairs of corresponding components are equal.

Exercises

1. Let $\mathbf{a} = (1, 3, 5, 7)$ and $\mathbf{b} = (2, -3, 0, 9)$. Verify that:

(a) $\mathbf{a} + \mathbf{b} = (3, 0, 5, 16)$
$\mathbf{a} - \mathbf{b} = (-1, 6, 5, -2)$
$4\mathbf{a} = (4, 12, 20, 28)$

(b) $-7\mathbf{b} = (-14, 21, 0, -63)$
$(0)\mathbf{a} = (0, 0, 0, 0)$
$(-1)\mathbf{b} = (-2, 3, 0, -9)$

2. Find the vectors $\mathbf{u}$ and $\mathbf{v}$, given that $\mathbf{u} + \mathbf{v} = (1, 0)$ and $\mathbf{u} - \mathbf{v} = (0, 1)$.

3. Solve the following vector equations:

(a) $\begin{pmatrix} x \\ y \end{pmatrix} + \begin{pmatrix} 3 \\ 4 \end{pmatrix} = \begin{pmatrix} 5 \\ 7 \end{pmatrix}$ (b) $\begin{pmatrix} x \\ 4 \end{pmatrix} + \begin{pmatrix} 3 \\ y \end{pmatrix} = \begin{pmatrix} 3 \\ -3 \end{pmatrix}$

4. Given the system

$$\begin{aligned} 3x + y &= 48 \\ 3x + 4y &= 120 \end{aligned}$$

Verify that we can express this system in vector form as

$$x\begin{pmatrix} 3 \\ 3 \end{pmatrix} + y\begin{pmatrix} 1 \\ 4 \end{pmatrix} = \begin{pmatrix} 48 \\ 120 \end{pmatrix}$$

5. The vector $\mathbf{v} = (2, 4, 6)$ is derived from $\mathbf{w} = \begin{pmatrix} 1 \\ 2 \\ 3 \end{pmatrix}$. How can we express the relationship between $\mathbf{v}$ and $\mathbf{w}$?

6. *True or false?*

(a) For any vector $\mathbf{x}$, $[(\mathbf{x})']' = \mathbf{x}$.
(b) $\mathbf{a} - (\mathbf{b} - \mathbf{c}) = \mathbf{a} - \mathbf{b} + \mathbf{c}$, for any vectors $\mathbf{a}$, $\mathbf{b}$, and $\mathbf{c}$.

7. Express the following systems as combinations of column vectors.

(a)
$$\begin{aligned} x + 2y + 3z &= b_1 \\ 4x + 5y + 6z &= b_2 \\ 7x + 8y + 9z &= b_3 \end{aligned}$$

(b)
$$\begin{aligned} x + 2y + 3z &= b_1 \\ 4x + 5y + 6z &= b_2 \end{aligned}$$

(c)
$$\begin{aligned} a_{11}x_1 + a_{12}x_2 + a_{13}x_3 &= b_1 \\ a_{21}x_1 + a_{22}x_2 + a_{23}x_3 &= b_2 \\ a_{31}x_1 + a_{32}x_2 + a_{33}x_3 &= b_3 \\ a_{41}x_1 + a_{42}x_2 + a_{43}x_3 &= b_4 \end{aligned}$$

7.2 THE PRODUCT OF TWO VECTORS

Given two n-dimensional vectors, $\mathbf{x} = (x_1, \ldots, x_n)$ and $\mathbf{y} = (y_1, \ldots, y_n)$; their *product* is defined to be the number $x_1y_1 + x_2y_2 + \cdots + x_ny_n$. This product is sometimes called the *inner product*, *scalar product*, or *dot product* of $\mathbf{x}$ and $\mathbf{y}$ and is denoted by $\mathbf{x} \cdot \mathbf{y}$. Thus we write

$$\mathbf{x} \cdot \mathbf{y} = x_1y_1 + \cdots + x_ny_n$$

Whenever two vectors have the same dimension, their product is defined whether they are both row vectors, both column vectors, or mixed. To illustrate, let $\mathbf{a} = (1, 2, 3)$, $\mathbf{b} = (4, -3, 2)$, and $\mathbf{c} = \begin{pmatrix} 0 \\ 1 \\ -2 \end{pmatrix}$. Then $\mathbf{a} \cdot \mathbf{b} = 1(4) + 2(-3) + 3(2) = 4$; $\mathbf{a} \cdot \mathbf{c} = 1(0) + 2(1) + 3(-2) = -4$; and $\mathbf{b} \cdot \mathbf{c} = -7$.

We will often need to calculate the product of two vectors where one is given as the row vector $\mathbf{r} = (a_1, a_2, \ldots, a_n)$, the other as the column vector

$$\mathbf{k} = \begin{pmatrix} b_1 \\ b_2 \\ \cdot \\ \cdot \\ \cdot \\ b_n \end{pmatrix}$$

In this case we write $\mathbf{r} \cdot \mathbf{k}$ as follows:

$$\mathbf{r} \cdot \mathbf{k} = (a_1, a_2, \ldots, a_n) \begin{pmatrix} b_1 \\ b_2 \\ \cdot \\ \cdot \\ \cdot \\ b_n \end{pmatrix} = a_1b_1 + a_2b_2 + \cdots + a_nb_n$$

The row vector will always be written to the left of the column vector.

Two vectors are called *orthogonal* if their product is 0. For example, $\mathbf{u}_1 = (1, 0, 0)$ and $\mathbf{u}_3 = (0, 0, 1)$ are orthogonal, as are $\mathbf{v}_1 = (2, 1)$ and $\mathbf{v}_2 = (1, -2)$.

Recall that in Chapter 4 we focused attention on the distinct *unit columns* of a tableau. Obviously, any two distinct unit columns can be thought of as vectors, and, as such, they are orthogonal.

An intuitive way of remembering the definition of the product of two vectors is to think of a "bundle of purchases": Given $\mathbf{x} = (x_1, \ldots, x_n)$ and $\mathbf{y} = (y_1, \ldots, y_n)$. Think of the x's as amounts purchased and the y's as prices paid. Then the product $\mathbf{x} \cdot \mathbf{y} = x_1y_1 + \cdots + x_ny_n$ represents the "total purchase price."

Exercises

1. Let $\mathbf{v}_1 = (3, 7, 5)$ and $\mathbf{v}_2 = (5, 9, -3)$. Calculate:

(a) $\mathbf{v}_1 \cdot \mathbf{v}_1$ (b) $\mathbf{v}_2 \cdot \mathbf{v}_2$ (c) $\mathbf{v}_1 \cdot \mathbf{v}_2$
(d) $(3\mathbf{v}_1) \cdot \mathbf{v}_2$ (e) $\mathbf{v}_1 \cdot (3\mathbf{v}_2)$ (f) $3(\mathbf{v}_1 \cdot \mathbf{v}_2)$

2. Show that the following pairs of vectors are orthogonal:

(a) $\mathbf{v}_1 = (3, 7)$ and $\mathbf{v}_2 = (7, -3)$.
(b) $\mathbf{v}_1 = (a_1, a_2)$ and $\mathbf{v}_2 = (a_2, -a_1)$.
(c) $\mathbf{v}_1 = (-6, -3, 2)$ and $\mathbf{v}_2 = (-2, 6, 3)$.

3. Verify that

$$(\mathbf{v}_1 + \mathbf{v}_2) \cdot \mathbf{v}_3 = \mathbf{v}_1 \cdot \mathbf{v}_3 + \mathbf{v}_2 \cdot \mathbf{v}_3$$

where $\mathbf{v}_1 = (a_1, a_2, a_3)$, $\mathbf{v}_2 = (b_1, b_2, b_3)$, and $\mathbf{v}_3 = (c_1, c_2, c_3)$.

4. *True or false?*

(a) The only four-dimensional vector orthogonal to itself is $\mathbf{0} = (0, 0, 0, 0)$.
(b) The only four-dimensional vector orthogonal to all other four-dimensional vectors is $\mathbf{0} = (0, 0, 0, 0)$.

5. Let $\mathbf{u}_1 = (1, 0, 0)$, $\mathbf{u}_2 = (0, 1, 0)$, $\mathbf{u}_3 = (0, 0, 1)$. For any vector $\mathbf{v}$ with three components, show that:

(a) $\mathbf{v} \cdot \mathbf{u}_1$ equals the *first* component of $\mathbf{v}$.
(b) $\mathbf{v} \cdot \mathbf{u}_2$ equals the *second* component of $\mathbf{v}$.
(c) $\mathbf{v} \cdot \mathbf{u}_3$ equals the *third* component of $\mathbf{v}$.

Generalize the above result to n-dimensional vectors.

6. Let $\mathbf{v} = (v_1, v_2, v_3)$, $\mathbf{k} = (k_1, k_2, k_3)$, and $\mathbf{u}_1$, $\mathbf{u}_2$, and $\mathbf{u}_3$ be as in Exercise 5. Verify that

$$\mathbf{v} \cdot \mathbf{k} = k_1(\mathbf{v} \cdot \mathbf{u}_1) + k_2(\mathbf{v} \cdot \mathbf{u}_2) + k_3(\mathbf{v} \cdot \mathbf{u}_3)$$

Generalize this result.

7.3 SIMPLEX MULTIPLIERS

Consider the following problem:

PROBLEM 7.1. Maximize $P = 5x_1 + 20x_2$

subject to $x_1 \geq 0, \quad x_2 \geq 0, \quad s_1 \geq 0, \quad s_2 \geq 0$

$$\begin{aligned} \text{and}\quad 3x_1 + x_2 + s_1 \quad &= 48 \\ 3x_1 + 4x_2 \quad + s_2 &= 120 \end{aligned}$$

The simplex solution of this program is

Tableau 1

	x_1	x_2	s_1	s_2		
s_1	3	1	1	0	48	(48)
s_2	3	④	0	1	120	←(30)
	5	20	0	0	$P - 0$	
		↑				

Tableau 2

	x_1	x_2	s_1	s_2	
s_1	$\frac{9}{4}$	0	1	$-\frac{1}{4}$	18
x_2	$\frac{3}{4}$	1	0	$\frac{1}{4}$	30
	-10	0	0	-5	$P - 600$

This tableau is terminal.

It will help clarify the discussion to follow if we draw attention to these points:

(1) In performing the simplex algorithm we pivot only on entries in the body of a tableau.

(2) Pivoting is a way of effecting elementary row operations. In performing a pivot operation, we multiply the pivot row by some constant and add a certain multiple of the pivot row to each of the other rows.

(3) The set of basic variables changes from tableau to tableau. Likewise, the set of corner, or nonbasic, variables changes from tableau to tableau.

(4) The set of slack variables does not change from tableau to tableau. More precisely, assume that the initial tableau contains a set of distinct unit columns, $s_1, s_2, \ldots, s_n$, ordered in the natural way, where the unit entry 1 of column s_j occurs in row j.

Henceforth let Tableau n (or in abbreviated form, Tbl n) denote the nth tableau in a sequence. By the very nature of pivoting, the last row of Tableau 2 above is obtained by adding a certain multiple of row 2 of Tableau 1 to row 3 of Tableau 1. Symbolically,

$$(\text{row 3 of Tbl 1}) + k\,(\text{row 2 of Tbl 1}) = (\text{row 3 of Tbl 2})$$

where k is the multiple in question. In Problem 7.1, $k = -5$. The reader can check this by noting that

$$\begin{aligned}(\text{row 3 of Tbl 1}) &= (5, 20, 0, 0, P - 0)\\ -5\,(\text{row 2 of Tbl 1}) &= -5(3, 4, 0, 1, 120)\\ &= (-15, -20, 0, -5, -600)\end{aligned}$$

Clearly, the sum is equal to row 3 of Tableau 2.

The fact that k must equal −5 can also be deduced as follows: In Tableau 1 the entries in the last row, under the s_1 and s_2 columns, are 0. Moreover, the only 1 in the s_1 column appears in row 1, and the only 1 in the s_2 column appears in row 2. Hence, the sum of row 3 of Tableau 1 and k times row 2 of Tableau 1 must exhibit k itself in the s_2 column. Symbolically,

	s_2		
$(\ldots,$	$0,$	$\ldots)$	Row 3 of Tbl 1
$+\,k(\ldots,$	$1,$	$\ldots)$	k (Row 2 of Tbl 1)
$=(\ldots,$	$k,$	$\ldots)$	Row 3 of Tbl 2

These are the operations which take us to row 3 of Tableau 2, where we observe −5 in the s_2 column. Hence $k = -5$.

For a more complex example, consider

PROBLEM 7.2. Maximize $P = 5x_1 + 6x_2$

where $x_1 \geq 0,\quad x_2 \geq 0,\quad s_1 \geq 0,\quad s_2 \geq 0$

and
$$\begin{aligned}3x_1 + x_2 + s_1 \quad &= 48\\ 3x_1 + 4x_2 \quad + s_2 &= 120\end{aligned}$$

The solution is

Tableau 1

	x_1	x_2	s_1	s_2		
s_1	3	1	1	0	48	(48)
s_2	3	④	0	1	120	←(30)
	5	6	0	0	$P - 0$	
		↑				

Tableau 2

	x_1	x_2	s_1	s_2		
s_1	$\textcircled{\frac{9}{4}}$	0	1	$-\frac{1}{4}$	18	←(8)
x_2	$\frac{3}{4}$	1	0	$\frac{1}{4}$	30	(40)
	$\frac{1}{2}$ ↑	0	0	$-\frac{3}{2}$	$P - 180$	

Tableau 3

	x_1	x_2	s_1	s_2	
x_1	1	0	$\frac{4}{9}$	$-\frac{1}{9}$	8
x_2	0	1	$-\frac{1}{3}$	$\frac{1}{3}$	24
	0	0	$-\frac{2}{9}$	$-\frac{13}{9}$	$P - 184$

The pivot operations performed above imply that the last row of Tableau 3 can be found by adding together the following three rows:

(a) Some multiple of row 1 of Tableau 1
(b) Some multiple of row 2 of Tableau 1
(c) Row 3 of Tableau 1

In other words, the following equation among row vectors must hold:

$$(\text{row 3 of Tbl 3}) = k_1(\text{row 1 of Tbl 1}) + k_2(\text{row 2 of Tbl 1}) + (\text{row 3 of Tbl 1})$$

where k_1 and k_2 are the multiples in question.

Just as in Problem 7.1, the values of k_1 and k_2 must appear in the objective row of Tableau 3. This is always the case, for the following reasons:

(a) $k_1(\text{row 1 of Tbl 1}) = k_1(3, 1, 1, 0, 48)$
(b) $k_2(\text{row 2 of Tbl 1}) = k_2(3, 4, 0, 1, 120)$
(c) $1(\text{row 3 of Tbl 1}) = 1\ (5, 6, 0, 0, P)$

The sum of these three row vectors is of the form

$$(\_, \_, k_1, k_2, \_)$$

But this vector must equal row 3 of Tableau 3, and hence $k_1 = -\frac{2}{9}$ and $k_2 = -\frac{13}{9}$.

We will soon generalize this result in several directions, but first it is helpful to change the format of our tableau. Instead of writing P in the lower right-hand corner of a tableau, we can introduce a new column, the $-P$ column, as shown below:

Tableau 1′

	x_1	x_2	s_1	s_2	$-P$	
s_1	3	1	1	0	0	48
s_2	3	4	0	1	0	120
$-P$	5	6	0	0	1	0

Tableau 1′ is evidently equivalent to Tableau 1 of Problem 7.2. We shall call the format of Tableau 1′ the format with "$-P$ upstairs." Here $-P$ is placed up with the variables labeling the columns, and because the entries in the column take the form of a unit vector it is also listed in the left-hand rim. In this format succeeding tableaux are

Tableau 2′

	x_1	x_2	s_1	s_2	$-P$	
s_1	$\frac{9}{4}$	0	1	$-\frac{1}{4}$	0	18
x_2	$\frac{3}{4}$	1	0	$\frac{1}{4}$	0	30
$-P$	$\frac{1}{2}$	0	0	$-\frac{3}{2}$	1	-180

Tableau 3′

	x_1	x_2	s_1	s_2	$-P$	
x_1	1	0	$\frac{4}{9}$	$-\frac{1}{9}$	0	8
x_2	0	1	$-\frac{1}{3}$	$\frac{1}{3}$	0	24
$-P$	0	0	$-\frac{2}{9}$	$-\frac{13}{9}$	1	-184

We are now ready to define $\mathbf{m}$, *the vector of simplex multipliers.* We assume that the initial tableau contains a set of unit columns arranged in natural order, like s_1, s_2, and $-P$ in Tableau 1′. *Then, by definition,*

$$\mathbf{m} = (\gamma_1, \gamma_2, 1)$$

where γ_1, γ_2, and 1 are the entries which appear in the objective row of the terminal tableau, in the columns which contained the unit vectors of Tableau 1′, in natural order. Thus, in Problem 7.2, $\mathbf{m} = (-\frac{2}{9}, -\frac{13}{9}, 1)$.

Let $\mathbf{k}_1, \mathbf{k}_2, \ldots, \mathbf{k}_6$ denote the columns of the initial tableau. *Then the "multiplier results" we have previously expressed in terms of "combinations of rows" can be reformulated using scalar products, as follows:* $\mathbf{m} \cdot \mathbf{k}_j =$ the element in the jth column of the terminal objective row.

To illustrate, in Problem 7.2, $\mathbf{m} = (-\frac{2}{9}, -\frac{13}{9}, 1)$, $\mathbf{k}_1 = \begin{pmatrix} 3 \\ 3 \\ 5 \end{pmatrix}$, and $\mathbf{k}_6 = \begin{pmatrix} 48 \\ 120 \\ 0 \end{pmatrix}$. Then $\mathbf{m} \cdot \mathbf{k}_1 = 0$ and $\mathbf{m} \cdot \mathbf{k}_6 = -184$.

These results can be generalized immediately. Suppose the initial and terminal tableaux of a problem are as indicated below:

Tableau 1 (Initial)

	x_1 x_2 $\cdots$ x_n	x_{n+1}	x_{n+2}	$\cdots$	x_{n+m}	$-P$	
x_{n+1}		1	0	$\cdots$	0	0	b_1
x_{n+2}		0	1	$\cdots$	0	0	b_2
$\cdot$		$\cdot$	$\cdot$		0	0	$\cdot$
$\cdot$		$\cdot$	$\cdot$				$\cdot$
x_{n+m}		0	0	$\cdots$	1	0	b_m
$-P$		0	0	$\cdots$	0	1	$-P_0$

Tableau t (Terminal)

	x_1 $\cdots\cdots$ x_n	x_{n+1}	x_{n+2}	$\cdots$	x_{n+m}	$-P$	
							B_1
							B_2
							$\cdot$
							$\cdot$
							$\cdot$
							B_m
		γ_1	γ_2	$\cdots$	γ_m	1	$-P_0 - (\Delta P) = -P^*$

Here P_0 equals the initial value of P (P_0 is usually 0), ΔP equals the *change* in P between the terminal and initial tableaux, and P^* is the terminal value of P. Clearly, if $P_0 = 0$, then $P^* = \Delta P$.

The vector of simplex multipliers is, by definition,

$$\mathbf{m} = (\gamma_1, \gamma_2, \ldots, 1)$$

Let $\mathbf{k}_1, \ldots, \mathbf{k}_{n+m}$, $-\mathbf{P}$ and $\mathbf{b} = (b_1, b_2, \ldots, b_m, -P_0)'$ denote the columns of Tableau 1. Then, by an analysis similar to that employed above, it follows that:

(a) $\mathbf{m} \cdot \mathbf{k}_j =$ the jth element in the objective row of Tableau t. In particular, if x_j is a basic variable at optimality, then $\mathbf{m}$ and $\mathbf{k}_j$ are orthogonal.

(b) $\mathbf{m} \cdot (-\mathbf{P}) = 1$

(c) $\mathbf{m} \cdot \mathbf{b} = -P^*$

These results constitute the scalar-product form of the *simplex-multiplier theorem.*

A good way to retain this theorem and form the required inner products is to replace the left rim of basic variables in Tableau 1 by the corresponding components of the optimum multiplier vector, as shown below:

Tableau 1

	x_1	x_2	$\cdots$	x_n	x_{n+1}	x_{n+2}	$\cdots$	x_{n+m}	$-P$	
γ_1					1	0	$\cdots$	0	0	b_1
γ_2					0	1	$\cdots$	0	0	b_2
.										.
.										.
.										.
γ_m					0	0	$\cdots$	1	0	b_m
1					0	0	$\cdots$	0	1	$-P_0$

The fact that the γ_i represent numbers makes it clear that the left rim of the above tableau contains multipliers, rather than basic variables. Using this format, we can form the inner product $\mathbf{m} \cdot \mathbf{k}_j$ simply by multiplying together corresponding elements in the left rim and in column j and then summing the products obtained.

Observe that part (c) of the simplex-multiplier theorem, when written out, reads:

$$-P^* = \mathbf{m} \cdot \mathbf{b} = (\gamma_1, \gamma_2, \ldots, \gamma_m, 1) \begin{pmatrix} b_1 \\ b_2 \\ \cdot \\ \cdot \\ \cdot \\ -P_0 \end{pmatrix} = \gamma_1 b_1 + \gamma_2 b_2 + \cdots + (-P_0)$$

and hence

$$\gamma_1 b_1 + \gamma_2 b_2 + \cdots + \gamma_m b_m = P_0 - P^* = -\Delta P \tag{1}$$

In Chapter 8 we shall analyze Eq. (1) and show when and why the numbers $-\gamma_i$ deserve the name *implicit prices.*

Exercises

1. The following tableau gives the optimal solution to a maximization problem whose slack variables are x_3 and x_4.

	x_1	x_2	x_3	x_4	$-P$	
x_1	1	0	1	−2	0	12
x_2	0	1	−2	6	0	24
$-P$	0	0	−2	−3	1	−200

(a) Describe in words the row operations needed to pass from the initial to the terminal tableau.

(b) Reconstruct the initial tableau.

2. Reread the illustrative problems of earlier chapters and verify the simplex-multiplier theorem.

3. The following tableau is terminal:

	x_1	x_2	x_3	x_4	$-P$	
x_3	0	$\frac{5}{2}$	1	$-\frac{1}{2}$	0	18
x_1	1	$\frac{3}{4}$	0	$\frac{1}{4}$	0	15
$-P$	0	$-\frac{1}{2}$	0	$-\frac{3}{2}$	1	−90

(a) If in the initial tableau x_3 and x_4 were basic, in this order, specify the optimal simplex multiplier.

(b) Can we specify the optimal simplex multiplier without knowing the order of appearance of the initial unit columns?

(c) By how much must the initial objective coefficient be increased for x_2 to be basic at optimality?

7.4 GENERALIZATIONS

In this section we give several more general reformulations of the multiplier theorem. As a first generalization, note that every result previously obtained still holds if we replace the phrase "*terminal* tableau" by "*current* tableau." To illustrate, consider Tableau 2′ of Problem 7.2. The multiplier vector of this tableau is recorded in the last row, in the s_1, s_2, and $-P$ columns, and reads

$$\mathbf{m} = (0, -\tfrac{3}{2}, 1)$$

To describe how the passage from Tableau 1′ to the objective row of Tableau 2′ was effected, replace the left rim of Tableau 1′ by the corresponding components of the multiplier vector of Tableau 2′, as shown below:

Tableau 1′

	x	y	s_1	s_2	$-P$	
0	3	1	1	0	0	48
$-\frac{3}{2}$	3	4	0	1	0	120
1	5	6	0	0	1	0

Thus the objective row of Tableau 2′ must have been formed by adding the following three row vectors:

$$0(3, 1, 1, 0, 0, \ 48)$$
$$-\tfrac{3}{2}(3, 4, 0, 1, 0, 120)$$
$$1(5, 6, 0, 0, 1, \ \ 0)$$

Below, we show how to describe all possible objective rows which can follow from a given initial tableau. To this end, consider the following problem:

PROBLEM 7.3. Maximize $P = 6x_1 + 7x_2$

where $x_1 \geq 0, \ x_2 \geq 0, \ s_1 \geq 0, \ s_2 \geq 0$

and $3x_1 + \ x_2 + s_1 + \quad = 200$

$3x_1 + 4x_2 \quad + s_2 = 120$

The initial tableau of this program reads

Tableau 1

	x_1	x_2	s_1	s_2	$-P$	
s_1	3	1	1	0	0	200
s_2	3	4	0	1	0	120
$-P$	6	7	0	0	1	0

Let $\mathbf{m} = (\gamma_1, \gamma_2, 1)$ be the multiplier vector belonging to *any* subsequent tableau, say, Tableau n. Then $\mathbf{m}$ summarizes the passage from Tableau 1 to the objective row of Tableau n, namely,

$$\gamma_1(\text{row 1 of Tbl 1}) + \gamma_2(\text{row 2 of Tbl 1}) + 1(\text{row 3 of Tbl 1}) = [\text{row 3 of Tbl n}]$$

If we replace the rim variables in Tableau 1 by the corresponding components of $\mathbf{m}$, we see immediately that the objective function of Tableau n must take the following form:

$$(3\gamma_1 + 3\gamma_2 + 6)x_1 + (\gamma_1 + 4\gamma_2 + 7)x_2 + \gamma_1 s_1 + \gamma_2 s_2 - P = 200\gamma_1 + 120\gamma_2$$

This equation can be called *the undetermined-multiplier form of the universal objective function.* Clearly, γ_1 and γ_2 are undetermined parameters and the objective function at any feasible corner can be determined by seeking γ_1 and γ_2 so that:

(1) The objective coefficients of the current basic variables become zero.

(2) The resulting values of the basic variables turn out nonnegative.

To illustrate, suppose past experience leads us to suspect that the optimum of Problem 7.3 occurs when s_1 and s_2 are corner variables, that is, when x_1 and x_2 are basic. We can test this belief immediately by calculating the values of γ_1

and y_2 which characterize x_1 and x_2 as basic variables. By condition (1), these values must satisfy the equations

$$\text{(I)} \qquad \begin{cases} 3y_1 + 3y_2 + 6 = 0 \\ y_1 + 4y_2 + 7 = 0 \end{cases}$$

The solution of this system is $y_1 = -\frac{1}{3}$ and $y_2 = -\frac{5}{3}$. Since these values of y_1 and y_2 are negative, it would seem that our original conjecture is correct and that $s_1 = s_2 = 0$ does define the optimal corner. *Nevertheless we must still test to see if this corner is feasible;* that is, we must solve the following system, derived from the original nontrivial constraints by setting $s_1 = s_2 = 0$:

$$\text{(II)} \qquad \begin{cases} 3x_1 + x_2 = 200 \\ 3x_1 + 4x_2 = 120 \end{cases}$$

But system (II) implies that $3x_2 = -80$, which contradicts feasibility. Hence the corner defined by $s_1 = s_2 = 0$ is *not feasible* (although the objective function at this corner has "optimal-type" coefficients).

Note the special relation between systems (I) and (II): The coefficients of the unknowns in a row of one system are identical to the coefficients of the unknowns in a column of the other system. For example, the coefficients 1 and 4 appear in row 2 of system (I) and in column 2 of system (II). As vectors, these lists of coefficients are transposes of each other. In succeeding chapters there will be many more instances where the rows of one system of equations are transposed to form the columns of another intimately related system.

Here, as an exercise, the reader should use the method of undetermined multipliers to show that, in Problem 7.3:

(1) The corner where x_1 and s_2 are nonbasic cannot be optimal.

(2) The corner where x_2 and s_2 are nonbasic satisfies all the conditions of optimality.

(3) The optimal multiplier vector is $\mathbf{m} = (0, -2, 1)$, leading to an optimal value $P^* = 240$.

For another illustration of the use of the undetermined-multiplier form of the universal objective function, we return to Problem 2.4. There we showed that this transportation problem can be reduced to the program

$$\text{Minimize} \quad C = 7x + 3y + 60$$

where all variables are nonnegative

$$\begin{array}{lrcl} \text{and} & x + y + s_1 & = 15 & (1) \\ & x + s_2 & = 10 & (2) \\ & y + s_3 & = 20 & (3) \\ & -x - y + s_4 & = -5 & (4) \end{array}$$

By setting any two of the six variables equal to 0 we specify a fundamental system. Clearly, when we set $x = y = 0$, the resulting system possesses a unique solution, namely, $s_1 = 15$, $s_2 = 10$, $s_3 = 20$, $s_4 = -5$. But, because s_4 is negative, this basic solution is nonfeasible. To find a basic feasible solution, we might set s_4 and another variable equal to 0. Thus, consider the fundamental system determined by setting $s_4 = x = 0$, that is,

$$
\begin{aligned}
y + s_1 \qquad\qquad &= 15 \qquad (1') \\
s_2 \qquad &= 10 \qquad (2') \\
y \qquad\qquad + s_3 &= 20 \qquad (3') \\
-y \qquad\qquad &= -5 \qquad (4')
\end{aligned}
$$

The solution to this system is readily seen to be unique and nonnegative, namely $y = 5$, $s_1 = 10$, $s_2 = 10$, $s_3 = 15$. Thus, by taking $x = s_4 = 0$, we have determined a basic feasible solution.

Now we determine the objective function at this corner. By putting our program in tableau form and focusing on the columns, or simply by examining the initial dictionary, we see that the universal objective function takes the form

$$
\begin{aligned}
[\gamma_1 + \gamma_2 + 0(\gamma_3) - \gamma_4 + 7]x + [\gamma_1 + 0(\gamma_2) + \gamma_3 - \gamma_4 + 3]y \\
+ \gamma_1 s_1 + \gamma_2 s_2 + \gamma_3 s_3 + \gamma_4 s_4 \\
= C + 15\gamma_1 + 10\gamma_2 + 20\gamma_3 - 5\gamma_4 - 60
\end{aligned}
$$

At the corner under study, the basic variables are y, s_1, s_2, and s_3, and hence the objective coefficients of these variables must be 0. This implies that $\gamma_1 = 0$, $\gamma_2 = 0$, $\gamma_3 = 0$. On substituting these values into the expression for the universal coefficient of y, we deduce immediately that $\gamma_4 = 3$. Therefore, the adjusted objective function must read

$$
\begin{aligned}
4x + 0(y) + 0(s_1) + 0(s_2) + 0(s_3) + 3s_4 \\
= C + 15(0) + 10(0) + 20(0) - 5(3) - 60
\end{aligned}
$$

or

$$C = 75 + 4x + 3s_4$$

Since the objective is to minimize C, the corner where x and s_4 are nonbasic must be the optimum.

Before turning to further generalizations of the multiplier theorem, we pause to emphasize that the method of undetermined coefficients can be viewed as an alternative way of pivoting, that is, a way of *diagonalizing*, or *solving*, a given system of equations. To clarify this remark, consider the system

$$
\text{(III)} \qquad \begin{cases} 3x + 4y + 6z = 60 & (1) \\ 2x + \ \ y + 3z = 30 & (2) \end{cases}
$$

Every consequence of this system can be obtained by adding together α times Eq. (1) and β times Eq. (2) to form

$$(3\alpha + 2\beta)x + (4\alpha + \beta)y + (6\alpha + 3\beta)z = 60\alpha + 30\beta \qquad (3)$$

Now we can express x in terms of z by solving system (IV) below for α and β:

$$\text{(IV)} \qquad \begin{cases} 3\alpha + 2\beta = 1 \\ 4\alpha + \beta = 0 \end{cases}$$

The solution to this system will put Eq. (3) in the form $1(x) + 0(y) + bz = k$. The solution is $\alpha = -\frac{1}{5}$, $\beta = \frac{4}{5}$. After substituting these values into Eq. (3), we obtain

$$1(x) + 0(y) + \tfrac{6}{5}z = 12$$

or

$$x = 12 - \tfrac{6}{5}z$$

Similarly, to express y in terms of z is equivalent to solving the following system for α and β:

$$\text{(V)} \qquad \begin{cases} 3\alpha + 2\beta = 0 \\ 4\alpha + \beta = 1 \end{cases}$$

The solution is $\alpha = \frac{2}{5}$, $\beta = -\frac{3}{5}$. Substituting these values in Eq. (3), we obtain

$$y = -\tfrac{3}{5}z + 6$$

Hence, system (III) is equivalent to the *diagonalized* system

$$x = 12 - \tfrac{6}{5}z$$
$$y = 6 - \tfrac{3}{5}z$$

We can easily obtain a further, slight generalization of the multiplier theorem by considering any tableau to be the initial tableau. For example, consider Tableau 2′ of Problem 7.2. The unit vectors of this tableau appear, in order, in the s_1, x_2, and $-P$ columns. By definition, this is the order in which the basic variables are listed in the left-hand rim. To specify the multiplier vector which describes how the rows of Tableau 2′ were combined to form the objective row of some subsequent Tableau m, all we need do is pick out the entries in the objective row of Tableau m which appear in those columns which contained the unit vectors of Tableau 2′, that is, the s_1, x_2, and $-P$ columns, in order. To illustrate, the vector $(-\frac{2}{9}, 0, 1)$ is the multiplier which records how the rows of Tableau 2′ were combined to form the objective row of Tableau 3′. It tells us that

$$-\tfrac{2}{9}(\text{row 1 of Tbl 2}') + 0(\text{row 2 of Tbl 2}') + 1(\text{row 3 of Tbl 2}') = [\text{row 3 of Tbl 3}']$$

Clearly, the same result can be visualized by listing the components of $(-\frac{2}{9}, 0, 1)$ in the left rim of Tableau 2′, as shown below:

Tableau 2′

	x_1	x_2	s_1	s_2	$-P$	
$-\frac{2}{9}$	$\frac{9}{4}$	0	1	$-\frac{1}{4}$	0	18
0	$\frac{3}{4}$	1	0	$\frac{1}{4}$	0	30
1	$\frac{1}{2}$	0	0	$-\frac{3}{2}$	1	-180

A third generalization of the multiplier theorem consists in associating a multiplier vector not only with the objective row but also with every other row of a tableau. To illustrate, assume that the initial tableau is as previously described, and that the current tableau reads

Tableau c (Current)

$x_1 \cdots x_n$	x_{n+1}	x_{n+2}	$\cdots$	x_{n+m}	$-P$	
	μ_{11}	μ_{12}	$\cdots$	μ_{1m}	0	β_1
	μ_{21}	μ_{22}	$\cdots$	μ_{2m}	0	β_2
					.	.
					.	.
					.	.
	μ_{m1}	μ_{m2}	$\cdots$	μ_{mm}	0	β_m
	$\mu_{m+1,1}$	$\mu_{m+1,2}$	$\cdots$	$\mu_{m+1,m}$	1	$-P_c$

The body of Tableau c has m rows. The vector of simplex multipliers $\boldsymbol{\mu}_{m+1}$ appears in the last, or $(m + 1)$th, row. With every row let us associate a multiplier vector as follows:

$$\begin{aligned}
\boldsymbol{\mu}_1 &= (\mu_{11}, \quad \mu_{12}, \quad \ldots, \mu_{1m}, \quad 0) \\
\boldsymbol{\mu}_2 &= (\mu_{21}, \quad \mu_{22}, \quad \ldots, \mu_{2m}, \quad 0) \\
&\qquad\qquad \cdots \\
\boldsymbol{\mu}_m &= (\mu_{m1}, \quad \mu_{m2}, \quad \ldots, \mu_{mm}, \quad 0) \\
\boldsymbol{\mu}_{m+1} &= (\mu_{m+1,1}, \mu_{m+1,2}, \ldots, \mu_{m+1,m}, 1)
\end{aligned}$$

In words, the current multiplier for row i is the vector whose components appear in row i of the current tableau under the same column headings which contained the unit columns of Tableau 1. The previous discussion of simplex multipliers was confined to $\boldsymbol{\mu}_{m+1}$.

As above, let $\mathbf{k}_j$ denote the jth column of any Tableau 1. Furthermore, let $a_{ij}^{(c)}$ denote the entry appearing in row i, column j of Tableau c. Then, on referring to Tableau c, it is clear that

$$a_{i,n+1}^{(c)} = \mu_{i1}$$

and, in general,

$$a_{i,n+s}^{(c)} = \mu_{is}$$

for $s = 1, 2, \ldots, m$.

Thus, by the same reasoning as used earlier, μ_{is} must be equal to the multiple of the initial sth *equation that has been added to the initial* ith *equation, directly or indirectly, in the process of obtaining the current* ith *equation.* Again, by the same reasoning as used in the previous discussion of $\boldsymbol{\mu}_{m+1}$, it is not difficult to show that

$$\boldsymbol{\mu}_i \cdot \mathbf{k}_j = a_{ij}^{(c)}$$

In other words, if $\mathbf{k}_j$ denotes the jth column of Tableau 1, and $\boldsymbol{\mu}_i$ denotes the ith row multiplier of Tableau c, then the inner product of these vectors is equal to the number appearing in row i, column j of Tableau c.

To understand why this result is valid, suppose that the body of Tableau c contains three rows; that is, $m = 3$. Then the product $\boldsymbol{\mu}_i \cdot \mathbf{k}_j$ equals

$$(\mu_{i1}, \mu_{i2}, \mu_{i3}, 0) \begin{pmatrix} k_{1j} \\ k_{2j} \\ k_{3j} \\ k_{4j} \end{pmatrix} = \mu_{i1}k_{1j} + \mu_{i2}k_{2j} + \mu_{i3}k_{3j}$$

A little thought shows that $\mu_{i1}k_{1j} + \mu_{i2}k_{2j} + \mu_{i3}k_{3j}$ is the exact history of how the element in row i, column j of Tableau c was obtained.

We can, of course, list the components of $\boldsymbol{\mu}_i$ in the left rim of Tableau 1 in place of the names of the corresponding basic variables and calculate $\boldsymbol{\mu}_i \cdot \mathbf{k}_j$ by focusing on both the left rim and on column j. The value of this inner product equals the coefficient in row i, column j of the current tableau, that is, the tableau from which $\boldsymbol{\mu}_i$ is taken.

The phrase "*simplex* multiplier" will always refer to the multiplier vector of an objective row. The phrase "*general* multiplier" refers to any row, but no ambiguity should arise if we omit either adjective.

In order to illustrate the general-multiplier theorem we should consider a tableau with at least three nontrivial constraints. The following problem serves this purpose:

PROBLEM 7.4. Maximize $P = 2x_1 + 4x_2 + 3x_3$

where $x_1 \geq 0, \quad x_2 \geq 0, \quad x_3 \geq 0$

and

$$\begin{aligned} x_1 + 2x_2 \qquad\quad &\leq 4 \\ x_2 + 3x_3 &\leq 6 \\ 2x_1 + x_2 + 2x_3 &\leq 10 \end{aligned}$$

We solve this problem below, using x_4, x_5, and x_6 to denote slack variables.

Tableau 1

	x_1	x_2	x_3	x_4	x_5	x_6	$-P$		
x_4	1	②	0	1	0	0	0	4	←(2)
x_5	0	1	3	0	1	0	0	6	(6)
x_6	2	1	2	0	0	1	0	10	(10)
$-P$	2	4	3	0	0	0	1	0	
		↑							

Tableau 2

	x_1	x_2	x_3	x_4	x_5	x_6	$-P$		
x_2	$\frac{1}{2}$	1	0	$\frac{1}{2}$	0	0	0	2	
x_5	$-\frac{1}{2}$	0	③	$-\frac{1}{2}$	1	0	0	4	$\leftarrow(\frac{4}{3})$
x_6	$\frac{3}{2}$	0	2	$-\frac{1}{2}$	0	1	0	8	(4)
$-P$	0	0	3	-2	0	0	1	-8	
			↑						

Tableau 3

	x_1	x_2	x_3	x_4	x_5	x_6	$-P$		
x_2	$\frac{1}{2}$	1	0	$\frac{1}{2}$	0	0	0	2	(4)
x_3	$-\frac{1}{6}$	0	1	$-\frac{1}{6}$	$\frac{1}{3}$	0	0	$\frac{4}{3}$	
x_6	$\left(\frac{11}{6}\right)$ (circled)	0	0	$-\frac{1}{6}$	$-\frac{2}{3}$	1	0	$\frac{16}{3}$	$\leftarrow(\frac{32}{11})$
$-P$	$\frac{1}{2}$	0	0	$-\frac{3}{2}$	-1	0	1	-12	
	↑								

Tableau 4

	x_1	x_2	x_3	x_4	x_5	x_6	$-P$	
x_2	0	1	0	$\frac{6}{11}$	$\frac{2}{11}$	$-\frac{3}{11}$	0	$\frac{6}{11}$
x_3	0	0	1	$-\frac{2}{11}$	$\frac{3}{11}$	$\frac{1}{11}$	0	$\frac{20}{11}$
x_1	1	0	0	$-\frac{1}{11}$	$-\frac{4}{11}$	$\frac{6}{11}$	0	$\frac{32}{11}$
$-P$	0	0	0	$-\frac{16}{11}$	$-\frac{9}{11}$	$-\frac{3}{11}$	1	$-13\frac{5}{11}$

To corroborate the general-multiplier theorem, consider the passage from Tableau 1 to Tableau 3. The multiplier vectors of Tableau 3 are

$$\boldsymbol{\mu}_1 = (\tfrac{1}{2}, 0, 0, 0)$$
$$\boldsymbol{\mu}_2 = (-\tfrac{1}{6}, \tfrac{1}{3}, 0, 0)$$
$$\boldsymbol{\mu}_3 = (-\tfrac{1}{6}, -\tfrac{2}{3}, 1, 0)$$
$$\boldsymbol{\mu}_4 = (-\tfrac{3}{2}, -1, 0, 1)$$

Consider the products $\boldsymbol{\mu}_i \cdot \mathbf{k}_2$, where $\mathbf{k}_2$ is the second column of Tableau 1, for $i = 1, 2, 3, 4$:

$$\boldsymbol{\mu}_1 \cdot \mathbf{k}_2 = (\tfrac{1}{2}, 0, 0, 0) \begin{pmatrix} 2 \\ 1 \\ 1 \\ 4 \end{pmatrix} = 1$$

$$\boldsymbol{\mu}_2 \cdot \mathbf{k}_2 = 0$$
$$\boldsymbol{\mu}_3 \cdot \mathbf{k}_2 = 0$$
$$\boldsymbol{\mu}_4 \cdot \mathbf{k}_2 = 0$$

These four products must equal the entries in the successive rows of column 2, Tableau 3.

Now form $\boldsymbol{\mu}_i \cdot \mathbf{b}$, for $i = 1, 2, 3, 4$, where $\mathbf{b}$ is the stub of Tableau 1.

$$\boldsymbol{\mu}_1 \cdot \mathbf{b} = (\tfrac{1}{2}, 0, 0, 0) \begin{pmatrix} 4 \\ 6 \\ 10 \\ 0 \end{pmatrix} = 2$$

$$\boldsymbol{\mu}_2 \cdot \mathbf{b} = -\tfrac{4}{6} + 2 = \tfrac{4}{3}$$
$$\boldsymbol{\mu}_3 \cdot \mathbf{b} = -\tfrac{4}{6} - 4 + 10 = \tfrac{16}{3}$$
$$\boldsymbol{\mu}_4 \cdot \mathbf{b} = -12$$

These four products equal the entries in the stub of Tableau 3.

The ultimate generalization of the multiplier methods involves *matrix algebra.* A *matrix* is nothing more than a rectangular table of numbers, usually enclosed in parentheses and treated as a mathematical object, subject to appropriate laws of "addition," "subtraction," and "multiplication." For example,

$$M = \begin{pmatrix} 1 & 2 & 3 \\ 4 & 5 & 9 \end{pmatrix}$$

is a matrix with two rows and three columns.

Observe that each row and column of a matrix can be considered to be a vector. It is this fact which lies at the heart of the "ultimate" generalization, which will not be explored in this text.

The next problem explains how to read off multipliers from a tableau containing artificial variables and M coefficients.

PROBLEM 7.5. Maximize $P = 19x_1 + 6x_2$

where $x_1 \geq 0, \quad x_2 \geq 0$

and
$$3x_1 + x_2 \leq 48$$
$$3x_1 + 4x_2 \geq 120$$

In the tableaux below s_1 and s_2 represent slack variables, A_2 represents an artificial variable, and we place both $-P$ and $-P_M$ upstairs. The column heading $-P$ corresponds to ordinary profits, the column heading $-P_M$ corresponds to M profits, and the use of these two columns allows us to associate a multiplier with each of the four rows of our tableaux.

Tableau 0

	x_1	x_2	s_1	s_2	A_2	$-P$	$-P_M$	
s_1	3	1	1	0	0	0	0	48
A_2	3	4	0	−1	①	0	0	120
$-P$	19	6	0	0	0	1	0	0
$-P_M$	0	0	0	0	$-M$	0	1	0

Tableau 0 must first be adjusted (hence, it is not the *initial* tableau).

Tableau 1

	x_1	x_2	s_1	s_2	A_2	$-P$	$-P_M$		
s_1	3	1	1	0	0	0	0	48	(48)
A_2	3	(4)	0	-1	1	0	0	120	←(30)
$-P$	19	6	0	0	0	1	0	0	
$-P_M$	$3M$	$4M$	0	$-M$	0	0	1	$120M$	
		↑							

Tableau 1 is the initial tableau. The distinct unit columns bear the headings s_1, A_2, $-P$, and $-P_M$, in that order. Note that the initial value of $-P_M$ is not 0 but rather $120M$. If the present problem is not contradictory, no trace of M can appear in the optimum value of $-P_M$, in which case the inner product of the terminal simplex multiplier for row 4 and the initial stub must be $-120M$.

Tableau 2

	x_1	x_2	s_1	s_2	A_2	$-P$	$-P_M$		
s_1	$\left(\frac{9}{4}\right)$	0	1	$\frac{1}{4}$	$-\frac{1}{4}$	0	0	18	←(8)
x_2	$\frac{3}{4}$	1	0	$-\frac{1}{4}$	$\frac{1}{4}$	0	0	30	(40)
$-P$	$\frac{29}{2}$	0	0	$\frac{3}{2}$	$-\frac{3}{2}$	1	0	-180	
$-P_M$	0	0	0	0	$-M$	0	1	0	
	↑								

Phase I has been completed, but for illustrative purposes we carry the A_2 column in all subsequent tableaux.

Tableau 3

	x_1	x_2	s_1	s_2	A_2	$-P$	$-P_M$	
x_1	1	0	$\frac{4}{9}$	$\frac{1}{9}$	$-\frac{1}{9}$	0	0	8
x_2	0	1	$-\frac{1}{3}$	$-\frac{1}{3}$	$\frac{1}{3}$	0	0	24
$-P$	0	0	$-\frac{58}{9}$	$-\frac{1}{9}$	$\frac{1}{9}$	1	0	-296
$-P_M$	0	0	0	0	$-M$	0	1	0

Tableau 3 is terminal. As contended earlier, the inner product of the stub of Tableau 1 with the vector of simplex multipliers for the fourth row of Tableau 3 is equal to $-120M$.

The reader may find it interesting to reread the illustrative problems of Chapter 6 and carefully observe the M objective row. *Is it true that, once all artificial variables become nonbasic, each M objective coefficient remains the same in all subsequent tableaux?*

Exercises

1. Reread the illustrative problems of earlier chapters and verify the general-multiplier theorem.

2. Given the following tableau, solve for x_2 and x_4 in terms of x_1 and x_3 by the method of undetermined multipliers:

x_1	x_2	x_3	x_4	
2	3	4	5	-1
2	-3	-1	-1	-7

3. The following tableau is the terminal tableau in a sequence:

Tableau t

	x_1	x_2	x_3	x_4	$-P$	
x_2	0	1	$\frac{3}{2}$	$-\frac{1}{2}$	0	12
x_1	1	0	-2	1	0	24
$-P$	0	0	-1	-1	1	-168

In the initial tableau the x_3, x_4, and $-P$ columns formed a set of distinct unit columns in natural order, that is, the x_3 column contained 1 in row 1, the x_4 column contained 1 in row 2, and the $-P$ column contained 1 in row 3.

(a) Determine the initial tableau.
(b) What is the numerical value of t? (Is this question unambiguous?)

4. The following tableau is terminal:

	x_1	x_2	x_3	x_4	$-P$	
x_2	0	1	2	4	0	1
x_1	1	0	-1	7	0	1
$-P$	0	0	-6	-30	1	6

It is known that in the initial tableau the x_3, x_4, and $-P$ columns, in that order, formed a set of distinct unit columns.

(a) Determine the initial tableau.
(b) What is the value of t?

5. Suppose that you are given the terminal tableau of a linear-programming problem but no other information. From this tableau can you determine the original stub and nonbasic columns?

6. Consider the transportation problem whose data are

	D_1	D_2	
S_1	\$0	\$1	5
S_2	\$2	\$3	7
S_3	\$5	\$1	9
	10	11	

Let x_{ij} denote the amount shipped from S_i to D_j for $i = 1, 2, 3$, and $j = 1, 2$. It is known that the following is a basic feasible solution: $x_{11} = 1$, $x_{12} = 4$, $x_{21} = 0$, $x_{22} = 7$, $x_{31} = 9$, $x_{32} = 0$. Use the universal form of the objective function to test whether this solution is optimal.

8

POSTOPTIMALITY ANALYSIS

Very often the user of linear programming is interested not only in the optimum solution but also in how this solution would change if resources were available in greater or lesser amounts or if costs were to change, and so on. The aim of *postoptimality analysis* is to answer these questions with a minimum of calculation. In this chapter a unified theory of postoptimality analysis is presented, *unified* in the sense that all problems are treated by one fundamental method.

8.1 IMPLICIT PRICES

Let us examine more closely the multiplier equation

$$\mathbf{m} \cdot \mathbf{b} = -P_0 - \Delta P = -P^*$$

first encountered in Chapter 7. Recall that $\mathbf{m}$ is the vector of simplex multipliers belonging to the typical *terminal* tableau, $\mathbf{b}$ is the stub of the typical *initial* tableau, P_0 is the value of the objective function at the initial corner, P^* the value at the optimum, and ΔP the optimum profit less the initial profit. In a program with $r + 1$ rows, the multiplier equation reads

$$\begin{aligned} -P^* &= -P_0 - \Delta P \\ &= (\gamma_1, \gamma_2, \ldots, \gamma_r, 1) \begin{pmatrix} b_1 \\ b_2 \\ \cdot \\ \cdot \\ \cdot \\ b_r \\ -P_0 \end{pmatrix} = \gamma_1 b_1 + \gamma_2 b_2 + \cdots + \gamma_r b_r - P_0 \end{aligned}$$

from which it follows immediately that

$$-\Delta P = \gamma_1 b_1 + \gamma_2 b_2 + \cdots + \gamma_r b_r$$

or

$$\Delta P = -\gamma_1 b_1 - \gamma_2 b_2 - \cdots - \gamma_r b_r \tag{1}$$

Equation (1) shows that ΔP is a linear function of each b_i, and that if any b_i is increased by 1 unit, ΔP will be increased by $-\gamma_i$ units—*provided, of course, that the increase in b_i does not change the sequence of pivot steps needed to reach optimality.*

To understand this proviso better, observe that if any b_i changes by 1 we have a new program whose optimal simplex multiplier need no longer equal the previous **m**. Usually the change in b_i will be relatively small and the very same sequence of row operations which led to optimality before the change will also lead to optimality after the change. When this is the case, the only difference between the old optimal tableau and the new one is in the stub, since this is the only place where the old and new initial tableaux differ.

To illustrate, suppose we increase b_2 of Problem 7.1 from 120 to 121. The optimal tableau of this problem has already been obtained. Its multipliers record the history of solution. Except for some change in notation, this tableau reads

Optimum Tableau (Problem 7.1)

	x_1	x_2	x_3	x_4	$-P$	
x_3	$\frac{9}{4}$	0	1	$-\frac{1}{4}$	0	18
x_2	$\frac{3}{4}$	1	0	$\frac{1}{4}$	0	30
$-P$	-10	0	0	-5	1	-600

Hence $\boldsymbol{\mu}_3 = (0, -5, 1)$, so $-\gamma_2 = 5$. Thus, the optimal profit in the new problem will be 5 units more than in the old problem, *provided the new problem can be solved by the same sequence of row operations as the old problem.* If this condition is met, both problems will have the same set of optimal row multipliers, namely,

$$\boldsymbol{\mu}_1 = (1, -\tfrac{1}{4}, 0)$$
$$\boldsymbol{\mu}_2 = (0, \ \tfrac{1}{4}, 0)$$
$$\boldsymbol{\mu}_3 = (0, -5, 1)$$

To calculate the new optimal stub we need only multiply the new initial stub by the above multiplier vectors. The new initial stub is $\boldsymbol{\beta} = \begin{pmatrix} 48 \\ 121 \\ 0 \end{pmatrix}$ and hence

$$\boldsymbol{\mu}_1 \cdot \boldsymbol{\beta} = (1, -\tfrac{1}{4}, 0) \begin{pmatrix} 48 \\ 121 \\ 0 \end{pmatrix} = 17\tfrac{3}{4}$$

$$\boldsymbol{\mu}_2 \cdot \boldsymbol{\beta} = 30\tfrac{1}{4}$$

$$\boldsymbol{\mu}_3 \cdot \boldsymbol{\beta} = -605$$

Since $\boldsymbol{\mu}_1 \cdot \boldsymbol{\beta}$ and $\boldsymbol{\mu}_2 \cdot \boldsymbol{\beta}$ are nonnegative, the optimal tableau of the new problem must read

	x_1	x_2	x_3	x_4	$-P$	
x_3	$\frac{9}{4}$	0	1	$-\frac{1}{4}$	0	$17\frac{3}{4}$
x_2	$\frac{3}{4}$	1	0	$\frac{1}{4}$	0	$30\frac{1}{4}$
$-P$	-10	0	0	-5	1	-605

As an exercise the reader should solve the new problem from the beginning.

If a resource level b_i is increased, then the increase in the optimum P^*, per unit increase in b_i, is called the *implicit value* of a unit increase in the b_i resource. This implicit value will be equal to $-\gamma_i$, provided that the change in resource level does not alter the set of optimal basic variables. (A small change in b_i will usually not alter this set.) Note that $-\gamma_i$ appears in the x_{n+i} column, the very column which measures slack against the b_i resource.

Synonyms often employed for implicit value are *incremental value*, *shadow price*, and *internal (accounting) price*. This last phrase is particularly suggestive—if we wish to investigate the purchase of additional resources, we should compare the market price to the internal price.

Observe that the optimal stub of the new problem above is equal to the sum of two columns appearing in the old optimal tableau, namely, (1) the stub of the old optimal tableau, and (2) the x_4, that is, the x_{n+i} column of the old optimal tableau (the column which measures slack against the particular b_i whose change is under study).

To see why this is so, let $\mathbf{b}$ denote the initial stub of the old problem, and $\boldsymbol{\beta}$ the initial stub of the new problem; let $\mathbf{b}^*$ denote the terminal stub of the old problem, $\boldsymbol{\beta}^*$ the terminal stub of the new problem; and, as usual, let $\boldsymbol{\mu}_i$ denote the terminal multiplier vectors of the old problem, for $i = 1, 2, 3$. Then

$$\boldsymbol{\beta} = \mathbf{b} + \begin{pmatrix} 0 \\ 1 \\ 0 \end{pmatrix}$$

and

$$\boldsymbol{\mu}_i \cdot \boldsymbol{\beta} = \boldsymbol{\mu}_i \cdot \mathbf{b} + \boldsymbol{\mu}_i \cdot \begin{pmatrix} 0 \\ 1 \\ 0 \end{pmatrix}$$

But $\boldsymbol{\mu}_i \cdot \boldsymbol{\beta}$ equals the ith coordinate of the new terminal stub, $\boldsymbol{\mu}_i \cdot \mathbf{b}$ equals the ith coordinate of the old terminal stub, and $\boldsymbol{\mu}_i \cdot \begin{pmatrix} 0 \\ 1 \\ 0 \end{pmatrix}$ is equal to the second coordinate of $\boldsymbol{\mu}_i$. *But the set of second coordinates of the* $\boldsymbol{\mu}_i$, *for* $i = 1, 2, 3$, *is nothing but the* x_{n+i} *column of the old optimal tableau.*

For another simple illustration of postoptimality analysis, consider

PROBLEM 8.1. Maximize $P = 2x_1 + 4x_2 + 3x_3$

$$\begin{aligned} \text{where} \quad & x_1 \geq 0, \quad x_2 \geq 0, \quad x_3 \geq 0 \\ \text{and} \quad & x_1 + 2x_2 \qquad \leq \mathbf{5} \\ & \qquad x_2 + 3x_3 \leq 6 \\ & 2x_1 + x_2 + 2x_3 \leq \mathbf{8} \end{aligned}$$

This problem differs from Problem 7.4 (the old problem) only in the stub.* In the new problem, b_1 is 5 rather than 4 and b_3 is 8 rather than 10. From the optimal tableau of the old problem we see that its multiplier vectors are

$$\begin{aligned} \boldsymbol{\mu}_1 &= \tfrac{1}{11}(\ \ 6, \ \ 2, -3, 0) \\ \boldsymbol{\mu}_1 &= \tfrac{1}{11}(\ -2, \ \ 3, \ \ 1, 0) \\ \boldsymbol{\mu}_3 &= \tfrac{1}{11}(\ -1, -4, \ \ 6, 0) \\ \boldsymbol{\mu}_4 &= \tfrac{1}{11}(-16, -9, -3, 11) \end{aligned}$$

Let $\boldsymbol{\beta}$ denote the stub of the problem at hand. Whenever $\boldsymbol{\mu}_1 \cdot \boldsymbol{\beta}$, $\boldsymbol{\mu}_2 \cdot \boldsymbol{\beta}$, and $\boldsymbol{\mu}_3 \cdot \boldsymbol{\beta}$ are all *nonnegative* we can quickly deduce the solution to the new problem. Accordingly, we calculate

$$\boldsymbol{\mu}_1 \cdot \boldsymbol{\beta} = \tfrac{1}{11}(6, 2, -3, 0) \begin{pmatrix} 5 \\ 6 \\ 8 \\ 0 \end{pmatrix} = \tfrac{18}{11}$$

$$\boldsymbol{\mu}_2 \cdot \boldsymbol{\beta} = \tfrac{16}{11}$$

$$\boldsymbol{\mu}_3 \cdot \boldsymbol{\beta} = \tfrac{19}{11}$$

Hence the optimal tableau of the new problem will differ from that of the old problem only in the stub. It still remains to calculate

$$\boldsymbol{\mu}_4 \cdot \boldsymbol{\beta} = -14\tfrac{4}{11}$$

Thus the optimal tableau of the new problem must read:

	x_1	x_2	x_3	x_4	x_5	x_6	$-P$	
x_2	0	1	0	$\frac{6}{11}$	$\frac{2}{11}$	$-\frac{3}{11}$	0	$\frac{18}{11}$
x_3	0	0	1	$-\frac{2}{11}$	$\frac{3}{11}$	$\frac{1}{11}$	0	$\frac{16}{11}$
x_1	1	0	0	$-\frac{1}{11}$	$-\frac{4}{11}$	$\frac{6}{11}$	0	$\frac{19}{11}$
$-P$	0	0	0	$-\frac{16}{11}$	$-\frac{9}{11}$	$-\frac{3}{11}$	1	$-14\frac{4}{11}$

* Since we will be comparing new problems with problems already solved throughout this chapter, the location of all changes will be indicated by boldface, for example, **8**.

Observe that the optimal stub of Problem 8.1 is equal to the sum of the following columns of Problem 7.4: (1) the optimal stub of the old problem; (2) the x_4 column of the optimal tableau of the old problem (corresponding to increasing b_1 by 1 unit); and (3) twice the negative of the x_6 column of the optimal tableau of the old problem (corresponding to decreasing b_3 by 2 units).

In what follows we show how the above *method of column transfer* can be generalized and used to construct a uniform theory of postoptimality analysis—"uniform" in the sense that all problems can be treated by one basic method.

This is the *fundamental principle* involved: Any change in a column x_j of the old initial tableau is of the form

$$\Delta\mathbf{x}_j = \begin{pmatrix} \delta_{1j} \\ \delta_{2j} \\ \cdot \\ \cdot \\ \cdot \\ \delta_{mj} \\ \delta_{m+1,j} \end{pmatrix} = \delta_{1j}\mathbf{u}_1 + \cdots + \delta_{m+1,j}\mathbf{u}_{m+1}$$

where $\mathbf{u}_1, \mathbf{u}_2, \ldots, \mathbf{u}_{m+1}$ are the $(m + 1)$ distinct unit vectors of dimension $(m + 1)$, taken in natural order. Accordingly, in order to determine the effect on $\Delta\mathbf{x}_j$ of the old optimal sequence of pivot operations we need only determine the effect of these operations on $\mathbf{u}_1, \ldots, \mathbf{u}_{m+1}$. But the effect on $\mathbf{u}_j$ of these operations is already recorded in the x_{m+j} column of the optimal tableau of the old problem.

The following set of problems will serve to clarify and illustrate how to handle problems of postoptimality analysis by transferring the appropriate columns of the old optimal tableau.

PROBLEM 8.2.

$$\begin{aligned} &\text{Maximize} \quad P = 5x_1 + 6x_2 \\ &\text{where} \quad x_1 \geq 0, \quad x_2 \geq 0 \\ &\text{and} \quad x_1 + 2x_2 \leq 48 \\ &\qquad 3x_1 + 4x_2 \leq 120 \end{aligned}$$

We shall use this problem as a basis from which to perform postoptimality analysis. Exhibited below is the simplex solution; x_3 and x_4 denote slack variables. (Throughout this chapter we employ the tableau format with $-P$ upstairs.)

Tableau 1

	x_1	x_2	x_3	x_4	$-P$		
x_3	1	②	1	0	0	48	←(24)
x_4	3	4	0	1	0	120	(30)
$-P$	5	6	0	0	1	0	
		↑					

Tableau 2

	x_1	x_2	x_3	x_4	$-P$		
x_2	$\frac{1}{2}$	1	$\frac{1}{2}$	0	0	24	(48)
x_4	①	0	-2	1	0	24	←(24)
$-P$	2	0	-3	0	1	-144	
	↑						

Tableau 3

	x_1	x_2	x_3	x_4	$-P$		
x_2	0	1	($\frac{3}{2}$)	$-\frac{1}{2}$	0	12	←
x_1	1	0	-2	1	0	24	
$-P$	0	0	1	-2	1	-192	
			↑				

Tableau 4

	x_1	x_2	x_3	x_4	$-P$	
x_3	0	$\frac{2}{3}$	1	$-\frac{1}{3}$	0	8
x_1	1	$\frac{4}{3}$	0	$\frac{1}{3}$	0	40
$-P$	0	$-\frac{2}{3}$	0	$-\frac{5}{3}$	1	-200

This tableau is optimal.

PROBLEM 8.3.

$$\text{Maximize} \quad P = 4x_1 + 6x_2$$
$$\text{where} \quad x_1 \geq 0, \quad x_2 \geq 0$$
$$\text{and} \quad x_1 + 2x_2 \leq 48$$
$$3x_1 + 4x_2 \leq 120$$

The only difference between Problems 8.2 and 8.3 is in the objective function. The x_1 column of the new tableau equals the old x_1 column plus the vector

$$\Delta\mathbf{x}_1 = \begin{pmatrix} 0 \\ 0 \\ -1 \end{pmatrix} = -\begin{pmatrix} 0 \\ 0 \\ 1 \end{pmatrix} = -\mathbf{u}_3$$

To see the effect on $\mathbf{u}_3$ of the old pivot operations we need only observe how the optimal tableau of Problem 8.2 reads under the $-P$ heading. There we see a unit vector with three components, having its 1 in row 3. In order to form a candidate for the optimal tableau of the new problem we need only add $-\mathbf{u}_3$ to the old optimal x_1 column. This new x_1 column is what we would have constructed from the new initial tableau, if we *had applied directly the old optimal sequence of pivot steps.* Let us henceforth refer to the *candidate* for the optimum of the new problem as Tableau t (t for *terminal candidate*). Thus Tableau t for Problem 8.3 must appear as follows.

Tableau t

	x_1	x_2	x_3	x_4	$-P$	
x_3	0	$\frac{2}{3}$	1	$-\frac{1}{3}$	0	8
x_1	①	$\frac{4}{3}$	0	$\frac{1}{3}$	0	40
$-P$	-1	$-\frac{2}{3}$	0	$-\frac{5}{3}$	1	-200

In constructing Tableau t we list in the left margin the old basic variables, that is, the same variables that appeared in the terminal tableau of Problem 8.2. (Just as the set of optimal multipliers is part of the record of how we obtain the optimal tableau of Problem 8.2, so is the optimal left margin.) The left margin of Tableau t shows that it needs to be adjusted, because the x_1 column no longer represents a basic variable. After adjusting, we obtain

Tableau (t + 1)

	x_1	x_2	x_3	x_4	$-P$		
x_3	0	($\frac{2}{3}$)	1	$-\frac{1}{3}$	0	8	←(12)
x_1	1	$\frac{4}{3}$	0	$\frac{1}{3}$	0	40	(30)
$-P$	0	$\frac{2}{3}$ ↑	0	$-\frac{4}{3}$	1	-160	

Tableau (t + 1) is not optimal. After pivoting as indicated, we obtain

Tableau (t + 2)

	x_1	x_2	x_3	x_4	$-P$	
x_2	0	1	$\frac{3}{2}$	$-\frac{1}{2}$	0	12
x_1	1	0	-2	1	0	24
$-P$	0	0	-1	-1	1	-168

We see immediately that Tableau (t + 2) describes the optimum solution of Problem 8.3 because it passes the following three necessary and sufficient tests: (1) It is *adjusted*; (2) the right-hand column indicates *feasibility*; (3) the objective row indicates *optimality*.

PROBLEM 8.4.

$$\text{Maximize } P = 5x_1 + 6x_2$$
$$\text{where } x_1 \geq 0, \quad x_2 \geq 0$$
$$\text{and } \tfrac{1}{2}x_1 + 2x_2 \leq 48$$
$$3x_1 + 4x_2 \leq 120$$

Problem 8.4 differs from Problem 8.2 in a technical coefficient, namely, the coefficient of x_1 in the first nontrivial constraint. The x_1 column of the initial tableau of Problem 8.4 equals the x_1 column of the initial tableau of Problem 8.2 plus the vector

$$\Delta \mathbf{x}_1 = \begin{pmatrix} -\frac{1}{2} \\ 0 \\ 0 \end{pmatrix} = -\tfrac{1}{2}\begin{pmatrix} 1 \\ 0 \\ 0 \end{pmatrix} = -\tfrac{1}{2}\mathbf{u}_1$$

To construct the relevant Tableau t we add $-\frac{1}{2}$ times the x_3 column of the old optimal tableau to its x_1 column. Thus we obtain

Tableau t

	x_1	x_2	x_3	x_4	$-P$	
x_3	$-\frac{1}{2}$	$\frac{2}{3}$	1	$-\frac{1}{3}$	0	8
x_1	①	$\frac{4}{3}$	0	$\frac{1}{3}$	0	40
$-P$	0	$-\frac{2}{3}$	0	$-\frac{5}{3}$	1	-200

Tableau t must first be adjusted, because the left rim indicates that x_1 should be a basic variable. After adjusting we obtain

Tableau (t + 1)

	x_1	x_2	x_3	x_4	$-P$	
x_3	0	$\frac{4}{3}$	1	$-\frac{1}{6}$	0	28
x_1	1	$\frac{4}{3}$	0	$\frac{1}{3}$	0	40
$-P$	0	$-\frac{2}{3}$	0	$-\frac{5}{3}$	1	-200

Tableau (t + 1) is clearly terminal.

PROBLEM 8.5.

$$\text{Maximize } P = 5x_1 + 6x_2$$
$$\text{where } x_1 \geq 0, \quad x_2 \geq 0, \quad x_3 \geq 0, \quad x_4 \geq 0$$
$$\text{and } x_1 + 2x_2 + x_3 = 48$$
$$3x_1 + 4x_2 + x_4 = \mathbf{180}$$

The initial stub of Problem 8.5 equals the initial stub of Problem 8.2 plus the vector

$$\Delta \mathbf{b} = \begin{pmatrix} 0 \\ 60 \\ 0 \end{pmatrix} = 60 \begin{pmatrix} 0 \\ 1 \\ 0 \end{pmatrix}$$

To form Tableau t of Problem 8.5 we must add 60 times the optimal x_4 column of Problem 8.2 to the optimal stub of Problem 8.2, keeping everything else as is. Thus we obtain

Tableau t

	x_1	x_2	x_3	x_4	$-P$	
x_3	0	$\frac{2}{3}$	1	$-\frac{1}{3}$	0	-12
x_1	1	$\frac{4}{3}$	0	$\frac{1}{3}$	0	60
$-P$	0	$-\frac{2}{3}$	0	$-\frac{5}{3}$	1	-300

We stress that the stub of this tableau indicates nonfeasibility. Below we continue this problem by elementary means, but here we pause to remark that there exists an algorithm, the so-called *dual simplex*, which can be used to convert the stub of Tableau t into a feasible solution. We shall not describe the dual simplex algorithm in this introductory text—not that it is difficult, but simply to draw a dividing line between elementary and less elementary material.* The logic of the dual simplex algorithm would instruct us to pivot on the entry in row 1, column 4 of Tableau t and thereby obtain

Tableau (t + 1)

	x_1	x_2	x_3	x_4	$-P$	
x_4	0	-2	-3	1	0	36
x_1	1	2	1	0	0	48
$-P$	0	-4	-5	0	1	-240

Tableau (t + 1) is terminal.

Figure 31 compares the regions of feasibility of Problems 8.2 and 8.5. The region for Problem 8.2 is the *quadrilateral OABC*; the region for Problem 8.5 is the *triangle OAD*. The change from Problem 8.2 to Problem 8.5 was so radical that it changed the shape of the region of feasibility.

We stress that it is possible to pass beyond Tableau t without having to invoke the dual simplex, by proceeding as follows:

(a) Multiply by (-1) all basic rows whose right-hand constants are negative. (This gives us too few unit columns.)

* The interested reader is referred to F. S. Hillier and G. J. Lieberman, *Introduction to Operations Research* (Holden-Day, San Francisco, 1967).

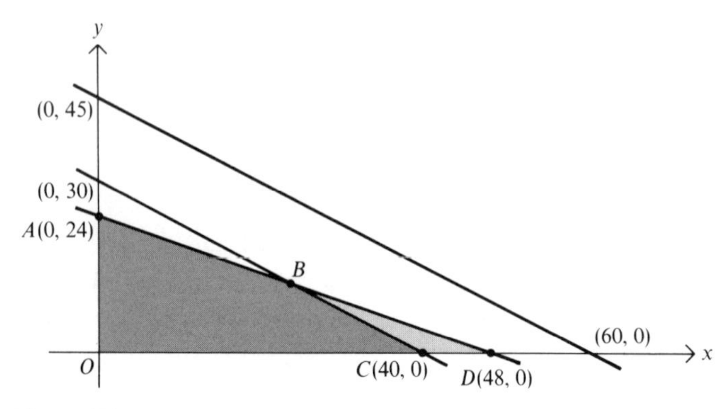

Figure 31

(b) Add artificial variables into all the basic rows which were multiplied in (a), and then drive them to 0 in the usual manner.

Thus we obtain

Tableau t′

	x_1	x_2	x_3	x_4	A	$-P$	
A	0	$-\frac{2}{3}$	-1	$\frac{1}{3}$	①	0	12
x_1	1	$\frac{4}{3}$	0	$\frac{1}{3}$	0	0	60
$-P$	0	$-\frac{2}{3}$	0	$-\frac{5}{3}$	$-M$	1	-300

We repeat that Tableau t′ is obtained from Tableau t by multiplying row 1 of Tableau t by -1 and then adjoining the artificial variable A with objective coefficient $-M$. *Tableau t′ is not adjusted*, but by pivoting on the entry in row 1 of the A column, we obtain the following adjusted tableau:

Tableau t″

	x_1	x_2	x_3	x_4	A	$-P$		
A	0	$-\frac{2}{3}$	-1	($\frac{1}{3}$)	1	0	12	←(36)
x_1	1	$\frac{4}{3}$	0	$\frac{1}{3}$	0	0	60	(180)
$-P$	0	$-\frac{2}{3}M - \frac{2}{3}$	$-M$	$\frac{1}{3}M - \frac{5}{3}$ ↑	0	1	$12M - 300$	

Continuing the simple algorithm, by pivoting as indicated above, we obtain

Tableau t‴

	x_1	x_2	x_3	x_4	A	$-P$	
x_4	0	-2	-3	1	3	0	36
x_1	1	2	1	0	-1	0	48
$-P$	0	-4	-5	0	$-M+5$	1	-240

Tableau t''' is terminal. Except for the A column, it is identical to Tableau (t + 1), previously obtained without much explanation.

The examples we have considered illustrate the following points:

(1) Whenever the passage from the old problem to the new one involves a change in the stub, the new candidate for optimality is always adjusted and always has an objective row of optimal form—the only difficulty is whether or not the new stub is feasible.

(2) Whenever the modification is in some column other than the stub the new candidate, that is, Tableau t, may or may not be adjusted. Furthermore, after adjustment, Tableau (t + 1) may not even be feasible.

The following problem illustrates some aspects of postoptimality analysis which arise when the two-phase method is employed.

PROBLEM 8.6. Maximize $P = -4x_1 - 15x_2$

where $x_1 \geq 0, \quad x_2 \geq 0$

and

$$4x_1 + 2x_2 \geq 75$$
$$3x_1 + 9x_2 \geq 100$$

We begin Phase I by maximizing the negative of the sum of the artificialities, namely, $-S = -x_5 - x_6$. In doing so, we also transform P, as shown below.

Tableau 0 (Phase I)

	x_1	x_2	x_3	x_4	x_5	x_6	$-P$	S	
x_5	4	2	−1	0	(1)	0	0	0	75
x_6	3	9	0	−1	0	(1)	0	0	100
$-P$	−4	−15	0	0	0	0	1	0	0
S	0	0	0	0	−1	−1	0	1	0

After adjusting, we obtain

Tableau 1 (Phase I)

	x_1	x_2	x_3	x_4	x_5	x_6	$-P$	S		
x_5	4	2	−1	0	1	0	0	0	75	$(\frac{75}{2})$
x_6	3	(9)	0	−1	0	1	0	0	100	←$(\frac{100}{9})$
$-P$	−4	−15	0	0	0	0	1	0	0	
S	7	11	−1	−1	0	0	0	1	175	
		↑								

Tableau 2 (Phase I)

	x_1	x_2	x_3	x_4	x_5	x_6	$-P$	S		
x_5	($\frac{10}{3}$)	0	−1	$\frac{2}{9}$	1	$-\frac{2}{9}$	0	0	$\frac{475}{9}$	←$(\frac{95}{6})$
x_2	$\frac{1}{3}$	1	0	$-\frac{1}{9}$	0	$\frac{1}{9}$	0	0	$\frac{100}{9}$	$(\frac{100}{3})$
$-P$	1	0	0	$-\frac{5}{3}$	0	$\frac{5}{3}$	1	0	$\frac{500}{3}$	
S	$\frac{10}{3}$	0	−1	$\frac{2}{9}$	0	$-\frac{11}{9}$	0	1	$\frac{475}{9}$	
	↑									

Tableau 3 (Phase I)

	x_1	x_2	x_3	x_4	x_5	x_6	$-P$	S	
x_1	1	0	$-\frac{3}{10}$	$\frac{1}{15}$	$\frac{3}{10}$	$-\frac{1}{15}$	0	0	$\frac{95}{6}$
x_2	0	1	$\frac{1}{10}$	$-\frac{2}{15}$	$-\frac{1}{10}$	$\frac{2}{15}$	0	0	$\frac{35}{6}$
$-P$	0	0	$\frac{3}{10}$	$-\frac{26}{15}$	$-\frac{3}{10}$	$\frac{26}{15}$	1	0	$\frac{905}{6}$
S	0	0	0	0	-1	-1	0	1	0

Tableau 3 marks the termination of Phase I. Upon discarding the columns of the artificial variables x_5 and x_6 and the row and column of S, we obtain

Tableau 1 (Phase II)

	x_1	x_2	x_3	x_4	$-P$		
x_1	1	0	$-\frac{3}{10}$	$\frac{1}{15}$	0	$\frac{95}{6}$	
x_2	0	1	$\boxed{\frac{1}{10}}$	$-\frac{2}{15}$	0	$\frac{35}{6}$	$\leftarrow(\frac{175}{3})$
$-P$	0	0	$\frac{3}{10}$	$-\frac{26}{15}$	1	$\frac{905}{6}$	
			$\uparrow$				

Tableau 1 is adjusted but nonterminal. After pivoting as indicated, we obtain

Tableau 2 (Phase II)

	x_1	x_2	x_3	x_4	$-P$	
x_1	1	3	0	$-\frac{1}{3}$	0	$\frac{100}{3}$
x_3	0	10	1	$-\frac{4}{3}$	0	$\frac{175}{3}$
$-P$	0	-3	0	$-\frac{4}{3}$	1	$\frac{400}{3}$

This tableau is terminal.

Now suppose we wish to solve the following modification of Problem 8.6:

PROBLEM 8.7.

$$\text{Maximize} \quad P = -4x_1 - 15x_2$$

$$\text{where} \quad x_1 \geq 0, \quad x_2 \geq 0$$

$$\text{and} \quad 4x_1 + 2x_2 \geq 75$$

$$3x_1 + 9x_2 \geq \mathbf{99}$$

Clearly, Problem 8.7 has a larger region of feasibility than Problem 8.6. We pass from Problem 8.6 to Problem 8.7 by adding to the stub of Problem 8.6 the vector

$$\Delta\mathbf{b} = -\begin{pmatrix} 0 \\ 1 \\ 0 \end{pmatrix} = -\mathbf{u}_2$$

To solve Problem 8.7 we must first locate the second unit vector in Tableau 0 or 1 (Phase I). A moment's thought shows that we can consider the negatives of the x_3 and x_4 columns as the first two unit columns. Accordingly, the effect of all the pivot operations of Problem 8.6 on the second unit vector must equal the negative of what appears in the x_4 column of the old optimal tableau, that is, Tableau 2, Phase II. Thus the candidate for the optimal stub of Problem 8.7 is

$$\begin{pmatrix} \frac{100}{3} \\ \frac{175}{3} \\ \frac{400}{3} \end{pmatrix} - \begin{pmatrix} \frac{1}{3} \\ \frac{4}{3} \\ \frac{4}{3} \end{pmatrix} = \begin{pmatrix} 33 \\ 57 \\ 132 \end{pmatrix}$$

This gives us

Tableau t

	x_1	x_2	x_3	x_4	$-P$	
x_1	1	3	0	$-\frac{1}{3}$	0	33
x_3	0	10	1	$-\frac{4}{3}$	0	57
$-P$	0	-3	0	$-\frac{4}{3}$	1	132

Tableau t is clearly feasible and is the optimal solution to Problem 8.7.

Exercises

1. Given the linear program:

$$\text{Maximize} \quad P = 5x_1 + 7x_2$$

where all variables are nonnegative

$$\text{and} \quad 3x_1 + x_2 + x_3 = 48 \qquad (1)$$

$$3x_1 + 4x_2 + x_4 = 120 \qquad (2)$$

Show, by direct methods or otherwise, that the optimal solution is given by the following tableau:

	x_1	x_2	x_3	x_4	$-P$	
x_3	$\frac{9}{4}$	0	1	$-\frac{1}{4}$	0	18
x_2	$\frac{3}{4}$	1	0	$\frac{1}{4}$	0	30
$-P$	$-\frac{1}{4}$	0	0	$-\frac{7}{4}$	1	-210

2. Modify Exercise 1 by changing the constant in Eq. (1) from 48 to 49. Deduce the solution to the modified problem with as little calculation as possible.

3. Solve the following program with as few steps as possible:

$$\text{Maximize} \quad P = 5x_1 + 7x_2$$
$$\text{where} \quad x_1 \geq 0, \quad x_2 \geq 0$$
$$\text{and} \quad 3x_1 + x_2 \leq 49$$
$$3x_1 + 4x_2 \leq 122$$

4. From the optimal tableau of Exercise 1 deduce the optimal tableaux of the following programs:

$$\text{Maximize} \quad P = \gamma x_1 + \delta x_2$$
$$\text{where} \quad x_1 \geq 0, \quad x_2 \geq 0$$
$$\text{and} \quad 3x_1 + x_2 \leq 48$$
$$3x_1 + 4x_2 \leq 120$$

and (a) $\gamma = 5, \delta = 5$; (b) $\gamma = 2, \delta = 7$; (c) $\gamma = 15, \delta = 4$.

5. Given the program

$$\text{Maximize} \quad P = 5x_1 + 6x_2$$
$$\text{where} \quad x_1 \geq 0, \quad x_2 \geq 0$$
$$\text{and} \quad 3x_1 + x_2 \leq 48$$
$$3x_1 + 4x_2 \leq 120$$

show, by working with as many of Exercises 1 to 4 as you can, that the optimal solution to this program is given in the tableau

	x_1	x_2	x_3	x_4	$-P$	
x_1	1	0	$\frac{4}{9}$	$-\frac{1}{9}$	0	8
x_2	0	1	$-\frac{1}{3}$	$\frac{1}{3}$	0	24
$-P$	0	0	$-\frac{2}{9}$	$-\frac{13}{9}$	1	-184

6. From the optimal tableau of Exercise 4(a), deduce the optimal solution to Problem 8.2.

7. From the optimal tableau of Problem 8.2 deduce the solution to the following program:

$$\text{Maximize} \quad P = 5x_1 + 6x_2$$
$$\text{where} \quad x_1 \geq 0, \quad x_2 \geq 0$$
$$\text{and} \quad 2x_1 + x_2 \leq 48$$
$$3x_1 + 4x_2 \leq 120$$

Then verify the answer by direct calculation.

8. Given the program

$$\begin{aligned} &\text{Maximize} \quad P = 5x_1 + 7x_2 \\ &\text{where} \quad x_1 \geq 0, \quad x_2 \geq 0 \\ &\text{and} \quad 3x_1 + x_2 \leq 148 \\ &\qquad\quad 3x_1 + 4x_2 \leq 120 \end{aligned}$$

show that the optimal solution to this program has a set of basic variables different from those of Exercise 1; then compare the graphs of their regions of feasibility.

9. Solve Problem 2.5 (portfolio problem) by the simplex algorithm. How does the optimum change if we require:

(a) At least \$350 appreciation over the next year, all other requirements remaining the same?

(b) At least \$300 in income next year, all other requirements remaining the same?

(c) \$11,000 appreciation over the next decade, as well as both requirements (a) and (b)?

8.2 ADJOINING NEW ROWS AND COLUMNS

In the following program we add a new constraint to an old problem. In doing this we implicitly assume that we are not simultaneously adding any new variables other than slack and artificial variables. Note that the first thing to do when a new constraint is imposed is to check whether the old optimal solution is still feasible—for if it is, it must still be optimal.

PROBLEM 8.8.

$$\begin{aligned} &\text{Maximize} \quad P = 5x_1 + 6x_2 \\ &\text{where} \quad x_1 \geq 0, \quad x_2 \geq 0 \\ &\text{and} \quad x_1 + 2x_2 \leq 48 \\ &\qquad\quad 3x_1 + 4x_2 \leq 120 \\ &\qquad\quad \mathbf{2x_1 + 3x_2 \leq 82} \end{aligned}$$

We passed from Problem 8.2 to Problem 8.8 by adjoining the third nontrivial constraint above. At the old optimum, $x_1 = 40$ and $x_2 = 0$, and hence the old optimum is still feasible.

To construct the optimal tableau of Problem 8.8 we append the new constraint, with new slack variable x_5, to the optimal tableau of Problem 8.2. Thus we form

Tableau t

	x_1	x_2	x_3	x_4	x_5	$-P$	
x_5	2	3	0	0	1	0	82
x_3	0	$\frac{2}{3}$	1	$-\frac{1}{3}$	0	0	8
x_1	①	$\frac{4}{3}$	0	$\frac{1}{3}$	0	0	40
$-P$	0	$-\frac{2}{3}$	0	$-\frac{5}{3}$	0	1	-200

The first row of this tableau contains the new constraint, and the other rows contain the terminal tableau of Problem 8.2. The natural choice of basic variables for Tableau t is x_5, x_3, and x_1 (corresponding to the slack in the new constraint and the set of basic variables belonging to the old optimal tableau). Clearly, Tableau t is not adjusted. After adjusting the x_1 column, we obtain

Tableau (t + 1)

	x_1	x_2	x_3	x_4	x_5	$-P$	
x_5	0	$\frac{1}{3}$	0	$-\frac{2}{3}$	1	0	2
x_3	0	$\frac{2}{3}$	1	$-\frac{1}{3}$	0	0	8
x_1	1	$\frac{4}{3}$	0	$\frac{1}{3}$	0	0	40
$-P$	0	$-\frac{2}{3}$	0	$-\frac{5}{3}$	0	1	-200

Tableau (t + 1) is optimal. Here the old solution is still feasible in the new problem (and hence still optimal). Usually, when a new constraint is added the old optimum is no longer feasible. For example, consider Problem 8.8′:

PROBLEM 8.8′. Maximize $P = 5x_1 + 6x_2$

where all variables are nonnegative

and

$$x_1 + 2x_2 + x_3 = 48$$

$$3x_1 + 4x_2 + x_4 = 120$$

$$\mathbf{2x_1 + 3x_2 + x_5 = 66}$$

We formed Problem 8.8′ by adjoining to Problem 8.2 the new constraint

$$2x_1 + 3x_2 \leq 66$$

Thus we obtain the following candidate for the optimal solution to Problem 8.8′:

Tableau t

	x_1	x_2	x_3	x_4	x_5	$-P$	
x_5	2	3	0	0	1	0	66
x_3	0	$\frac{2}{3}$	1	$-\frac{1}{3}$	0	0	8
x_1	①	$\frac{4}{3}$	0	$\frac{1}{3}$	0	0	40
$-P$	0	$-\frac{2}{3}$	0	$-\frac{5}{3}$	0	1	-200

After adjusting, we obtain

Tableau (t + 1)

	x_1	x_2	x_3	x_4	x_5	$-P$	
x_5	0	$\frac{1}{3}$	0	$\left(-\frac{2}{3}\right)$	1	0	-14
x_3	0	$\frac{2}{3}$	1	$-\frac{1}{3}$	0	0	8
x_1	1	$\frac{4}{3}$	0	$\frac{1}{3}$	0	0	40
$-P$	0	$-\frac{2}{3}$	0	$-\frac{5}{3}$	0	1	-200

The stub of Tableau (t + 1) indicates nonfeasibility (the new constraint must have cut away the old optimum corner). However, if we pivot in row 1, column 4, we obtain

Tableau (t + 2)

	x_1	x_2	x_3	x_4	x_5	$-P$	
x_4	0	$-\frac{1}{2}$	0	1	$-\frac{3}{2}$	0	21
x_3	0	$\frac{1}{2}$	1	0	$-\frac{1}{2}$	0	15
x_1	1	$\frac{3}{2}$	0	0	$\frac{1}{2}$	0	33
$-P$	0	$-\frac{3}{2}$	0	0	$-\frac{5}{2}$	1	-165

This tableau is terminal.

As an exercise, the reader should effect the passage from Tableau (t + 1) to Tableau (t + 2) by employing the methods of this text, that is, by adding artificial variable(s), as previously explained in Problem 8.5.

Before studying the next problem the reader should reconsider Problem 8.2 and think of it as a product-mix problem, with the first nontrivial constraint referring to some resource, say time on machine I, and the second referring to time on machine II. Then suppose a new product is proposed. One unit of the proposed product requires 3 units of time on machine I and 4 units of time on machine II and makes a contribution to net profit of \$7 per unit. *All other things being equal*, should we produce the new product?

One way to answer this question is to let x_0 denote the number of units of the new product, and then solve the following problem directly:

PROBLEM 8.9. Maximize $P = \mathbf{7x_0} + 5x_1 + 6x_2$

where $\mathbf{x_0 \geq 0}$, $x_1 \geq 0$, $x_2 \geq 0$

and
$$\begin{aligned} \mathbf{3x_0} + x_1 + 2x_2 &\leq 48 \\ \mathbf{4x_0} + 3x_1 + 4x_2 &\leq 120 \end{aligned}$$

Observe that the initial x_0 column of Problem 8.9, in the usual tableau format, reads

$$\begin{pmatrix}3\\4\\7\end{pmatrix} = 3\begin{pmatrix}1\\0\\0\end{pmatrix} + 4\begin{pmatrix}0\\1\\0\end{pmatrix} + 7\begin{pmatrix}0\\0\\1\end{pmatrix}$$

Hence, to pass immediately from Problem 8.2 to Tableau t of Problem 8.9 we form the x_0 column of Tableau t by adding together the following columns of the optimal tableau of Problem 8.2: (a) 3 times the x_3 column, (b) 4 times the x_4 column, (c) 7 times the $(-P)$ column. Thus we obtain

Tableau t (Problem 8.9)

	x_0	x_1	x_2	x_3	x_4	$-P$		
x_3	$\textcircled{\frac{5}{3}}$	0	$\frac{2}{3}$	1	$-\frac{1}{3}$	0	8	$\leftarrow(\frac{24}{5})$
x_1	$\frac{4}{3}$	1	$\frac{4}{3}$	0	$\frac{1}{3}$	0	40	(30)
	$\frac{1}{3}$ ↑	0	$-\frac{2}{3}$	0	$-\frac{5}{3}$	1	-200	

We can increase the value of the objective function by pivoting in the column of x_0. This gives

Tableau (t + 1)

	x_0	x_1	x_2	x_3	x_4	$-P$	
x_0	1	0	$\frac{2}{5}$	$\frac{3}{5}$	$-\frac{1}{5}$	0	$\frac{24}{5}$
x_1	0	1	$\frac{4}{5}$	$-\frac{4}{5}$	$\frac{3}{5}$	0	$33\frac{3}{5}$
$-P$	0	0	$-\frac{4}{5}$	$-\frac{1}{5}$	$-\frac{8}{5}$	1	$-201\frac{3}{5}$

This tableau is terminal. It shows, among other things, *that the new product should be produced.*

Exercises

1. A toy manufacturer makes two types of model airplanes, type X and type Y. These products are subjected to three processes, namely, molding, sanding and painting, and assembling, with requirements as shown in the following table:

	Product		
Process	*Toy X*	*Toy Y*	*Time available*
Molding	5 hr	4 hr	40 hr
Sanding and painting	2 hr	3 hr	25 hr
Assembly	1 hr	2 hr	10 hr
Contribution to profit and overhead (per unit)	\$10	\$13	

(a) Determine the optimal production plan.

(b) Find the implicit value of having available 5 more hours of molding time (all other resource capacities being held constant).

(c) Find the implicit value of having 10 more hours of sanding and painting time.

(d) Find the implicit value of having 5 more hours of assembly time.

2. In connection with Exercise 1, if a new toy is proposed whose contribution to profit and overhead is \$15 but which utilizes 4 hours of holding time, 3 hours of sanding and painting time, and 3 hours of assembly time, would its production be profitable, all other things being equal? If so, what will the new optimal production be?

3. If you had solved a linear program and were then asked to exhibit the *next best solution* for comparison with the optimal solution, how would you find it? Do so for Problems 8.2 and 8.5 in this chapter.

4. *True or false?*

(a) Adding more variables to a linear program may improve the optimum value of the objective function.

(b) Adding a new, nonredundant constraint may improve the optimum value of the objective function.

(c) Adding more variables does not increase the number of basic variables in a tableau, nor the number of nonbasic variables.

(d) Adding a new constraint does not increase the number of basic variables.

8.3 DEGENERACY AND POSTOPTIMALITY ANALYSIS

The reader should not believe that the order in which he has learned a subject is necessarily *the natural order*. To warn against this *prejudice* we now use quite elementary tools to perform a postoptimality analysis of changes in the limiting value b_i of a nontrivial constraint. To illustrate, consider the task of passing from Problem 8.2 to Problem 8.10.

PROBLEM 8.10.

$$\begin{aligned} &\text{Maximize} \quad P = 5x_1 + 6x_2 \\ &\text{where} \quad x_1 \geq 0, \quad x_2 \geq 0 \\ &\text{and} \quad x_1 + 2x_2 \leq \mathbf{49} \\ &\qquad\quad 3x_1 + 4x_2 \leq 120 \end{aligned}$$

The first nontrivial constraint in Problem 8.2, that is,

$$x_1 + 2x_2 \leq 48$$

becomes

$$x_1 + 2x_2 \leq 49$$

in Problem 8.10. In our previous study of these problems, we used the variable x_3 to denote slack in both these inequalities—but x_3 really has a different meaning in each problem. To see why, let x_3 measure slack against 48 and X_3 measure slack against 49. Then we have

$$x_1 + 2x_2 + x_3 = 48$$

$$x_1 + 2x_2 + X_3 = 49$$

Clearly,

$$x_3 = X_3 - 1 = 48 - x_1 - 2x_2$$

Below we show how to pass from the optimal dictionary of Problem 8.2 to that of Problem 8.10 by making the *change of variable* $x_3 = (X_3 - 1)$. We obtain

$$x_3 = X_3 - 1 = 8 - \tfrac{2}{3}x_2 + \tfrac{1}{3}x_4 \quad (1)$$

$$x_1 = 40 - \tfrac{4}{3}x_2 - \tfrac{1}{3}x_4 \quad (2)$$

$$P = 200 - \tfrac{2}{3}x_2 - \tfrac{5}{3}x_4 \quad (3)$$

Equation (1) implies that

$$X_3 = 9 - \tfrac{2}{3}x_2 + \tfrac{1}{3}x_4 \quad (1')$$

Equations (1′), (2), and (3) constitute an *optimal dictionary for Problem 8.10.*

For a less trivial example of postoptimality analysis by change of variable, consider

PROBLEM 8.11. Maximize $P = 5x_1 + 6x_2$

where $x_1 \geq 0, \quad x_2 \geq 0$

and $x_1 + 2x_2 \leq 48$

$3x_1 + 4x_2 \leq \mathbf{121}$

Problem 8.11 differs from Problem 8.2 only in the second nontrivial constraint. Let x_4 and X_4 denote the slack in these constraints, that is,

$$3x_1 + 4x_2 + x_4 = 120$$

$$3x_1 + 4x_2 + X_4 = 121$$

Then $x_4 = X_4 - 1$, and if we make this substitution in the optimal dictionary of Problem 8.2 we obtain

$$x_3 = 8 - \tfrac{2}{3}x_2 + \tfrac{1}{3}(X_4 - 1) = \tfrac{23}{3} - \tfrac{2}{3}x_2 + \tfrac{1}{3}X_4 \quad (1)$$

$$x_1 = 40 - \tfrac{4}{3}x_2 - \tfrac{1}{3}(X_4 - 1) = \tfrac{121}{3} - \tfrac{4}{3}x_2 - \tfrac{1}{3}X_4 \quad (2)$$

$$P = 200 - \tfrac{2}{3}x_2 - \tfrac{5}{3}(X_4 - 1) = \tfrac{605}{3} - \tfrac{2}{3}x_2 - \tfrac{5}{3}X_4 \quad (3)$$

Clearly, Eqs. (1), (2), and (3) constitute the optimal dictionary of Problem 8.11.

For a final example of postoptimality analysis by change of variable, *turn back and reread Problem 8.5*. Let the slack variables x_4 of Problem 8.2 and X_4 of Problem 8.5 be defined through the equations,

$$3x_1 + 4x_2 + x_4 = 120$$

$$3x_1 + 4x_2 + X_4 = 180$$

Obviously, $x_4 = X_4 - 60$. As we observed earlier, adding 60 times the optimal

x_4 column of Problem 8.2 to its optimal stub gives a new stub which is infeasible. We can, however, add 24 times the x_4 column to the optimal stub without violating feasibility. In other words, by the usual techniques of postoptimality analysis we can quickly pass from the solution of Problem 8.2 to the solution of the following program.

PROBLEM 8.12. Maximize $P = 5x_1 + 6x_2$

where all variables are nonnegative

$$\text{and} \quad \begin{aligned} x_1 + 2x_2 + x_3 \qquad &= 48 \\ 3x_1 + 4x_2 \qquad + W_4 &= 144 \end{aligned}$$

Clearly,

$$x_4 = W_4 - 24$$
$$W_4 = X_4 - 36$$

Now add 24 times the optimal x_4 column of Problem 8.2 to the optimal stub, obtaining

Tableau t

	x_1	x_2	x_3	W_4	$-P$	
x_3	0	$\frac{2}{3}$	1	$\boxed{-\frac{1}{3}}$	0	0
x_1	1	$\frac{4}{3}$	0	$\frac{1}{3}$	0	48
$-P$	0	$-\frac{2}{3}$	0	$-\frac{5}{3}$	1	-240

This tableau gives the optimal solution to Problem 8.12. It is degenerate and hence we cannot add a positive multiple of the W_4 column to the stub without violating feasibility—this means that the substitution $W_4 = X_4 - 36$ will not take us from the optimal dictionary of Problem 8.12 to that Problem 8.5. (Our objective is to pass from Problem 8.2 to Problem 8.5 by elementary means.)

However, precisely because Tableau t is degenerate, we can interchange the roles of x_3 and W_4. We make W_4 basic by pivoting on the *negative* entry in row 1, column 4. Thus we obtain

Tableau (t + 1)

	x_1	x_2	x_3	W_4	$-P$	
W_4	0	-2	-3	1	0	0
x_1	1	2	1	0	0	48
$-P$	0	-4	-5	0	1	-240

Tableaux t and (t + 1) give different descriptions of the unique optimal solution to Problem 8.12, but because W_4 is basic in Tableau (t + 1) we can now make the substitution

$$W_4 = X_4 - 36$$

without violating feasibility. Thus

$$W_4 = X_4 - 36 = 0 + 2x_2 + 3x_3$$

and hence

$$X_4 = 36 + 2x_2 + 3x_3 \tag{1}$$

Now replace row 1 of Tableau (t + 1) by Eq. (1). We obtain

	x_1	x_2	x_3	X_4	$-P$	
X_4	0	−2	−3	1	0	36
x_1	1	2	1	0	0	48
$-P$	0	−4	−5	0	1	−240

This tableau gives the optimal solution to Problem 8.5, thus completing the task we set.

The objective of the next problem is not only to find the optimum P^* but also to study *how P^* varies with the parameter b_1*.

PROBLEM 8.13.

$$\begin{aligned} &\text{Maximize} \quad P = 5x_1 + 6x_2 \\ &\text{where} \quad x_1 \geq 0, \quad x_2 \geq 0 \\ &\text{and} \quad 3x_1 + x_2 \leq b_1 \\ &\qquad\quad 3x_1 + 4x_2 \leq 120 \end{aligned}$$

Since we have already studied several particular cases of Problem 8.13, especially the case where $b_1 = 48$, let us relate the general problem to this case. To this end, let $b_1 = 48 + \lambda$. Now turn to the optimal tableau already obtained when $b_1 = 48$. This tableau reads

Optimal Tableau ($\lambda = 0$)

	x_1	x_2	x_3	x_4	$-P$	
x_1	1	0	$\frac{4}{9}$	$-\frac{1}{9}$	0	8
x_2	0	1	$-\frac{1}{3}$	$\frac{1}{3}$	0	24
$-P$	0	0	$-\frac{2}{9}$	$-\frac{13}{9}$	1	−184

x_3 and x_4 designate the slack variables in the initial constraints, namely,

$$\begin{aligned} 3x_1 + x_2 + x_3 \qquad &= 48 \\ 3x_1 + 4x_2 \qquad + x_4 &= 120 \end{aligned}$$

The "rate of change" of the above tableau with respect to λ is, within certain limits, the x_3 column itself. These limits are easily found by determining the multiples of the x_3 column which, when added to the stub above, lead to degeneracy. Thus the new stub, as a function of λ, is

$$\begin{pmatrix} 8 \\ 24 \\ -184 \end{pmatrix} + \lambda \begin{pmatrix} \frac{4}{9} \\ -\frac{1}{3} \\ -\frac{2}{9} \end{pmatrix} = \begin{pmatrix} 8 + \frac{4}{9}\lambda \\ 24 - \frac{1}{3}\lambda \\ -184 - \frac{2}{9}\lambda \end{pmatrix}$$

Clearly, the right-hand side above is feasible only if $-18 \leq \lambda \leq 72$. Accordingly, the following tableau is optimal within the indicated range:

Optimal Tableau ($-18 \leq \lambda \leq 72$)

	x_1	x_2	x_3	x_4	$-P$	
x_1	1	0	$\frac{4}{9}$	$-\frac{1}{9}$	0	$8 + \frac{4}{9}\lambda$
x_2	0	1	$-\frac{1}{3}$	$\frac{1}{3}$	0	$24 - \frac{1}{3}\lambda$
$-P$	0	0	$-\frac{2}{9}$	$-\frac{13}{9}$	1	$-184 - \frac{2}{9}\lambda$

Now suppose that $-48 < \lambda < -18$. (We assume that $b_1 > 0$, and hence $\lambda > -48$). Then the above optimal tableau is no longer feasible because the entry in row 1 of the stub is now strictly negative. In this case we can restore feasibility and continue the simplex algorithm by multiplying row 1 by -1 and adjoining an artificial column, with appropriate objective coefficient, as follows:

Tableau t ($-48 < \lambda < -18$)

	x_1	x_2	x_3	x_4	A_1	$-P_0$	$-P_M$	
A_1	-1	0	$-\frac{4}{9}$	$\frac{1}{9}$	(1)	0	0	$-8 - \frac{4}{9}\lambda$
x_2	0	1	$-\frac{1}{3}$	$\frac{1}{3}$	0	0	0	$24 - \frac{1}{3}\lambda$
$-P_0$	0	0	$-\frac{2}{9}$	$-\frac{13}{9}$	0	1	0	$-184 - \frac{2}{9}\lambda$
$-P_M$	0	0	0	0	$-M$	0	1	

Note that A_1 is now a basic variable and hence Tableau t is not adjusted. Upon adjusting, we obtain

Tableau (t + 1) ($-48 < \lambda < -18$)

	x_1	x_2	x_3	x_4	A_1	$-P_0$	$-P_M$		
A_1	-1	0	$-\frac{4}{9}$	($\frac{1}{9}$)	1	0	0	$-8 - \frac{4}{9}\lambda$	←$(-72 - 4\lambda)$
x_2	0	1	$-\frac{1}{3}$	$\frac{1}{3}$	0	0	0	$24 - \frac{1}{3}\lambda$	$(72 - \lambda)$
$-P_0$	0	0	$-\frac{2}{9}$	$-\frac{13}{9}$	0	1	0	$-184 - \frac{2}{9}\lambda$	
$-P_M$	$-M$	0	$-\frac{4}{9}M$	$\frac{1}{9}M$	0	0	1	$-8M - \frac{4}{9}\lambda M$	
				↑					

Tableau (t + 1) is not optimal. We are motivated to increase x_4, obtaining

Optimal Tableau ($-48 < \lambda \leq -18$)

	x_1	x_2	x_3	x_4	$-P$	
x_4	-9	0	-4	1	0	$-72 - 4\lambda$
x_2	3	1	1	0	0	$48 + \lambda$
$-P$	-13	0	-6	0	1	$-288 - 6\lambda$

Now let us return to the optimal tableau obtained for $-18 \leq \lambda \leq 72$. Suppose that $\lambda > 72$; then this tableau is no longer feasible. It is left as an exercise to show that, when $\lambda > 72$, we obtain the following solution:

Optimal Tableau ($\lambda > 72$)

	x_1	x_2	x_3	x_4	$-P$	
x_1	1	$\frac{4}{3}$	0	$\frac{1}{3}$	0	40
x_3	0	-3	1	-1	0	$-72 + \lambda$
$-P$	0	$-\frac{2}{3}$	0	$-\frac{15}{9}$	1	-200

As a last exercise, the reader should show that when $\lambda < -48$, the region of feasibility is empty.

In Figure 32 the optimum value of P is graphed as a function of λ. This function is *piecewise linear* and is defined as follows:

$$\begin{aligned} P(\lambda) &= 288 + 6\lambda && \text{for } -48 \leq \lambda \leq -18 \\ &= 184 + \tfrac{2}{9}\lambda && \text{for } -18 \leq \lambda \leq 72 \\ &= 200 && \text{for } \lambda > 72 \end{aligned}$$

Note that as λ increases, the rate of change of $P(\lambda)$ with respect to λ decreases. Furthermore, $P(\lambda)$ is *concave* in the sense that, as λ increases, the rate

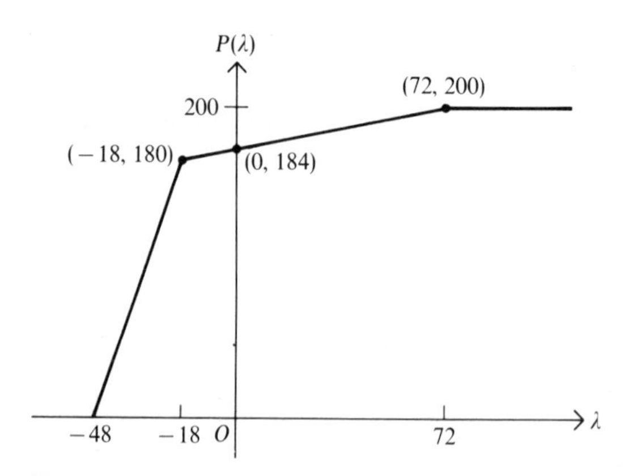

Figure 32

of change of $P(\lambda)$ remains the same or decreases, but never increases. This is in accord with the "law of diminishing returns."

The graph of $P(\lambda)$ as a piecewise linear function shows vividly that the *implicit value of a change in* b_i *depends on both the value of* b_i *from which the change is initiated and on the size of the change.* To illustrate, consider what happens at a point where two of the pieces of $P(\lambda)$ come together, say at $\lambda = -18$, that is, $b_1 = 30$. The corresponding optimal tableau reads

Optimal Tableau ($b_1 = 30$)

	x_1	x_2	x_3	x_4	$-P$	
x_4	-9	0	-4	1	0	0
x_2	3	1	1	0	0	30
$-P$	-13	0	-6	0	1	-180

The vector of simplex multipliers is $\mathbf{m} = (-6, 0, 1)$, but $-(-6)$ is not the implicit value of increasing b_1 by 1 unit, since such a change necessarily leads to a new optimal corner, that is, a terminal tableau with a different vector of simplex multipliers. (However, if we *decrease* b_1 from 30 to 29, optimal profit does decrease by 6. Why?)

In general, if an optimal tableau is degenerate, with $b_i = 0$, the simplex multiplier corresponding to x_{n+i} is a *one-sided shadow price.*

For emphasis, we repeat that the methods of this chapter enable us to manipulate a former optimum tableau to construct the new optimum solution whenever problem conditions change. The key idea is that any change in the original problem is a linear combination of unit columns; and the *imprint*, or record, of all the previously optimal pivot operations is contained in the appropriate columns of the formerly optimal tableau.

Postoptimality analysis is often referred to as *sensitivity analysis.* This is because the methods of postoptimality analysis can be employed to measure the sensitivity of the optimum to changes in the parameters. If the optimum is sensitive to changes in certain parameters, these parameters should be estimated with great care.

It is important to distinguish between "the sensitivity of the solution" and "the sensitivity of the objective function." It can often happen that a minor modification in a parameter can lead to a drastic change in the solution; for example, the optimum can jump to a different corner of the region of feasibility. But this does not necessarily mean that the optimum value of the objective has also changed drastically.

Exercises

1. Solve Problems 8.3 and 8.4 by change of variable in Problem 8.2.

2. Given the following program:

$$\begin{aligned} &\text{Minimize} \quad C = 3x_1 + 2x_2 \\ &\text{where} \quad x_1 \geq 0, \quad x_2 \geq 0 \\ &\text{and} \quad x_1 + x_2 \leq 8 \\ &\qquad 6x_1 + 4x_2 \geq \lambda \\ &\qquad 5x_1 + 8x_2 = \gamma \end{aligned}$$

(a) Find the optimum tableau when $\lambda = 12$ and $\gamma = 20$.

(b) Show that if $\gamma = 20$ and λ is decreased from 12 to to 11, then the optimum value of C drops by $\frac{1}{2}$.

(c) Find the optimum when $\lambda = 12$ and $\gamma = 21$.

(d) Compare graphically the regions of feasibility of the programs in (a) to (c).

3. Consider the following program:

$$\begin{aligned} &\text{Maximize} \quad P = 2x_1 + 3x_2 + x_3 \\ &\text{subject to} \quad x_1 \geq 0, \quad x_2 \geq 0, \quad x_3 \geq 0 \\ &\text{and} \quad 2x_1 + x_2 \leq 5 && (1) \\ &\qquad 3x_1 + 3x_2 + x_3 \leq 9 && (2) \\ &\qquad x_1 + 2x_2 + 2x_3 \leq 4 && (3) \end{aligned}$$

(a) If x_4, x_5, and x_6 denote the slack variables corresponding to inequalities (1), (2), (3), respectively, verify that the following tableau is optimal:

	x_1	x_2	x_3	x_4	x_5	x_6	$-P$	
x_4	0	0	1	1	-1	1	0	0
x_1	1	0	$-\frac{4}{3}$	0	$\frac{2}{3}$	-1	0	2
x_2	0	1	$\frac{5}{3}$	0	$-\frac{1}{3}$	1	0	1
$-P$	0	0	$-\frac{4}{3}$	0	$-\frac{1}{3}$	-1	1	-7

(b) If we increase the right-hand side of inequality (2) by 1, will P increase by the value of the optimal x_5 simplex multiplier? Explain.

(c) Verify that the following tableau is also optimal:

	x_1	x_2	x_3	x_4	x_5	x_6	$-P$	
x_1	1	0	$-\frac{2}{3}$	$\frac{2}{3}$	0	$-\frac{1}{3}$	0	2
x_5	0	0	-1	-1	1	-1	0	0
x_2	0	1	$\frac{4}{3}$	$-\frac{1}{3}$	0	$\frac{2}{3}$	0	1
$-P$	0	0	$-\frac{5}{3}$	$-\frac{1}{3}$	0	$-\frac{4}{3}$	1	-7

(d) Explain the "paradox": The given program does not possess more than one optimal corner, yet it has two different sets of optimal simplex multipliers.

4. Discuss:

(a) The surest method of performing postoptimality analysis is to "start from scratch."

(b) There are limitations on the use of shadow prices. They must be interpreted with care.

(c) If, in the application of the simplex algorithm, a pivot column is obtained which has the property that all its entries are small in comparison with the other entries in the tableau, then beware of the final answer.

(d) A linear-programming solution cannot improve the reliability of its input.

9

SELECTED APPLICATIONS

Linear programming is applicable to a wide range of business and industrial problems. In this chapter only a few problems are discussed, with the aim of giving the reader some idea of the diversity of applications. In Section 9.1, some simple one-dimensional problems of best fit are treated by elementary techniques not involving linear programming. The same problems are then reexamined in Section 9.2 using the techniques of linear programming. In Section 9.3, more complex, multidimensional problems of best approximation are studied. Section 9.4 treats the cutting-stock and discrete-loading problems, and their interrelations. In Section 9.5, the reader is introduced to the fractional-assignment problem and to the celebrated traveling-salesman problem.

The examples of this chapter are, of course, only toy problems. A clear outline of the relevant ideas is more helpful to the beginner than a complex set of equations and involved explanations. Nevertheless, we emphasize that linear programming is actually useful in the real world, precisely because it is a method for analyzing the behavior of a complex system by building a model of the interrelations among its component parts.

It is the author's opinion that the need for realistic examples should be met facally, *viva voce*. The reader must become familiar with the local computer locilities and know which computer programs are available for solving linear programs and how to use them. The instructor should present in detail some case study of mutual interest. Only then does it make sense to examine the appropriateness of the model and the quality of the data.*

In this chapter we do not give all calculations in detail. The reader should fill in all the necessary steps. Since computer codes are available for solving linear programs, the difficulty of hand calculation need not prevent the reader from checking the author's findings. We stress again that the reader need only formulate the problem correctly and prepare the input data according to the specifications for the computer program he is using.

9.1 ONE-DIMENSIONAL PROBLEMS OF BEST FIT

Consider the following example.

PROBLEM 9.1. Given five points on a straight line, with coordinates 1, 2, 5, 8, and 9. Find a point X on the line, with coordinate x, such that the *sum* of the distances from X to the five given points is minimum.

This is the skeletal form of a word problem which appears in several standard statistics texts: A concessionaire has five refreshment stands at various points on the (straight-line) boardwalk of the popular beach at Ocean City. He wants to locate a central supply station at some point on the boardwalk so that the total distance between the stands and the supply station is as small as possible.

In order to translate into mathematics the phrase "distance between the points with coordinates x and a" we use the symbol for *absolute value*, defined as follows:

$$\begin{aligned} |x - a| &= \text{absolute value of } (x - a) \\ &= \text{distance between the points with coordinates } x \text{ and } a \\ &= x - a, \qquad \text{if } x \geq a \\ &= -(x - a), \qquad \text{if } x < a \end{aligned}$$

In particular, if $a = 0$, then

$$|x| = \begin{cases} x, & \text{if } x \geq 0 \\ -x, & \text{if } x < 0 \end{cases}$$

* For the reader who is working alone, we recommend:

(1) S. Vadja, *Readings in Mathematical Programming* (Wiley, New York, 1962), 2nd ed. This beautifully written little book is self-contained and treats more than two dozen toy problems.

(2) George Dantzig, *Linear Programming and Extensions* (Princeton University Press, Princeton, N.J., 1962). This is the "bible." Chapter 3 is devoted to the formulation of linear-programming models.

(3) N. J. Driebeck, *Applied Linear Programming* (Addison-Wesley, New York, 1969). Driebeck deemphasizes proofs and algorithms in favor of detailed descriptions of applications. His text makes an excellent companion to the one at hand.

(4) E. H. Bowman and R. B. Fetter, eds., *Analyses of Industrial Operations* (R. D. Irwin, Homewood, Illinois, 1959). This book is a collection of empirical studies of industrial operating problems, many of which use linear programming.

To illustrate, if $x = 7$ and $a = 5$, then $|7 - 5| = 2$. If $x = 3$ and $a = 5$, then $|3 - 5| = |-2| = 2$ and $|-5| = |5| = 5$. Accordingly, to solve Problem 9.1 we must minimize the function

$$d(x) = |x - 1| + |x - 2| + |x - 5| + |x - 8| + |x - 9|$$

One way to minimize $d(x)$ is to graph it and then visually determine the low point. But how can this function be graphed? We might, of course, draw up a table of values of x and $d(x)$, for example,

x	0	1	2	3	4	5	6	7	8	9
$d(x)$	25	20	17	16	15	14*	15	16	17	20

This table suggests that $d(x)$ is minimum when $x = 5$. But it is impossible to be sure, because not all values of x have been considered.

We give below a way of graphing $d(x)$ which enables us to examine implicitly all possible x's. This is accomplished by breaking up the domain of definition of $d(x)$ into six subdomains, and renaming $d(x)$ in simpler terms over each subdomain. Thus, let us consider the following cases.

Case (a). Suppose $x \geq 9$. In this case, $|x - 9| = x - 9$; $|x - 8| = x - 8$; $|x - 5| = x - 5$; $|x - 2| = x - 2$; and $|x - 1| = x - 1$. Hence, for $x \geq 9$,

$$d(x) = (x - 1) + (x - 2) + (x - 5) + (x - 8) + (x - 9) = 5x - 25$$

a linear function of x. It is easy to graph this function for every value of $x \geq 9$.

Case (b). $8 \leq x < 9$. In this case, $|x - 9| = -(x - 9) = 9 - x$; $|x - 8| = x - 8$; $|x - 5| = x - 5$; $|x - 2| = x - 2$; and $|x - 1| = x - 1$. Hence

$$\begin{aligned} d(x) &= |x - 1| + |x - 2| + |x - 5| + |x - 8| + |x - 9| \\ &= (x - 1) + (x - 2) + (x - 5) + (x - 8) + (9 - x) \\ &= 3x - 7 \end{aligned}$$

The two subdomains of definition stipulated in cases (a) and (b) abut at the point $x = 9$. At $x = 9$, both expressions $5x - 25$ and $3x - 7$ assume the same value. Hence the graphs of the linear functions defined in cases (a) and (b), that is,

$$\begin{aligned} d(x) &= 5x - 25, \qquad \text{for } x \geq 9 \\ d(x) &= 3x - \ \ 7, \qquad \text{for } 8 \leq x < 9 \end{aligned}$$

join together without any gap at $x = 9$, as shown in Figure 33.

Case (c). $5 \leq x < 8$. Here $|x - 9| = 9 - x$, $|x - 8| = 8 - x$; $|x - 5| = x - 5$; $|x - 2| = x - 2$; $|x - 1| = x - 1$. Hence, over this subdomain,

$$\begin{aligned} d(x) &= (x - 1) + (x - 2) + (x - 5) + (8 - x) + (9 - x) \\ &= x + 9 \end{aligned}$$

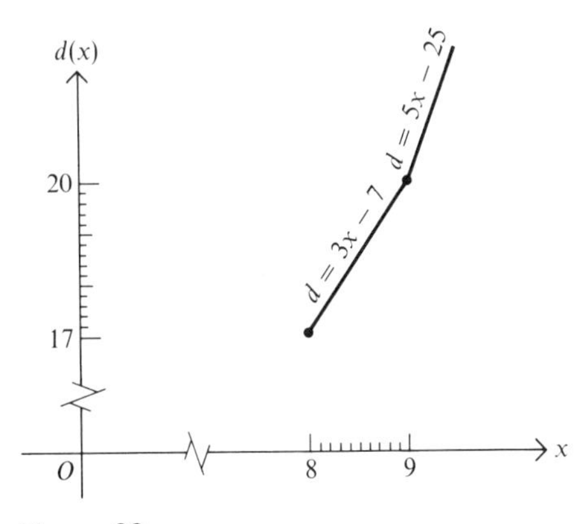

Figure 33

Note that, when $x = 8$, the linear expressions in cases (b) and (c), namely, $3x - 7$ and $x + 9$, are both equal to 17.

Case (d). $2 \leq x < 5$. Here $|x - 9| = 9 - x$; $|x - 8| = 8 - x$; $|x - 5| = 5 - x$; $|x - 2| = x - 2$; $|x - 1| = x - 1$; and

$$d(x) = (x - 1) + (x - 2) + (5 - x) + (8 - x) + (9 - x)$$
$$= -x + 19$$

We point out that, when $x = 5$, the expressions $x + 9$ and $(-x + 19)$ are equal.

Case (e). $1 \leq x < 2$. In this case, $|x - 9| = 9 - x$; $|x - 8| = 8 - x$; $|x - 5| = 5 - x$; $|x - 2| = 2 - x$; $|x - 1| = x - 1$; and hence

$$d(x) = -3x + 23$$

When $x = 2$, both $-x + 19$ and $-3x + 23$ are equal to 17.

Case (f). $x < 1$. The reader can easily show that, in this last case, $d(x) = -5x + 25$.

To sum up, in order to gain insight into the graph of the function $d(x) = |x - 1| + |x - 2| + |x - 5| + |x - 8| + |x - 9|$, it has been represented in six portions, with six different names over the six different sections of the x axis determined by the five fixed points; these are

(1) When $x < 1$, $d(x) = -5x + 25$
(2) When $1 \leq x < 2$, $d(x) = -3x + 23$
(3) When $2 \leq x < 5$, $d(x) = -x + 19$
(4) When $5 \leq x < 8$, $d(x) = x + 9$
(5) When $8 \leq x < 9$, $d(x) = 3x - 7$
(6) When $9 \leq x$, $d(x) = 5x - 25$

Over each of the subdomains of definition, $d(x)$ is a linear function. Over the whole x axis, we say that $d(x)$ is a *piecewise linear* function. This function is well defined for any x, and its graph exhibits no gaps. We stress that a function need not be defined by any one simple, explicit algebraic formula. All that matters is that, given any value of the independent variable x, there exists a well-defined, unambiguous rule for determining the value of the dependent variable $d(x)$.

Figure 34 shows the graph of $d(x) = |x - 1| + |x - 2| + |x - 5| + |x - 8| + |x - 9|$. By inspection, we see that when $x = 5$, $d(x)$ achieves its minimum value of 14.

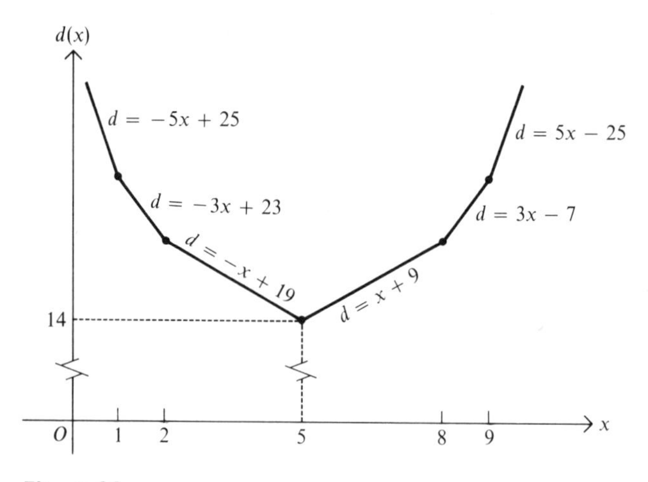

Figure 34

The study of particular cases sometimes provides insight into the general case, sometimes not. For instance, what characterizes values of x for which $d(x)$ is minimum? In Problem 9.1, the optimal x appears to be the *average* of the coordinates of the fixed points which represent the refreshment stands. But this is not always true. Our aim is to study the general case; nevertheless we begin by considering problems much simpler than Problem 9.1. We hope to understand the general case not by graphing $d(x)$ but rather by somehow capturing the underlying structure of the problem at hand.

PROBLEM 9.2. Minimize $d(x) = |x - 1| + |x - 3|$

Here the fixed points representing stands have coordinates 1 and 3. As indicated in Figure 35, no points *outside* the interval $1 \leq x \leq 3$ can render $d(x)$ minimum, because any point to the left of $x = 1$ makes $d(x)$ larger than does $x = 1$, while any point to the right of $x = 3$ likewise makes $d(x)$ larger than does $x = 3$. Therefore, in seeking to minimize $d(x)$, only values of x lying between 1 and 3 inclusive need be considered. Evidently, any choice of x

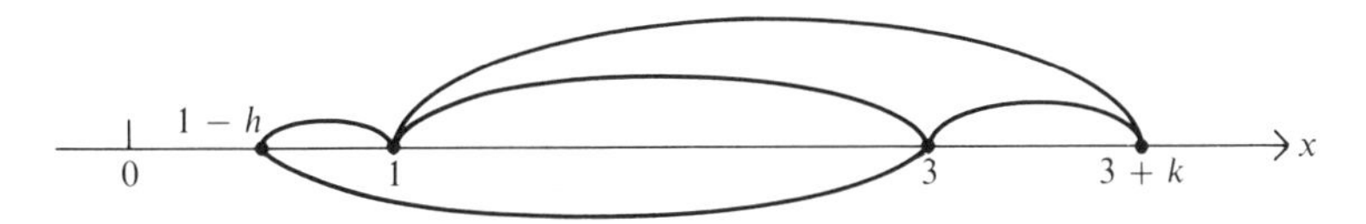

Figure 35

between 1 and 3 inclusive is as good as any other, since all such points make $d(x) = 2$. Mathematically, if $1 \leq x \leq 3$, then

$$\begin{aligned} d(x) &= |x - 1| + |x - 3| \\ &= (x - 1) + (3 - x) = 2 \end{aligned}$$

Now consider a somewhat more complex situation.

PROBLEM 9.3. Minimize $d(x) = |x - 1| + |x - 4| + |x - 5|$

In this case the fixed points have coordinates 1, 4, and 5, as shown in Figure 36. For the same reasons as in Problem 9.2, only values of x lying between the extremes of 1 and 5 need be considered.

For any point X with coordinate x between 1 and 5 inclusive, the sum of the distances from X to the extreme points 1 and 5 is always constant. Figure

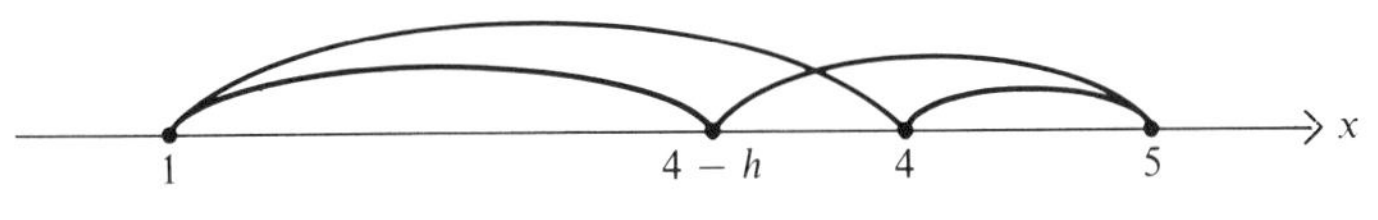

Figure 36

36 shows this pictorially. Mathematically, the fact that x lies between 1 and 5 implies that

$$|x - 1| + |x - 5| = (x - 1) + (5 - x) = 4$$

Thus, for $1 \leq x \leq 5$,

$$\begin{aligned} d(x) &= |x - 1| + |x - 4| + |x - 5| \\ &= 4 + |x - 4| \end{aligned}$$

In order to make $d(x)$ minimum we need only minimize the variable part $|x - 4|$; it is minimal when $x = 4$. Hence Problem 9.3 is completely solved. (The best x in this problem is *not* the average of the given coordinates 1, 4, and 5.)

With our experience in Problems 9.2 and 9.3, let us return to Problem 9.1 and see if it can be solved without the aid of a graph. Recall that the objective is to minimize $d(x) = |x - 1| + |x - 2| + |x - 5| + |x - 8| + |x - 9|$. First, for the same reason as in Problems 9.2 and 9.3, we need only consider values of

x which lie between the extremes of 1 and 9. Furthermore, for any point within this interval, the sum of its distances to the extreme points, with coordinates 1 and 9, is constant. Hence, when $1 \leq x \leq 9$, the total distance can be rewritten

$$d(x) = 8 + |x - 2| + |x - 5| + |x - 8|$$

Problem 9.1 can therefore be solved by minimizing $f(x) = |x - 2| + |x - 5| + |x - 8|$. But by an argument now familiar, no x outside the range $2 \leq x \leq 8$ can be the minimum of $f(x)$. In other words, we need only study $f(x)$ over the range $2 \leq x \leq 8$. Now, for x between 2 and 8, $|x - 2| + |x - 8| = (x - 2) + (8 - x)$. Hence, for $2 \leq x \leq 8, f(x) = |x - 5| + 6$. The variable part of $f(x)$ is as small as possible when $x = 5$, and so $d(x)$ is itself minimum when $x = 5$.

Problems 9.1 to 9.3 suggest that the optimal location(s) of the supply station is at the *central value(s)* of the refreshment stands. To be more precise, let $a_1, a_2, \ldots, a_n$ be a sequence of numbers arranged in ascending order. We say that M is a *median* of these a's if the number of a's less than or equal to M is the same as the number of a's greater than or equal to M. According to this definition, if there are an odd number of distinct a's, there is always a middle item whose value is the unique median. For example, the median of the five numbers $\{1, 8, 9, 2, 6\}$ is 6, as can easily be verified by rearranging these numbers in ascending order.

If there are an even number of distinct a's, there is never a middle item but rather two central ones, say a_k and a_{k+1}. In this case, any number x satisfying the closed inequality $a_k \leq x \leq a_{k+1}$ is said to be a *median* of the a's. For example, for the six numbers $\{1, 2, 5, 8, 9, 12\}$ any number between 5 and 8 inclusive is a median.

Problems 9.2 and 9.3 can be thought of as the *canonical* cases for the refreshment-stand problem, canonical in the sense that they contain the germ of all ideas needed to elucidate the general case.

The reader should now be able to demonstrate the following theorem: Given a set of points on an axis with coordinates $a_1, a_2, \ldots, a_n$. The function $d(x) = |x - a_1| + |x - a_2| + \cdots + |x - a_{n-1}| + |x - a_n|$ takes on its minimum value at the median(s) of the a_i.

The appropriate criterion for deciding how to choose a particular point, such as the point representing the central supply station in the preceding examples, evidently depends on the nature of the underlying problem and the penalties involved in making the wrong choice. The criterion in the following problem is to make the *largest* of the deviations between the supply station and the different refreshment stands as small as possible. This is the so-called *Tchebycheff criterion*, or *minimax rule*.

PROBLEM 9.4. Given three points on a straight line, with coordinates 1, 3, and 7. Find the point X, with coordinate x, which makes the largest of the deviations, namely $|x - 1|$, $|x - 3|$, and $|x - 7|$, as small as possible.

The solution to this problem is $x = 4$, as can be verified by considering what happens when we move away from this point.

In general, if the given points have coordinates $a_1, a_2, \ldots, a_n$, arranged in ascending order, then the objective function

$$t(x) = \text{maximum of } (|x - a_1|, |x - a_2|, \ldots, |x - a_n|)$$

takes on its *minimum* value at the average of the extreme a's, that is, when $x = \frac{1}{2}(a_1 + a_n)$. In practical terms, the supply station is located halfway between the furthest outlying refreshment stands.

Another yardstick that might be used for choosing X in such a problem is the *least-squares criterion*. According to this criterion, we minimize the following quadratic function of x:

$$q(x) = (x - a_1)^2 + (x - a_2)^2 + \cdots + (x - a_n)^2$$

that is, the sum of the squares of the distances from supply point to stands.

The minimum of $q(x)$ occurs when x equals the average of the a_i, that is, when $x = (a_1 + \cdots + a_n)/n$. For example, the minimum of $q(x) = (x - 1)^2 + (x - 3)^2 + (x - 7)^2$ is at $x = \frac{1}{3}(1 + 3 + 7)$. The reader who knows calculus can find the minimum of $q(x)$ by studying its derivative. The following is a more elementary demonstration: Let us determine the value of x that minimizes

$$q(x) = (x - a_1)^2 + (x - a_2)^2 + \cdots + (x - a_n)^2 \tag{1}$$

By performing all indicated multiplications and then collecting like powers of x, we obtain

$$\begin{aligned} q(x) &= nx^2 - 2(a_1 + a_2 + \cdots + a_n)x + (a_1^2 + a_2^2 + \cdots + a_n^2) \\ &= nx^2 - 2sx + t \end{aligned} \tag{2}$$

where $s = a_1 + \cdots + a_n$ and $t = a_1^2 + \cdots + a_n^2$. Equation (2) can be rewritten as

$$\begin{aligned} q(x) &= n\left(x - \frac{s}{n}\right)^2 + \left(t - \frac{s^2}{n}\right) \\ &= n(x - \bar{a})^2 + w \end{aligned} \tag{3}$$

where $\bar{a} = s/n$ and $w = t - s^2/n$. [The reader can check the equivalence of Eqs. (2) and (3) by expanding Eq. (3).] From Eq. (3) it is clear that when $x = \bar{a}$, $q(x) = w$ and that $q(x) > w$ for any other value of x.

In passing, we have obtained two expressions for w, namely,

$$w = t - \frac{s^2}{n} = (a_1^2 + \cdots + a_n^2) - n\bar{a}^2$$

$$w = q\bar{a} = (\bar{a} - a_1)^2 + (\bar{a} - a_2)^2 + \cdots + (\bar{a} - a_n)^2$$

(see Exercise 4).

Exercises

1. Consider the function $d(x) = |x - 1| + |x - 2| + |x - 6|$.

   (a) Show that $d(x)$ can be redefined as follows:

$$\begin{aligned} x < 1, &\quad d(x) = -3x + 9 \\ 1 \le x < 2, &\quad d(x) = -x + 7 \\ 2 \le x < 6, &\quad d(x) = x + 3 \\ 6 \le x\ \ \ , &\quad d(x) = 3x - 9 \end{aligned}$$

   (b) Graph $d(x)$ against x.
   (c) Determine the minimum value of $d(x)$ and the value of x when $d(x)$ is minimal.

2. Given the six points on the x axis with coordinates $x = 1, 2, 5, 8, 9$, and 10.77. *Minimize* the following functions:

   (a) $d(x) = |x - 1| + |x - 2| + |x - 5| + |x - 8| + |x - 9| + |x - 10.77|$
   (b) $q(x) = (x - 1)^2 + (x - 2)^2 + (x - 5)^2 + (x - 8)^2 + (x - 9)^2 + (x - 10.77)^2$
   (c) $T(x) = \text{maximum of } \{|x - 1|, |x - 2|, |x - 5|, |x - 8|, |x - 9|, |x - 10.77|\}$

3. Add the point with coordinate $x = 100$ to Exercise 2. Then minimize $d(x)$, $q(x)$, and $T(x)$.

4. *True or false?* Given any sequence of numbers $a_1, a_2, \ldots, a_n$, let $\bar{a} = (a_1 + a_2 + \cdots + a_n)/n$. Then

$$(a_1^2 + a_2^2 + \cdots + a_n^2) = (a_1 - \bar{a})^2 + (a_2 - \bar{a})^2 + \cdots + (a_n - \bar{a})^2 + n\bar{a}^2$$

5. Consider the following problem:

$$\begin{aligned} &\text{Maximize} \quad P = x_2 \\ &\text{where} \quad x_1 \ge 0, \quad x_2 \ge 0 \\ &\text{and} \quad 2x_1 + 3x_2 \le 9 \\ &\qquad\quad |x_1 - 2| \le 1 \end{aligned}$$

   (a) Solve graphically.
   (b) Reformulate as a linear program.
   (c) Solve by the simplex algorithm.

6. Briefly describe the advantages and disadvantages of:

   (a) The minimax criterion
   (b) Minimization of the sum of the absolute deviations
   (c) Least squares

9.2 THE BEST-FIT PROBLEMS REEXAMINED

This section serves as a bridge between Sections 9.1 and 9.3. Here methods will be developed to solve the simple problems of Section 9.1 as linear programs; these methods will be carried over to the more complicated problems of Section 9.3.

Consider again Problem 9.2, namely,

$$\text{Minimize} \quad d(x) = |x - 1| + |x - 3|$$

In order to solve this problem by linear programming, we first introduce a new variable z which serves as an upper bound to $d(x)$, defined by the inequality

$$d(x) = |x - 1| + |x - 3| \leq z$$

We need only consider the following three cases:

Case (a). $x \geq 3$.

Case (b). $3 \geq x \geq 1$.

Case (c). $1 \geq x$.

In case (a), $d(x) = (x - 1) + (x - 3) = 2x - 4$, and hence one constraint on x and z is

$$2x - 4 \leq z \tag{1}$$

In case (b), $|x - 1| = x - 1$, $|x - 3| = 3 - x$, and $d(x) = 2$. Another constraint is therefore

$$2 \leq z \tag{2}$$

Similarly, case (c) implies that

$$-2x + 4 \leq z \tag{3}$$

Furthermore, the variable z is by its very nature nonnegative, and a moment's thought shows that we can also assume $x \geq 0$.

Problem 9.2 is therefore equivalent to the following linear program.

PROBLEM 9.5.

$$\begin{aligned} &\text{Minimize} \quad z \\ &\text{where} \quad x \geq 0, \quad z \geq 0 \\ &\text{and} \quad 2x - z \leq 4 && (4) \\ &\quad -2x - z \leq -4 && (5) \\ &\quad -z \leq -2 && (6) \end{aligned}$$

Inequalities (4), (5), and (6) follow immediately from (1), (2), and (3), respectively, by transposing all variables to the left and all constants to the right.

One way to solve this problem is to enumerate all feasible corners. As an exercise, the reader should solve it by the simplex algorithm. He will find that the minimum value of z is 2 and that multiple optima are indicated.

It can be useful to realize that the above linear program can be handled quite easily, without artificial variables, by making the following change of variables: Let $z = 1/z'$ and $x = x'/z'$. Then constraints (1), (2), and (3) become, respectively,

$$2(x'/z') - 4 \leq 1/z' \tag{1'}$$

$$-2(x'/z') + 4 \leq 1/z' \tag{2'}$$

$$2 \leq 1/z' \tag{3'}$$

Multiplying both sides of each of these inequalities by z', we obtain

$$2x' - 4z' \leq 1$$
$$-2x' + 4z' \leq 1$$
$$2z' \leq 1$$

These transformations are only valid if the minimum of z is not zero. Under this assumption, Problem 9.6 below is equivalent to Problem 9.5.

PROBLEM 9.6. Maximize z'

$$\text{where} \quad x' \geq 0, \quad z' \geq 0$$
$$\text{and} \quad 2x' - 4z' \leq 1$$
$$-2x' + 4z' \leq 1$$
$$2z' \leq 1$$

The reader should solve Problem 9.6 in detail.

Now consider a *minimax* problem solved by linear programming.

PROBLEM 9.7. Minimize the maximum of $\{|x - 3|, |x - 7|, |x - 1|\}$.

Define a new variable z which satisfies the inequalities

$$|x - 3| \leq z, \qquad |x - 7| \leq z, \qquad |x - 1| \leq z$$

These three inequalities are equivalent to the system of six algebraic inequalities

$$x - 3 \leq z \qquad x - 7 \leq z \qquad x - 1 \leq z$$
$$-x + 3 \leq z \qquad -x + 7 \leq z \qquad -x + 1 \leq z$$

The given minimax problem is thus equivalent to the linear program

Minimize z

$$\text{where} \quad x \geq 0, \quad z \geq 0$$

$$\text{and} \quad x - z \leq 3 \tag{1}$$
$$-x - z \leq -3 \tag{2}$$
$$x - z \leq 7 \tag{3}$$
$$-x - z \leq -7 \tag{4}$$
$$x - z \leq 1 \tag{5}$$
$$-x - z \leq -1 \tag{6}$$

Observe that if inequality (5) holds, then so do inequalities (1) and (3). Similarly, inequality (4) implies inequalities (2) and (4). Hence, by discarding these redundancies, Problem 9.7 can be reformulated as

Minimize z

$$\text{where} \quad x \geq 0, \quad z \geq 0$$
$$\text{and} \quad x - z \leq 1$$
$$-x - z \leq -7$$

The solution to this program is easily seen to be $x = 4$, $z = 3$.

Exercises

1. Consider the problem of minimizing $d(x) = |x - 1| + |x - 3|$.

(a) Show that this problem is equivalent to

$$\begin{aligned} &\text{Minimize} \quad S = z_1 + z_2 \\ &\text{where} \quad |x - 1| \leq z_1 \\ &\text{and} \quad |x - 3| \leq z_2 \end{aligned}$$

(b) Show that the problem is equivalent to the following program, where s_1, s_2, s_3, and s_4 are appropriately defined slack variables:

	x	z_1	z_2	s_1	s_2	s_3	s_4	
s_1	1	−1	0	1	0	0	0	1
s_2	−1	−1	0	0	1	0	0	−1
s_3	1	0	−1	0	0	1	0	3
s_4	−1	0	−1	0	0	0	1	−3
	0	−1	−1	0	0	0	0	$-S$

(c) Solve the program by introducing artificial variables wherever needed.

2. Given three points on a line with coordinates 1, 3, and 7, let

$$d(x) = \text{maximum}\left(\left|\frac{x-1}{1}\right|, \left|\frac{x-3}{3}\right|, \left|\frac{x-7}{7}\right|\right)$$

(a) Show that $d(x)$ can be minimized by solving the program

$$\begin{aligned} &\text{Minimize} \quad z \\ &\text{where} \quad x \geq 0, \quad z \geq 0 \\ &\text{and} \quad x - 1 \leq z \\ &\qquad -x + 1 \leq z \\ &\qquad x - 3 \leq 3z \\ &\qquad -x + 3 \leq 3z \\ &\qquad x - 7 \leq 7z \\ &\qquad -x + 7 \leq 7z \end{aligned}$$

(b) Solve this program, after first making the change in variables: $x' = x/z$, $z' = 1/z$.

3. Given three points on a line with coordinates 1, 3, and 7, let

$$d(x) = \text{maximum}\left(\left|\frac{x-1}{x}\right|, \left|\frac{x-3}{x}\right|, \left|\frac{x-7}{x}\right|\right)$$

(a) Show that the minimum of $d(x)$ can be found by solving the linear program

$$\begin{aligned}
&\text{Minimize} \quad z \\
&\text{where} \quad u \geq 0, \quad z \geq 0 \\
&\text{and} \qquad 1 - u \leq z \\
&\qquad\quad -1 + u \leq z \\
&\qquad\quad 1 - 3u \leq z \\
&\qquad\quad -1 + 3u \leq z \\
&\qquad\quad 1 - 7u \leq z \\
&\qquad\quad -1 + 7u \leq z
\end{aligned}$$

where $u = 1/x$.

(b) Solve the program.

4. Consider the problem:

$$\begin{aligned}
&\text{Maximize} \quad |x_1| + x_2 + x_3 \\
&\text{where } x_1 \text{ is free,} \quad x_2 \geq 0, \quad x_3 \geq 0 \\
&\text{and} \quad x_1 + 2x_3 + 3x_3 = 12 \\
&\qquad 2x_1 + x_2 + x_3 \geq 10 \\
&\qquad x_1 - x_2 + x_3 \leq 6
\end{aligned}$$

(a) Formulate this problem as a linear program.
(b) Solve.

5. At what stage in the simplex algorithm must a simple minimax problem be solved?

9.3 MULTIDIMENSIONAL PROBLEMS OF APPROXIMATION

In this section we generalize the material discussed in the previous sections. The first problem is concerned with a line of best fit in the plane.

PROBLEM 9.8. Given three points in the plane, $P_1(1, 2)$, $P_2(3, 4)$, and $P_3(4, 7)$, as shown in Figure 37. Call these the *observed points*. The objective is to find the straight line, with equation $y_p = b_0 + b_1 x$, which minimizes the largest of the absolute deviations between observed values of y and those predicted by the equation.

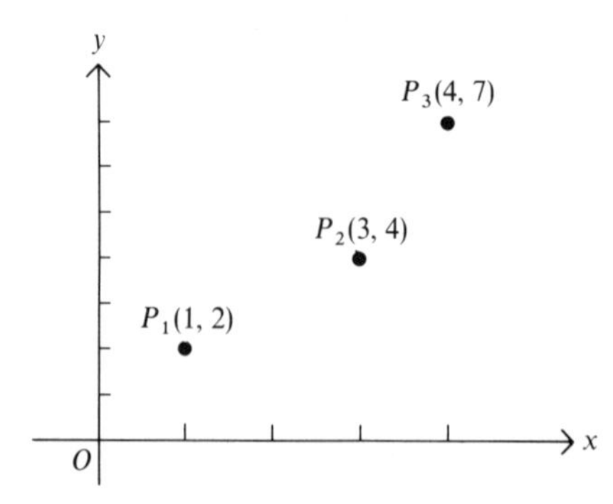

Figure 37

The following table describes the problem in greater detail:

Observed x	*Observed y*	$y_p = b_0 + b_1(x)$	*Absolute deviation*
1	2	$b_0 + b_1(1)$	$\lvert b_0 + b_1 - 2\rvert$
3	4	$b_0 + b_1(3)$	$\lvert b_0 + 3b_1 - 4\rvert$
4	7	$b_0 + b_1(4)$	$\lvert b_0 + 4b_1 - 7\rvert$

We seek b_0 and b_1 so as to render as small as possible the maximum of $\{|b_0 + b_1 - 2|, |b_0 + 3b_1 - 4|, |b_0 + 4b_1 - 7|\}$.

Let z be an upper bound to these three absolute deviations; that is, suppose

$$|b_0 + b_1 - 2| \leq z$$
$$|b_0 + 3b_1 - 4| \leq z$$
$$|b_0 + 4b_1 - 7| \leq z$$

Clearly, if we can make z as small as possible, the objective will be achieved. From Section 9.2, it would appear that the problem of minimizing z is equivalent to the linear program

Minimize z

subject to the trivial inequalities $b_0 \geq 0, \quad b_1 \geq 0, \quad z \geq 0$

and

$$\begin{aligned} b_0 + b_1 - z &\leq 2 \\ -b_0 - b_1 - z &\leq -2 \\ b_0 + 3b_1 - z &\leq 4 \\ -b_0 - 3b_1 - z &\leq -4 \\ b_0 + 4b_1 - z &\leq 7 \\ -b_0 - 4b_1 - z &\leq -7 \end{aligned}$$

This happens to be an incorrect formulation of Problem 9.8. Both b_0 and b_1 are constrained to be nonnegative in this formulation, thus forcing a prediction line with too particular an inclination. This stipulation can produce a larger minimax absolute deviation than would result from allowing the line to have *any* inclination. Under the arbitrary condition that both b_0 and b_1 must be nonnegative, the line that fits the given points best according to the minimax, or Tchebycheff, criterion is

$$y_p = b_0 + b_1x = 0 + \tfrac{11}{7}x = \tfrac{11}{7}x$$

The maximum absolute deviation corresponding to this line is $\frac{5}{7}$:

x	*y observed*	$y_p = \frac{11}{7}x$	*Absolute deviation*
1	2	$\frac{11}{7}$	$\frac{3}{7}$
3	4	$\frac{33}{7}$	$\frac{5}{7}$
4	7	$\frac{44}{7}$	$\frac{5}{7}$

To reformulate the above program and remove the nonnegativity restrictions on b_0 and b_1, we need only express b_0 and b_1 as the difference between nonnegative variables. Having done this, we can minimize z subject to a revised set of constraints. Accordingly, let

$$b_0 = a_0 - c_0$$

$$b_1 = a_1 - c_1$$

where $a_0 \geq 0$, $c_0 \geq 0$, $a_1 \geq 0$, $c_1 \geq 0$. Then

$$y_p = (a_0 - c_0) + (a_1 - c_1)x$$

The solution to the program which allows b_0 and b_1 to be unrestricted in sign is

$$a_0 = 0, \quad c_0 = \tfrac{1}{3}, \quad a_1 = \tfrac{5}{3}, \quad c_1 = 0$$

$$y_p = -\tfrac{1}{3} + \tfrac{5}{3}x$$

Then the deviations become

x	*y observed*	$y_p = -\frac{1}{3} + \frac{5}{3}x$	*Absolute deviation*
1	2	$\frac{4}{3}$	$\frac{2}{3}$
3	4	$\frac{14}{3}$	$\frac{2}{3}$
4	7	$\frac{19}{3}$	$\frac{2}{3}$

The value of the minimax deviation is clearly smaller when b_0 and b_1 are unconstrained: $\frac{2}{3} < \frac{5}{7}$.

The next problem requires minimization of the sum of the deviations of all observations.

PROBLEM 9.9. Given the observed points in Problem 9.8, find a straight line

$$y_p = b_0 + b_1 x$$

so as to minimize the *total* of all the absolute deviations between observed y's and those predicted by the equation. The following table shows these deviations

x	*Observed y*	*Predicted y*	*Absolute deviation*
1	2	$b_0 + b_1$	$\lvert b_0 + b_1 - 2\rvert$
3	4	$b_0 + 3b_1$	$\lvert b_0 + 3b_1 - 4\rvert$
4	7	$b_0 + 4b_1$	$\lvert b_0 + 4b_1 - 7\rvert$

The objective is to find b_0 and b_1 so as to minimize the sum

$$S = |b_0 + b_1 - 2| + |b_0 + 3b_1 - 4| + |b_0 + 4b_1 - 7|$$

One way of proceeding is to define the variable z to be an upper bound to S, that is, $S \leq z$, and then minimize z subject to the appropriate constraints. This would lead to a program with eight nontrivial constraints which are not immediately obvious.

A better procedure is to define the three variables z_1, z_2, and z_3, according to the following conditions:

$$|b_0 + b_1 - 2| \leq z_1$$
$$|b_0 + 3b_1 - 4| \leq z_2$$
$$|b_0 + 4b_1 - 7| \leq z_3$$

The linear program which will solve Problem 9.9 can now be formulated as follows:

$$\text{Minimize} \quad z = z_1 + z_2 + z_3$$
$$\text{where} \quad z_1 \geq 0, \quad z_2 \geq 0, \quad z_3 \geq 0$$
$$\begin{aligned} \text{and} \quad b_0 + b_1 - z_1 \qquad\qquad & \leq 2 \\ -b_0 - b_1 - z_1 \qquad\qquad & \leq -2 \\ b_0 + 3b_1 \qquad - z_2 \qquad & \leq 4 \\ -b_0 - 3b_1 \qquad - z_2 \qquad & \leq -4 \\ b_0 + 4b_1 \qquad\qquad - z_3 & \leq 7 \\ -b_0 - 4b_1 \qquad\qquad - z_3 & \leq -7 \end{aligned}$$

Observe that both b_0 and b_1 are assumed to be free variables. The best approximating line turns out to be

$$y_p = \tfrac{1}{3} + \tfrac{5}{3}x$$

The associated table of absolute deviations is

x	*y observed*	$y_p = \frac{1}{3} + \frac{5}{3}x$	*Absolute deviation*
1	2	2	0
3	4	$\frac{16}{3}$	$\frac{4}{3}$
4	7	7	0

For a different example of approximation, let us now consider how to find the "best approximate solution" to an inconsistent system of equations. Such a problem often arises when we take measurements of interrelated phenomena. For example, suppose that O represents a bench mark and that Z_1, Z_2, and Z_3 are three points whose elevation above O are to be determined. From direct observation we determine that

(1) Z_1 is 10 units above O.
(2) Z_2 is 7 units above Z_1.
(3) Z_2 is 18 units above O.
(4) Z_2 is 9 units above Z_3.
(5) Z_1 is 2 units above Z_3.

Let z_i equal the elevation of Z_i above O, for $i = 1, 2, 3$. To determine the z_i we have to "solve" the following inconsistent system of five equations in three unknowns,

$$z_1 = 10 \quad (1)$$

$$-z_1 + z_2 = 7 \quad (2)$$

$$z_2 = 18 \quad (3)$$

$$z_2 - z_3 = 9 \quad (4)$$

$$z_1 - z_3 = 2 \quad (5)$$

We assume no equation is correct, and let d_i represent the absolute value of the difference between the left and right sides of Eq. (i), for $i = 1, 2, 3, 4, 5$. Then $d_1 = |z_1 - 10|$, $d_2 = |-z_1 + z_2 - 7|$, and so on. These d_i are called *discrepancies*, *errors*, or *residuals*. Some possible criteria for choosing the best approximate solution are:

(1) *Least-squares criterion:* Minimize $(d_1^2 + d_2^2 + \cdots + d_n^2)$.
(2) *Sum of absolute deviations:* Minimize $(d_1 + d_2 + \cdots + d_n)$.
(3) *Tchebycheff criterion:* Minimize the maximum of $(d_1, d_2, \ldots, d_n)$.

Many practical problems of approximation reduce to one of the above problems or sometimes to a sequence of such problems. Of course, which criterion to use depends on the nature of the problem at hand.

PROBLEM 9.10. Given the inconsistent system

$$x - y = 7$$

$$2x + 3y = 5$$

$$3x + y = -1$$

find the best approximate Tchebycheff solution under the constraint that both coordinates of the solution are nonnegative.

Clearly, this problem reduces to minimizing d, subject to $|x - y - 7| \le d$, $|2x + 3y - 5| \le d$, $|3x + y + 1| \le d$, and $x \ge 0, y \ge 0$. As a linear program, the objective is to

Minimize d

where all variables are nonnegative

and
$$\begin{aligned} x - y - d &\le 7 \\ -x + y - d &\le -7 \\ 2x + 3y - d &\le 5 \\ -2x - 3y - d &\le -5 \\ 3x + y - d &\le -1 \\ -3x - y - d &\le 1 \end{aligned}$$

The best nonnegative Tchebycheff solution to Problem 9.10 is $x = \frac{3}{2}$, $y = 0$, $d = \frac{11}{2}$.

If we allow x and y to be free, that is, not necessarily nonnegative, the best solution becomes $x = \frac{3}{2}$, $y = -\frac{7}{8}$, $d = \frac{37}{8}$.

PROBLEM 9.11. Given the inconsistent system

$$\begin{aligned} x - y &= 7 \\ 2x + 3y &= 5 \\ 3x + y &= -1 \end{aligned}$$

find the approximate solution which minimizes the sum of all three discrepancies and satisfies the further condition that the coordinates of the solution are nonnegative. This problem is equivalent to minimizing the sum $S = d_1 + d_2 + d_3$, where

$$\begin{aligned} d_1 &= |x - y - 7| \\ d_2 &= |2x + 3y - 5| \\ d_3 &= |3x + y + 1| \end{aligned}$$

As usual, S can be minimized by solving the following linear program involving six nontrivial constraints:

Minimize $S = d_1 + d_2 + d_3$

where all variables are nonnegative

and

$$\begin{aligned} x - y - d_1 &= 7 \\ -x + y - d_1 &= -7 \\ 2x + 3y - d_2 &= 5 \\ -2x - 3y - d_2 &= -5 \\ 3x + y - d_3 &= -1 \\ -3x - y - d_3 &= 1 \end{aligned}$$

The optimal solution of this program is $x = 0$, $y = \frac{5}{3}$, $d_1 = \frac{26}{3}$, $d_2 = 0$, $d_3 = \frac{8}{3}$, which gives $\frac{34}{3}$ as the sum of the absolute errors.

If we remove the nonnegativity restriction on x and y, the best approximate solution becomes $x = -\frac{8}{7}$, $y = \frac{17}{7}$, $d_1 = \frac{74}{7}$, $d_2 = 0$, $d_3 = 0$. The corresponding absolute errors are

$$\begin{aligned} |x - y - 7| &= |-\tfrac{8}{7} - \tfrac{17}{7} - 7| = \tfrac{74}{7} \\ |2x + 3y - 5| &= |-\tfrac{16}{7} + 3(\tfrac{17}{7}) - 5| = 0 \\ |3x + y + 1| &= |3(-\tfrac{8}{7}) + \tfrac{17}{7} + 1| = 0 \end{aligned}$$

The sum of these absolute errors is $\frac{74}{7}$. As is usual, the sum of the discrepancies decreased when x and y were allowed to be free variables.

Exercises

1. Solve Problem 9.8 under the additional constraint that $b_1 \geq 2$.

2. Given $P_1(2, 3)$, $P_2(4, 5)$, $P_3(6, 15)$, find $y_p = b_0 + b_1 x$, so as to minimize the maximum value of $|(y_p - y_0)/y_0|$, where y_0 denotes the second coordinate of the observed points.

3. Given $P_1(0, 0)$, $P_2(1, 0)$, $P_3(2, 1)$, show that the line which best fits these points in the Tchebycheff sense is $y_p = -\frac{1}{4} + \frac{1}{2}x$.

4. Given the points $P_1(1, 0)$, $P_2(2, 1)$, $P_3(6, 2)$, show that the Tchebycheff line is $y_p = -\frac{1}{10} + \frac{2}{5}x$.

5. Find the best Tchebycheff solution to the system

$$\begin{aligned} x + y + z &= 10 \\ 2x + 3y + 4z &= 12 \\ 5x + 6y + 7z &= 20 \\ 8x + 9y + 10z &= 30 \end{aligned}$$

6. The values of a function $f(x)$ at $x = 0, 1, 2,$ and 3 are

x	0	1	2	3
$f(x)$	3	-1	2	5

Let $q(x)$ represent a polynomial of degree 2, that is, $q(x) = a_0 + a_1 x + a_2 x^2$. It is desired to determine the coefficients of $q(x)$ so as to minimize the maximum absolute difference between $f(x)$ and $q(x)$ at the points $x = 0, 1, 2,$ and 3.

(a) Formulate this minimization problem as a linear program.
(b) Solve.

9.4 COLUMN GENERATION AND THE CUTTING-STOCK PROBLEM *

Suppose that some product of interest, say sheet metal or paper, is produced in *standard rolls*, all of the same width. (By the width of a roll we mean the width of the unrolled sheet, that is, the actual height of the roll when it rests on its circular base.) Assume that the width of the standard roll is 10 feet, but that customers require smaller, nonstandard rolls of widths 2, 3, and 4 feet, in minimum amounts of 250, 60, and 180 rolls, respectively.

To meet these demands, a certain number of standard rolls will have to be cut into smaller rolls. The following table describes several ways of cutting the standard roll:

	Method											
Width	X_1	X_2	X_3	X_4	X_5	X_6	X_7	X_8	X_9	X_{10}	X_{11}	X_{12}
2 ft	5	3	3	2	1	0	1	0	0	1	0	0
3 ft	0	1	0	2	0	3	1	2	0	0	1	0
4 ft	0	0	1	0	2	0	1	1	2	0	0	1
Total width cut	10	9	10	10	10	9	9	10	8	2	3	4
Standard width	10	10	10	10	10	10	10	10	10	10	10	10
Trim loss	0	1	0	0	0	1	1	0	2	8	7	6

* The material in this section is somewhat difficult and may be omitted on a first reading.

Each column in the table represents a different pattern of cutting. For example, column 2 corresponds to cutting a standard roll into 3 rolls of width 2 feet and 1 roll of width 3 feet, which leaves a trim loss of 1 foot. The first eight columns describe patterns which are *efficient*, in that the associated trim loss is always strictly less than the smallest desired width. The remaining columns correspond to inefficient cutting patterns. For example, the pattern in column 9 is inefficient because the trim loss can be used to cut other smaller rolls.

We distinguish between $\mathbf{x}_i$ and x_i as follows: $\mathbf{x}_i$ represents the ith cutting pattern, and x_i denotes the number of times a standard roll is cut according to the ith pattern. Thus we shall write $\mathbf{x}_i = \begin{pmatrix} 5 \\ 0 \\ 0 \end{pmatrix}$ to denote the first cutting pattern.

Note that we can always meet customer requirements by using patterns $\mathbf{x}_9$, $\mathbf{x}_{10}$, and $\mathbf{x}_{11}$ often enough, but this choice of patterns may cause us to cut too many standard rolls. When we speak of *the "cutting-stock" problem* we assume that our objective is to meet customer demands and *minimize the number of standard rolls that are cut.*

Let us treat the given problem according to this cutting-stock criterion. First, note that we need only consider *all efficient patterns*, but for reasons of mathematical convenience we also include pattern $\mathbf{x}_9$. The program we must study is

PROBLEM 9.12.

$$\begin{aligned}
&\text{Minimize} \quad C = x_1 + x_2 + \cdots + x_9 \\
&\text{where} \quad x_i \geq 0 \text{ for } i = 1, \ldots, 9 \\
&\text{and} \quad 5x_1 + 3x_2 + 3x_3 + 2x_4 + x_5 + x_7 \geq 250 \\
&\qquad\qquad x_2 + 2x_4 + 3x_6 + x_7 + 2x_8 \geq 60 \\
&\qquad\qquad x_3 + 2x_5 + x_7 + x_8 + 2x_9 \geq 180
\end{aligned}$$

Each nontrivial constraint is derived by adding the number of rolls of a particular width that would result from x_i of each type of cut. The optimal solution to this program is shown below, where s_1, s_2, and s_3 represent the surplus variables that must be subtracted from the nontrivial constraints, respectively, in order to convert them into equalities.

Optimal Tableau (Problem 9.12)

	x_1	x_2	x_3	x_4	x_5	x_6	x_7	x_8	x_9	s_1	s_2	s_3	C	
x_1	1	$\frac{2}{5}$	$\frac{1}{2}$	0	0	$-\frac{3}{5}$	$-\frac{1}{10}$	$-\frac{1}{2}$	$-\frac{1}{5}$	$-\frac{1}{5}$	$\frac{1}{5}$	$\frac{1}{10}$	0	20
x_4	0	$\frac{1}{2}$	0	1	0	$\frac{3}{2}$	$\frac{1}{2}$	1	0	0	$-\frac{1}{2}$	0	0	30
x_5	0	0	$\frac{1}{2}$	0	1	0	$\frac{1}{2}$	$\frac{1}{2}$	1	0	0	$-\frac{1}{2}$	0	90
C	0	$-\frac{1}{10}$	0	0	0	$-\frac{1}{10}$	$-\frac{1}{10}$	0	$-\frac{1}{5}$	$-\frac{1}{5}$	$-\frac{3}{10}$	$-\frac{2}{5}$	1	140

Here 140 standard rolls are cut. Twenty rolls are cut according to pattern 1, 30 according to pattern 4, and 90 according to pattern 5. (This solution holds the "total trim loss" to zero; see the following Exercise 6.)

Note that when we formulated the cutting-stock problem we implicitly assumed that all x_i were integers. This is true in the optimal tableau above, but it will not always be true unless we extend the simplex algorithm in the appropriate fashion. This extension is more suited to an advanced course in linear programming. In this course we treat the cutting-stock problem as a linear program—if the optimal x_i's are nonintegral we shall be content to round off to the nearest integer.

Of course, rounding off can create nonfeasible solutions. To meet this objection, assume that a limited stock of nonstandard rolls is kept in inventory. Now if rounding off should lead to a cutting plan which does not satisfy the minimum demands, we can dip into inventory to create a feasible solution. Experience shows that by proceeding in this fashion practical, near-optimal solutions are obtained.

The number of efficient patterns depends on the relationship between the standard width and the desired smaller widths. In Problem 9.12 we were able to list all efficient setups without too much difficulty. However, in many cutting-stock problems, the standard roll is relatively large and the required rolls are of numerous widths; as a result the task of actually enumerating all possible efficient setups can be quite formidable. Later, after having studied *the loading problem*, we will then solve a cutting-stock problem without listing all the feasible setups.

The terminology of the loading problem is derived from the situation in which a hold of capacity b is to be packed with n types of objects of utility $c_1, c_2, \ldots, c_n$ and size $a_1, a_2, \ldots, a_n$, respectively, in such a manner that the total amount packed has maximum value (see Problem 3.6).

It is helpful to distinguish between three types of loading problems:

(1) The continuous-cargo loading problem
(2) The discrete-cargo loading problem
(3) The knapsack problem

The *continuous-cargo loading problem* is a linear program of the form

$$\begin{aligned} &\text{Maximize} \quad V = c_1x_1 + c_2x_2 + \cdots + c_nx_n \\ &\text{where all } x_i \geq 0, \text{ for } i = 1, \ldots, n \\ &\text{and} \quad a_1x_1 + a_2x_2 + \cdots + a_nx_n \leq b \end{aligned}$$

The x_i here can take on any nonnegative values whatsoever. But if we add the requirement that the x_i must be integers, the problem becomes a *discrete-cargo loading problem.*

The *knapsack problem* is a discrete loading problem where each x_i is allowed to assume only one of the two values, 0 and 1. This name is suggestive of how a knapsack is packed, say for a camping trip—either we take one transistor radio or none at all, and so on.

To help clarify these distinctions, consider

PROBLEM 9.13. Maximize $V = 10x_1 + 8x_2 + 7x_3$

where $x_1 \geq 0, \quad x_2 \geq 0, \quad x_3 \geq 0$

and $8x_1 + 7x_2 + 5x_3 \leq 34$

If we view this problem as a linear program, that is, a continuous loading problem, then it is easy to see that the optimal solution is $x_1 = 0$, $x_2 = 0$, $x_3 = 6.8$, at which point $V = 47.6$. [In general, to solve a continuous loading problem it suffices to find the maximum of $(c_1/a_1, c_2/a_2, \ldots, c_n/a_n)$. Let k denote the index for which this "relative value quotient" is maximum. Then the optimum occurs at the point where $x_k = b/a_k$ and all other coordinates are 0. To prove this result, pivot in column k and observe what happens.]

Note that if we round off the above continuous solution to the nearest feasible integer solution, we do not obtain the optimum of the analogous discrete loading problem. Thus $x_1 = 0$, $x_2 = 0$, $x_3 = 7$ is not a feasible point and hence $x_1 = 0$, $x_2 = 0$, $x_3 = 6$ is the "nearest" feasible integer solution to the continuous optimum. But there exist better solutions to the discrete problem, for example, $x_1 = 1$, $x_2 = 0$, $x_3 = 5$; and $x_1 = 3$, $x_2 = 0$, $x_3 = 2$.

The discrete loading problem can always be solved by enumeration of all feasible integer solutions. However, as n increases in size, the task of complete enumeration grows quite formidable if not practically impossible. A method of solving Problem 9.13, viewed as a discrete problem, is sketched below. This method is ad hoc and not the most efficient, but it is suggestive of a general line of attack—delimit the set of integer solutions so that complete enumeration can be replaced by *partial enumeration.*

To illustrate, consider the three integer solutions already encountered, namely,

Solution	*Value of* $V = 10x_1 + 8x_2 + 7x_3$
(i) $x_1 = 0, \quad x_2 = 0, \quad x_3 = 6$	42
(ii) $x_1 = 1, \quad x_2 = 0, \quad x_3 = 5$	45
(iii) $x_1 = 3, \quad x_2 = 0, \quad x_3 = 2$	44

The sum of the coordinates in each of these solutions is at least 5. The question then arises as to whether we can avoid examining any solution the sum of whose coordinates is 5 or less. To help find an answer, consider the following linear program, whose variables need not be discrete:

PROBLEM 9.14A. Maximize $V = 10x_1 + 8x_2 + 7x_3$

where $x_1 \geq 0, \quad x_2 \geq 0, \quad x_3 \geq 0$

and $8x_1 + 7x_2 + 5x_3 \leq 34 \qquad (1)$

$x_1 + x_2 + x_3 \leq 5 \qquad (2)$

(This is Problem 9.13 with the added constraint that the sum of the three variables is less than or equal to 5.) Let x_4 and x_5 denote slack in constraints (1) and (2). Then the reader can easily verify that the optimum solution reads as follows.

Optimal Tableau (Problem 9.14A)

	x_1	x_2	x_3	x_4	x_5	
x_1	1	$\frac{2}{3}$	0	$\frac{1}{3}$	$-\frac{5}{3}$	3
x_3	0	$\frac{1}{3}$	1	$-\frac{1}{3}$	$\frac{8}{3}$	2
	0	-1	0	-1	-2	$V-44$

The optimum solution of Problem 9.14A happens to be integral, but, in any case, we already have a better integral solution to Problem 9.13, namely, solution (ii) above. Hence, in working Problem 9.13, we need not examine any integer solution the sum of whose coordinates is 5 or less.

Let us now determine the effect of relaxing constraint (2) of Problem 9.14A. More precisely, consider

PROBLEM 9.14B. Maximize $V = 10x_1 + 8x_2 + 7x_3$

where $x_1 \geq 0, \quad x_2 \geq 0, \quad x_3 \geq 0$

and $8x_1 + 7x_2 + 5x_3 \leq 34$

$$x_1 + x_2 + x_3 \leq 6$$

By the usual techniques of postoptimality analysis, that is, by adding the optimal x_5 column of Problem 9.14A to the optimal stub of Problem 9.14A, we obtain

Optimal Tableau (Problem 9.14B)

	x_1	x_2	x_3	x_4	x_5	
x_1	1	$\frac{2}{3}$	0	$\frac{1}{3}$	$-\frac{5}{3}$	$\frac{4}{3}$
x_3	0	$\frac{1}{3}$	1	$-\frac{1}{3}$	$\frac{8}{3}$	$\frac{14}{3}$
	0	-1	0	-1	-2	$V-46$

Hence, to solve Problem 9.13, *we should limit our examination to integer solutions the sum of whose coordinates is* 6 *or more*. We list all such points below:

Case (a). $x_3 = 6$. There is only one solution: $x_1 = x_2 = 0, x_3 = 6$; $V = 42$.
Case (b). $x_3 = 5$. There are only two relevant solutions:

(1) $x_1 = 0, x_2 = 1, x_3 = 5$; $V = 43$

(2) $x_1 = 1, x_2 = 0, x_3 = 5$; $V = 45$

Case (*c*). $x_3 = 4$. There is only one relevant solution: $x_1 = 0$, $x_2 = 2$, $x_3 = 4$; $V = 44$.

These are the only cases which give solutions whose coordinates add up to at least 6 and which do not violate the nontrivial constraint. Hence the optimum integer solution to Problem 9.13 is $V = 45$, attained when $x_1 = 1$, $x_2 = 0$, $x_3 = 5$.

The above solution to a discrete loading problem is only meant to illustrate some difficulties and some ways of overcoming them. One of the most efficient ways of handling such problems is through *dynamic programming*.*

Let us now return to cutting-stock problems and describe a method of solution which does not require the enumeration of all possible efficient setups. Since our purpose is pedagogical, we use an example which can be checked by considering the efficient setups.

PROBLEM 9.15. The standard roll is 130 inches wide. We require at least 30 rolls of width 40 inches, at least 60 rolls of width 35 inches, and at least 60 rolls of width 25 inches.

Let $\mathbf{x}_i = \begin{pmatrix} u \\ v \\ w \end{pmatrix}$ represent the setup in which we cut u rolls of width 40, v rolls of width 35, and w rolls of width 25. It is convenient to let $\mathbf{x}_i$ denote any pattern, not necessarily efficient. In linear-program form, we have to

$$\text{Minimize} \quad C = x_1 + x_2 + \cdots + x_n$$

$$\text{subject to} \quad x_i \geq 0, \text{ for } i = 1, \ldots, n$$

$$\text{and} \quad x_1\mathbf{x}_1 + x_2\mathbf{x}_2 + \cdots + x_n\mathbf{x}_n \geq \begin{pmatrix} 30 \\ 60 \\ 60 \end{pmatrix}$$

First we point out that the components of the typical setup must satisfy the inequality

$$40u + 35v + 25w \leq 130$$

or, equivalently,

$$8u + 7v + 5w \leq 26 \tag{1}$$

From inequality (1) we can deduce the following "pure," but not necessarily efficient, patterns:

$$\mathbf{x}_1 = \begin{pmatrix} 3 \\ 0 \\ 0 \end{pmatrix}, \quad \mathbf{x}_2 = \begin{pmatrix} 0 \\ 3 \\ 0 \end{pmatrix}, \quad \mathbf{x}_3 = \begin{pmatrix} 0 \\ 0 \\ 5 \end{pmatrix}$$

* For the elements of dynamic programming, see F. Hillier and G. Lieberman, *Introduction to Operations Research* (Holden-Day, San Francisco, 1967); for a dynamic programming treatment of the discrete loading problem, see E. Bowman and R. Fetter, *Analysis for Production Management* (R. D. Irwin, Homewood, Illinois, 1961), rev. ed.

Now think of Problem 9.15 as schematized below:

Tableau 0

	x_1	x_2	x_3	$\cdots$	x_i	$\cdots$	s_1	s_2	s_3	C	
x_1	③	0	0	$\cdots$	u	$\cdots$	-1	0	0	0	30
x_2	0	③	0	$\cdots$	v	$\cdots$	0	-1	0	0	60
x_3	0	0	⑤	$\cdots$	w	$\cdots$	0	0	-1	0	60
C	-1	-1	-1	$\cdots$	-1	$\cdots$	0	0	0	1	0

Note that Tableau 0 is not yet adjusted, but that we can convert the first three columns into distinct unit vectors by pivoting, in any order, on the circled entries in row i, column i, for $i = 1, 2, 3$. Thus we obtain a tableau of the form:

Tableau 1

	x_1	x_2	x_3	$\cdots$	x_i	$\cdots$	s_1	s_2	s_3	C	
x_1	1	0	0	$\cdots$	$\frac{1}{3}u$	$\cdots$	$-\frac{1}{3}$	0	0	0	10
x_2	0	1	0	$\cdots$	$\frac{1}{3}v$	$\cdots$	0	$-\frac{1}{3}$	0	0	20
x_3	0	0	1	$\cdots$	$\frac{1}{5}w$	$\cdots$	0	0	$-\frac{1}{5}$	0	12
C	0	0	0	$\cdots$	π_1	$\cdots$	$-\frac{1}{3}$	$-\frac{1}{3}$	$-\frac{1}{5}$	1	42

Three pivot operations had to be performed to pass from Tableau 0 to Tableau 1. As a consequence, π_1, the objective coefficient of the typical x_i in Tableau 1, is a function of u, v, and w, whose value is easily seen to be

$$\pi_1 = \tfrac{1}{3}u + \tfrac{1}{3}v + \tfrac{1}{5}w - 1$$

Note further that the surplus-variable columns in any tableau can be formed by fixing the entries in its general x_i column at their appropriate levels. For example, as Tableau 0 shows, the s_1 column can be formed by taking $u = -1$, $v = 0$, $w = 0$. On substituting these values into the body of the x_i column of Tableau 1, we obtain the body of its s_1 column. As for the objective coefficients of the surplus-variable columns, observe that they are 0 (and not -1) in Tableau 0, and thus their values in Tableau 1 can be found by substituting the appropriate values of u, v, and w into $\pi_1 + 1 = \frac{1}{3}u + \frac{1}{3}v + \frac{1}{5}w$.

Clearly, the choice of a pivot column in Tableau 1 can be reduced to the study of

SUBPROBLEM 9.15A.

$$\text{Maximize} \quad \pi_1 = \tfrac{1}{3}u + \tfrac{1}{3}v + \tfrac{1}{5}w - 1$$

$$\text{where} \quad u, v, \text{ and } w \text{ are nonnegative integers}$$

$$\text{such that} \quad 8u + 7v + 5w \leq 26$$

Subproblem 9.15A is a discrete loading problem. As indicated earlier, there exist several algorithms for determining the maximum of π_1, but for the present we stress that, in order to pivot profitably in Tableau 1, we need not actually find the maximum of π_1; any admissible solution to Subproblem 9.15A for which $\pi_1 > 0$ will determine a suitable pivot column. Note further that $\pi_1 = 0$ for any cutting pattern corresponding to a variable which is currently basic, for example, $u = 0$, $v = 3$, $w = 0$. The reader can easily check that the solution $u = 0$, $v = 3$, $w = 1$ makes $\pi_1 = \frac{1}{5} > 0$.

Let us therefore replace some old basic pattern by $\mathbf{x}_4 = \begin{pmatrix} 0 \\ 3 \\ 1 \end{pmatrix}$. In other words, in addition to the three pure patterns originally considered, we now imagine a fourth pattern spelled out explicitly in Tableau 0, namely, $\mathbf{x}_4$. As for Tableau 1, the x_4 column now takes the form specified by substituting $u = 0$, $v = 3$, $w = 1$ into its general x_i column, as indicated below:

	x_1	x_2	x_3	$\cdots$	x_i	x_4	$\cdots$	s_1	s_2	s_3	C		
x_1	1	0	0	$\cdots$	$\frac{1}{3}u$	0	$\cdots$	$-\frac{1}{3}$	0	0	0	10	
x_2	0	1	0	$\cdots$	$\frac{1}{3}v$	①	$\cdots$	0	$-\frac{1}{3}$	0	0	20	←(20)
x_3	0	0	1	$\cdots$	$\frac{1}{5}w$	$\frac{1}{5}$	$\cdots$	0	0	$-\frac{1}{5}$	0	12	(60)
C	0	0	0	$\cdots$	π_1	$\frac{1}{5}$ ↑	$\cdots$	$-\frac{1}{3}$	$-\frac{1}{3}$	$-\frac{1}{5}$	1	42	

After pivoting on the circled entry in the x_4 column, we obtain

Tableau 2

	x_1	x_2	x_3	$\cdots$	x_i	x_4	$\cdots$	
x_1	1	0	0	$\cdots$	$\frac{1}{3}u$	0		10
x_4	0	1	0	$\cdots$	$\frac{1}{3}v$	1		20
x_3	0	$-\frac{1}{5}$	1	$\cdots$	$\frac{1}{5}w - \frac{1}{15}v$	0		8
C	0	$-\frac{1}{5}$	0	$\cdots$	π_2	0		38

Note that the new x_i column represents the general column of Tableau 2 and that

$\pi_2 = \pi_1 - \frac{1}{15}v = \frac{1}{3}u + \frac{4}{15}v + \frac{1}{5}w - 1$. Observe further that we list little extraneous information in Tableau 2. (We show later how to recover the surplus-variable columns, if needed.)

To specify the pivot column of Tableau 2 we must now consider the following discrete loading problem.

SUBPROBLEM 9.15B.

$$\begin{aligned}&\text{Maximize} \quad \pi_2 = \tfrac{1}{3}u + \tfrac{4}{15}v + \tfrac{1}{5}w - 1\\&\text{where} \quad u,\ v, \text{ and } w \text{ are nonnegative integers}\\&\text{such that} \quad 8u + 7v + 5w \leq 26\end{aligned}$$

One solution which makes π_2 positive is $u = 2$, $v = 0$, $w = 2$, for which $\pi_2 = \frac{1}{15}$. Whether or not this solution furnishes the maximum of π_2, we can take $\mathbf{x}_5 = \begin{pmatrix}2\\0\\2\end{pmatrix}$ to be the new basic pattern. Observe that $\mathbf{x}_5$ corresponds to a column in Tableau 0 but that the current column in Tableau 2 whose heading is x_5 is found by substituting the values $u = 2$, $v = 0$, $w = 2$ into the general column of this tableau. Thus we must consider the following tableau, with choice of pivot as indicated:

	x_1	x_2 $\cdots$	x_i	x_5	$\cdots$		
x_1	1		$\frac{1}{3}u$	$\left(\frac{2}{3}\right)$	$\cdots$	10	←(15)
x_4	0		$\frac{1}{3}v$	0		20	
x_3	0		$\frac{1}{5}w - \frac{1}{15}v$	$\frac{2}{5}$		8	(20)
C	0		π_2	$\frac{1}{15}$ ↑		38	

After pivoting we obtain the general tableau

Tableau 3

	x_1 x_2 $\cdots$	x_i	x_5 $\cdots$	
x_5		$\frac{1}{2}u$	1	15
x_4		$\frac{1}{3}v$	0	20
x_3		$\frac{1}{5}w - \frac{1}{15}v - \frac{1}{5}u$	0	2
C		π_3		37

It is easy to see that

$$\pi_3 = \pi_2 - \tfrac{1}{30}u = \tfrac{3}{10}u + \tfrac{4}{15}v + \tfrac{1}{5}w - 1$$

To choose the next pivot column we must consider

SUBPROBLEM 9.15C. Maximize $\pi_3 = \frac{3}{10}u + \frac{4}{15}v + \frac{1}{5}w - 1$

where u, v, and w are nonnegative integers

subject to $8u + 7v + 5w \leq 26$

Observe that $\pi_3 = 0$ for the current basic variables x_5, x_4, and x_3, corresponding to the patterns $\mathbf{x}_5 = (2, 0, 2)'$ $\mathbf{x}_4 = (0, 3, 1)'$, $\mathbf{x}_3 = (0, 0, 5)'$.

For an ad hoc proof that the maximum of π_3 above is 0, note that

$$30\pi_3 = (9u + 8v + 6w) - 30$$

However, each setup must satisfy the constraint

$$8u + 7v + 5w \leq 26$$

Hence $30\pi_3 = (8u + 7v + 5w) + (u + v + w) - 30$ will surely be negative if $u + v + w < 4$. Since π_3 must be nonnegative for pivoting in the x_i column to be profitable, we need only examine solutions to Subproblem 9.15C for which $u + v + w \geq 4$:

	Solution	*Value of* $30\pi_3 = 9u + 8v + 6w - 30$
$w = 0$	No relevant solution	. . .
$w = 1$	$u = 0,\quad v = 3,\quad w = 1$	0
$w = 2$	$u = 2,\quad v = 0,\quad w = 2$ $u = 1,\quad v = 1,\quad w = 2$ $u = 0,\quad v = 2,\quad w = 2$	0 Negative Negative
$w = 3$	$u = 1,\quad v = 0,\quad w = 3$ $u = 0,\quad v = 1,\quad w = 3$	Negative Negative
$w = 4$	$u = 0,\quad v = 0,\quad w = 4$	Negative
$w = 5$	$u = 0,\quad v = 0,\quad w = 5$	0

It still remains to be seen whether or not we need to recover the columns of Tableau 3 corresponding to the surplus variables. To recover the body of the current s_1 column, for example, substitute $u = -1$, $v = 0$, $w = 0$ into the body of the x_i column. To recover the objective coefficient for s_1, make this substitution into the general expression for $\pi_3 + 1$ (not into π_3). Thus we see that the current objective coefficient of s_1 is $-\frac{3}{10}$. By similar reasoning, it can be shown that the objective coefficients for s_2 and s_3 are also negative, which proves that Tableau 3 is terminal. (If some objective coefficient were positive we would have to recover its column completely and continue pivoting.)

As an exercise, the reader should solve Problem 9.15 directly, by starting with a tableau where all possible efficient setups are listed explicitly. We stress again that the "implicit tableau" method is meant to illustrate a column-generating procedure to be used whenever a cutting-stock problem is so complex that it is impossible to enumerate all relevant patterns.

To consolidate ideas, consider one more example.

PROBLEM 9.16. The standard roll is 10 feet wide. Customer demands require that we cut at least 30 rolls of width 2 feet, at least 20 rolls of width 3 feet, and at least 45 rolls of width 4 feet.

Let $\mathbf{x}_i = \begin{pmatrix} u \\ v \\ w \end{pmatrix}$ denote a typical setup, where u, v, and w represent the number of rolls of widths 2, 3, and 4 feet, respectively, in pattern $\mathbf{x}_i$. Clearly, the components of this typical setup must satisfy the inequality

$$2u + 3v + 4w \leq 10 \tag{1}$$

Inequality (1) gives the following "pure" patterns:

$$\mathbf{x}_1 = \begin{pmatrix} 5 \\ 0 \\ 0 \end{pmatrix}, \quad \mathbf{x}_2 = \begin{pmatrix} 0 \\ 3 \\ 0 \end{pmatrix}, \quad \mathbf{x}_3 = \begin{pmatrix} 0 \\ 0 \\ 2 \end{pmatrix}$$

Hence we can think of Problem 9.16 as depicted below:

Tableau 0

	x_1	x_2	x_3	$\cdots$	x_i	$\cdots$	s_1	s_2	s_3	C	
x_1	⑤	0	0	$\cdots$	u	$\cdots$	-1	0	0	0	30
x_2	0	③	0	$\cdots$	v	$\cdots$	0	-1	0	0	20
x_3	0	0	②	$\cdots$	w	$\cdots$	0	0	-1	0	45
C	-1	-1	-1	$\cdots$	-1	$\cdots$	0	0	0	1	0

Tableau 0 can be adjusted by pivoting on the circled entries in row i, column i, for $i = 1, 2, 3$. Thus we obtain

Tableau 1

	x_1	x_2	x_3	$\cdots$	x_i	$\cdots$	
x_1	1	0	0	$\cdots$	$\frac{1}{5}u$	$\cdots$	6
x_2	0	1	0	$\cdots$	$\frac{1}{3}v$	$\cdots$	$\frac{20}{3}$
x_3	0	0	1	$\cdots$	$\frac{1}{2}w$	$\cdots$	$\frac{45}{2}$
	0	0	0	$\cdots$	π_1	$\cdots$	

Observe that $\pi_1 = \frac{1}{5}u + \frac{1}{3}v + \frac{1}{2}w - 1$.

The first discrete loading problem to study is therefore

SUBPROBLEM 9.16A. Maximize $\pi_1 = \frac{1}{5}u + \frac{1}{3}v + \frac{1}{2}w - 1$

where u, v, and w are nonnegative integers

such that $2u + 3v + 4w \leq 10$

Rather than seek to maximize π_1, we follow the ad hoc procedure of finding any solution to Subproblem 9.16A for which $\pi_1 > 0$. One such solution is $u = 1$, $v = 0$, $w = 2$. Thus, in Tableau 0 we can list the worthwhile pattern $\mathbf{x}_4 = \begin{pmatrix} 1 \\ 0 \\ 2 \end{pmatrix}$ and in Tableau 1 we substitute the numerical values of u, v, and w into $\mathbf{x}_i$, thereby specifying the new pivot column, that is, the x_4 column shown below:

Tableau 1′

	x_1	x_2	x_3	$\cdots$	x_i	x_4	$\cdots$		
x_1	1	0	0	$\cdots$	$\frac{1}{5}u$	$\frac{1}{5}$		6	(30)
x_2	0	1	0	$\cdots$	$\frac{1}{3}v$	0		$\frac{20}{3}$	
x_3	0	0	1	$\cdots$	$\frac{1}{2}w$	①		$\frac{45}{2}$	$\leftarrow(\frac{45}{2})$
	0	0	0	$\cdots$	π_1	$\frac{1}{5}$ ↑			

After pivoting we obtain

Tableau 2

	$\cdots$	x_i	x_4	
x_1	$\cdots$	$\frac{1}{5}u - \frac{1}{10}w$	0	$\frac{3}{2}$
x_2	$\cdots$	$\frac{1}{3}v$	0	$\frac{20}{3}$
x_4	$\cdots$	$\frac{1}{2}w$	1	$\frac{45}{2}$
		π_2	0	

We repeat that the x_4 column of Tableau 1′ was defined by substituting $u = 1$, $v = 0$, $w = 2$ into the general x_i column of Tableau 1. The x_i column of Tableau 2 is clearly different from that of Tableau 1; observe that x_4 has replaced x_3 as a basic variable and that

$$\pi_2 = \pi_1 - \tfrac{1}{10}w = \tfrac{1}{5}u + \tfrac{1}{3}v + \tfrac{2}{5}w - 1$$

Thus the next loading problem to consider is

SUBPROBLEM 9.16B. Maximize $\pi_2 = \frac{1}{5}u + \frac{1}{3}v + \frac{2}{5}w - 1$

where u, v, and w are nonnegative integers

such that $2u + 3v + 4w \leq 10$

Again we content ourselves with noting that the solution $u = 0$, $v = 2$, $w = 1$ makes $\pi_2 = \frac{1}{15} > 0$. Thus by taking $u = 0$, $v = 2$, $w = 1$ we define a fifth pattern, $\mathbf{x}_5 = \begin{pmatrix} 0 \\ 2 \\ 1 \end{pmatrix}$. To form the current x_5 column of Tableau 2 we substitute $u = 0$, $v = 2$, $w = 1$ into the general column of this tableau and obtain

Tableau 2′

	x_1	x_2	x_4	$\cdots$	x_i	x_5	$\cdots$		
x_1	1	0	0	$\cdots$	$\frac{1}{5}u - \frac{1}{10}w$	$-\frac{1}{10}$		$\frac{3}{2}$	
x_2	0	1	0	$\cdots$	$\frac{1}{3}v$	$\textcircled{\frac{2}{3}}$		$\frac{20}{3}$	←(10)
x_4	0	0	1		$\frac{1}{2}w$	$\frac{1}{2}$		$\frac{45}{2}$	(45)
					π_2	$\frac{1}{15}$ ↑			

After pivoting as indicated above, we reach

Tableau 3

	x_1	x_4	$\cdots$	x_i	x_5	$\cdots$	
x_1	1	0		$\frac{1}{5}u + \frac{1}{20}v - \frac{1}{10}w$	0		$\frac{5}{2}$
x_5	0	0		$\frac{1}{2}v$	1		10
x_4	0	1		$\frac{1}{2}w - \frac{1}{4}v$	0		$\frac{35}{2}$
	0	0		π_3	0		

Observe that $\pi_3 = \pi_2 - \frac{1}{30}v$. We must now study

SUBPROBLEM 9.16C. Maximize $\pi_3 = \pi_2 - \frac{1}{30}v = \frac{1}{5}u + \frac{3}{10}v + \frac{2}{5}w - 1$

where u, v, and w are nonnegative integers

such that $2u + 3v + 4w \leq 10$

Observe that $u = 2$, $v = 2$, $w = 0$ is a solution to this loading problem for which $\pi_3 = 0$. If we use these values to specify a new pattern and pivot column, the objective function will be neither increased nor decreased but we will obtain a solution in which all basic variables are integral. Thus let us define a new pattern

$$\mathbf{x}_6 = \begin{pmatrix} 2 \\ 2 \\ 0 \end{pmatrix}$$

and examine the following tableau:

Tableau 3′

	x_1	x_5	x_4	x_i	x_6		
x_1	1	0	0	$\frac{1}{5}u + \frac{1}{20}v - \frac{1}{10}w$	($\frac{1}{2}$)	$\frac{5}{2}$	←(5)
x_5	0	1	0	$\frac{1}{2}v$	1	10	(10)
x_4	0	0	1	$\frac{1}{2}w - \frac{1}{4}v$	$-\frac{1}{2}$	$\frac{35}{2}$	
C	0	0	0	π_3	0 ↑	30	

By pivoting as indicated above, we obtain

Tableau 4

	x_5	x_4	x_i	x_6	$\cdots$ C	
x_6	0	0	$\frac{2}{5}u + \frac{1}{10}v - \frac{1}{5}w$	1	0	5
x_5	1	0	$-\frac{2}{5}u + \frac{2}{5}v + \frac{1}{5}w$	0	0	5
x_4	0	1	$\frac{1}{5}u - \frac{1}{5}v + \frac{2}{5}w$	0	0	20
C	0	0	π_4	0	1	30

Of course, $\pi_4 = \pi_3 = \frac{1}{5}u + \frac{3}{10}v + \frac{2}{5}w - 1$, and we obtain the same loading problem as before. It can be shown that the *maximum* value of π_4 is 0, and that the objective coefficients for s_1, s_2, and s_3 are negative. Hence, Tableau 4 furnishes an optimal solution to Problem 9.16. In other words, the minimum number of setups is 30, corresponding to 5 setups according to pattern $\mathbf{x}_6$, 5 setups according to $\mathbf{x}_5$, and 20 setups according to $\mathbf{x}_4$.

For an ad hoc proof that the maximum of π_3 is 0, observe that

$$10\pi_3 = (2u + 3v + 4w) - 10$$

But the constraint on any pattern happens to be

$$(2u + 3v + 4w) \leq 10$$

Hence $10\pi_3 \leq 10 - 10 = 0$.

As an exercise, the reader should recover completely the s_1, s_2, and s_3 columns of Tableau 4.

Exercises

1. A paper mill produces paper in reels of standard width equal to 215 inches. The following orders must be met:

Width, in.	*Minimum number of reels required*
64	180
60	90
35	96

The objective is to fill these orders so as to minimize the number of standard rolls which must be cut.

(a) Show that these are 10 efficient patterns, namely,

	Pattern									
Width	1	2	3	4	5	6	7	8	9	10
64	3	2	2	1	1	1	0	0	0	0
60	0	1	0	2	1	0	3	2	1	0
35	0	0	2	0	2	4	1	2	4	6
Trim loss	23	27	17	31	21	11	0	25	15	5

(b) Let x_i denote the number of standard reels to be processed according to pattern i. Show that the cutting-stock problem takes the following form:

Minimize $C = x_1 + x_2 + x_3 + \cdots + x_{10}$
where all $x_i \geq 0$ (and integers?)

$$\begin{aligned} \text{and}\quad 3x_1 + 2x_2 + 2x_3 + x_4 + x_5 + x_6 &\geq 180 \\ x_2 + 2x_4 + x_5 + 3x_7 + 2x_8 + x_9 &\geq 90 \\ 2x_3 + 2x_5 + 4x_6 + x_7 + 2x_8 + 4x_9 + 6x_{10} &\geq 90 \end{aligned}$$

(c) Explain the constraint "all $x_i \geq 0$ (and integers)."
(d) Solve explicitly.
(e) Solve by column generation.

2. Consider the discrete loading problem:

$$\begin{aligned} &\text{Maximize} \quad P = 10x_1 + 6x_2 + x_3 \\ &\text{where all} \quad x_i \geq 0 \quad \text{and integers} \\ &\text{and} \quad 3x_1 + 2x_2 + x_3 \leq 16 \end{aligned}$$

(a) Show that the solution to the analogous continuous loading problem is $x_1 = \frac{16}{3}$, $x_2 = 0$, $x_3 = 0$, for which $P = \frac{160}{3}$.

(b) Show that $x_1 = 5$, $x_2 = 0$, $x_3 = 1$ is a solution to the given discrete loading problem, for which $P = 41$; and that $x_1 = 4$, $x_2 = 2$, $x_3 = 0$ is a solution for which $P = 52$.

(c) Show that by rounding off the optimal solution to the continuous loading problem in (a) we do not obtain an optimal solution to the discrete problem.

(d) By solving the following linear program, show that any integer solution to the discrete problem which satisfies $x_1 + x_2 + x_3 \leq 5$ cannot be optimal:

$$\begin{aligned} &\text{Maximize} \quad P = 10x_1 + 6x_2 + x_3 \\ &\text{where} \quad x_1 \geq 0, \quad x_2 \geq 0, \quad x_3 \geq 0 \\ &\text{and} \quad 3x_1 + 2x_2 + x_3 \leq 16 \\ &\qquad\quad x_1 + x_2 + x_3 \leq 5 \end{aligned}$$

(e) Find the optimal solution to the given discrete loading problem.

3. Show, by linear programming, that in studying Subproblem 9.15C we can limit our search to solutions the sum of whose coordinates is 4 or more.

4. Solve Problem 9.12 implicitly, that is, by generating columns as needed.

5. A standard roll of paper is of width 200 inches, from which it is necessary to cut 30 rolls of width 60 inches, 60 rolls of width 55 inches, and 60 rolls of width 39 inches.

(a) List all efficient setups.

(b) Find a way of cutting that minimizes the total number of standard rolls needed, first solving implicitly, and then explicitly.

6. Turn back to Problem 9.12 and, instead of minimizing $C = x_1 + x_2 + \cdots + x_9$, minimize

(a) $W_1 = x_2 + x_6 + x_7$.

(b) $W_2 = x_2 + x_6 + x_7 + 2s_1 + 3s_2 + 4s_3$.

(c) When are W_1 and W_2 the appropriate objective functions?

(d) Show that, in general, the cutting-stock criterion is equivalent to minimizing an objective function, such as W_2 above, where total waste not only includes the waste associated with the efficient patterns but also includes all surplus exceeding customer demands.

9.5 THE FRACTIONAL ASSIGNMENT PROBLEM; THE TRAVELING-SALESMAN PROBLEM

Two male scientists, M_1 and M_2, and two female secretaries, W_1 and W_2, find themselves stranded. The secretaries have established the table of incompatibility ratings shown below, where the entry in row i, column j designates the degree of *incompatibility* between M_i and W_j.

	W_1	W_2
M_1	3	6
M_2	2	4

The scientists have been asked to determine the optimal percentage of time each man should spend with each woman, so as to minimize the total sum of incompatibilities.

Let f_{ij} denote the fraction of time M_i spends with W_j. The scientists must solve the following linear program:

PROBLEM 9.17. Minimize $I = (3f_{11} + 6f_{12}) + (2f_{21} + 4f_{22})$

where $f_{11} \geq 0,\ f_{12} \geq 0,\ f_{21} \geq 0,\ f_{22} \geq 0$

and

$$f_{11} + f_{12} = 1 \quad (1)$$

$$f_{21} + f_{22} = 1 \quad (2)$$

$$f_{11} + f_{21} = 1 \quad (3)$$

$$f_{12} + f_{22} = 1 \quad (4)$$

Equations (1) and (2) describe how the time of M_1 and M_2, respectively, is allocated; similarly, Eqs. (3) and (4) describe how the time of W_1 and W_2, respectively, is allocated. Observe that the sum of Eqs. (1) and (2) equals the sum of Eqs. (3) and (4). Furthermore, by performing subtraction of pairs of the above equations we can express all the variables in terms of any one single variable, say f_{11} (for example, $f_{12} = f_{21} = 1 - f_{11}$), as exhibited in the following table of allocations:

	W_1	W_2
M_1	f_{11}	$1 - f_{11}$
M_2	$1 - f_{11}$	f_{11}

Expressed in terms of f_{11}, Problem 9.17 reads

Minimize $I = [3f_{11} + 6(1 - f_{11})] + [2(1 - f_{11}) + 4f_{11}] = 8 - f_{11}$

where $0 \leq f_{11} \leq 1$

Clearly, the objective function will be minimum, that is, 7, when f_{11} assumes its maximum value of 1. In words, M_1 and W_1 should spend all available time together, as should M_2 and W_2. As a check, note (from the table of incompatibilities) that the total incompatibility associated with this set of assignments is $3 + 4 = 7$, which agrees with the optimal value of I above.

Before daring to make any generalizations as to incompatibilities and monogamy, let us consider the following somewhat more complex assignment problem. Suppose that M_3 and W_3 have joined the previous group and that the table of incompatibilities now reads

	W_1	W_2	W_3
M_1	3	6	7
M_2	2	4	3
M_3	1	5	7

The entry in row i, column j of this table represents the degree of incompatibility between M_i and W_j for i and j ranging from 1 to 3.

Again, the scientists have been challenged to determine the optimal percentage of time f_{ij} that each M_i and W_j should spend with each other, so as to minimize the total sum of incompatibilities. They must solve the following linear program:

PROBLEM 9.18.

Minimize $I = (3f_{11} + 6f_{12} + 7f_{13}) + (2f_{21} + 4f_{22} + 3f_{23}) + (f_{31} + 5f_{32} + 7f_{33})$
where $f_{ij} \geq 0$ for $i = 1, 2, 3,\ \ j = 1, 2, 3$
and

$$
\begin{aligned}
&f_{11} + f_{12} + f_{13} &= 1 \quad &(1)\\
&f_{21} + f_{22} + f_{23} &= 1 \quad &(2)\\
&f_{31} + f_{32} + f_{33} &= 1 \quad &(3)\\
&f_{11} + f_{21} + f_{31} &= 1 \quad &(4)\\
&f_{12} + f_{22} + f_{32} &= 1 \quad &(5)\\
&f_{13} + f_{23} + f_{33} &= 1 \quad &(6)
\end{aligned}
$$

The reader may have noticed that, in general, it is much easier to prove that the optimal solution to a linear program is actually optimal than to discover just which solution is optimal. In Chapter 15 it will be shown how to discover that the optimal solution in this case is $f_{12} = 1$, $f_{23} = 1$, $f_{31} = 1$, all other $f_{ij} = 0$. Below, we prove that this solution is indeed optimal by expressing the given problem in terms of the nonbasic variables f_{21}, f_{22}, f_{32}, and f_{33}, as indicated in the following table of fractional assignments:

	W_1	W_2	W_3
M_1	$-f_{21} + f_{32} + f_{33}$	$1 - f_{22} - f_{32}$	$f_{21} + f_{22} - f_{33}$
M_2	f_{21}	f_{22}	$1 - f_{21} - f_{22}$
M_3	$1 - f_{32} - f_{33}$	f_{32}	f_{33}

In terms of these nonbasic variables, Problem 9.18 reads

$$
\begin{aligned}
\text{Minimize}\quad I &= 3(f_{32} + f_{33} - f_{21}) + 6(1 - f_{22} - f_{32}) + 7(f_{21} + f_{22} - f_{33}) + 2f_{21}\\
&\quad + 4f_{22} + 3(1 - f_{21} - f_{22}) + (1 - f_{32} - f_{33}) + 5f_{32} + 7f_{33}\\
&= 10 + [3f_{21} + 2f_{22} + f_{32} + 2f_{33}]
\end{aligned}
$$

where $f_{21} \geq 0,\ \ f_{22} \geq 0,\ \ f_{32} \geq 0,\ \ f_{33} \geq 0$
and

$$
\begin{aligned}
f_{21} - f_{32} - f_{33} &\leq 0 \quad &(1)\\
f_{22} + f_{32} &\leq 1 \quad &(2)\\
-f_{21} - f_{22} + f_{33} &\leq 0 \quad &(3)\\
f_{21} + f_{22} &\leq 1 \quad &(4)\\
f_{32} + f_{33} &\leq 1 \quad &(5)
\end{aligned}
$$

Constraints (1) to (5) correspond, respectively, to affirming that $f_{11} \geq 0$, $f_{12} \geq 0, f_{13} \geq 0, f_{23} \geq 0, f_{31} \geq 0$. The form of I shows that if $f_{21} = f_{22} = f_{32} = f_{33} = 0$ is a feasible solution, it must be optimal. It is obviously feasible.

The solutions to Problems 9.17 and 9.18 are of the "all-or-nothing" type, that is, $f_{ij} = 0$ or $f_{ij} = 1$. This result is perfectly general—an $n \times n$ fractional assignment problem always possesses an optimal solution where each f_{ij} is 0 or 1. We can therefore reformulate the fractional assignment problem as a *combinatorial assignment problem*, as follows: The assignment problem is a type of allocation problem in which n items (men, persons, etc.) are distributed among n boxes (women, jobs, etc.), *one item to a box*, so that the cost of distribution is optimized. There are $n! = n(n - 1)(n - 2) \cdots (2)(1)$ different assignments. Thus if $n = 4$, then $n! = 4(3)(2)(1) = 24$ and we can minimize the objective function by examining each possible assignment. But, if $n = 6$, there are 720 possible assignments, and evaluating each combination is a formidable task. And if $n = 10$, there are more than 3.5 million combinations. (In Chapter 15 the assignment problem is studied further.)

A special type of assignment problem is the so-called *traveling-salesman problem:* A salesman must visit n cities during a trip. Given the distances between cities, the objective is to find the shortest path that visits each city once and returns to the point of departure only after each city has been visited.

To illustrate, consider the following network, with cities and distances as shown:

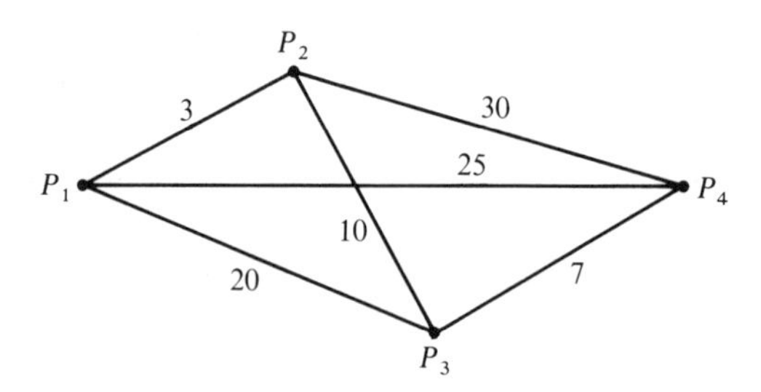

Note that only three tours are possible, beginning and ending at P_1, namely,

(1) $P_1\,P_2\,P_3\,P_4\,P_1$
(2) $P_1\,P_2\,P_4\,P_3\,P_1$
(3) $P_1\,P_3\,P_2\,P_4\,P_1$

A tour like $P_1\,P_3\,P_1$, followed by $P_1\,P_2\,P_4\,P_1$ is not allowed—the salesman cannot return home until he has visited all other cities. Furthermore, since a completed cycle is involved in each admissible tour, it makes no difference which city is called the starting point. Finally, in the present problem tours like $P_1\,P_2\,P_4\,P_3\,P_1$ and $P_1\,P_3\,P_4\,P_2\,P_1$ are considered equivalent. (Why?)

In viewing the problem as an assignment problem, the table of distances plays the role of the ratings between pairs of cities. Thus we have, for the above network,

Table 1

	P_1	P_2	P_3	P_4
P_1	0	3	20	25
P_2	3	0	10	30
P_3	20	10	0	7
P_4	25	30	7	0

This table is symmetric, but this need not hold in general. If any cities cannot be connected, we consider their mutual distance to be M, an extremely large positive constant.

Thinking of Table 1 as defining an assignment problem might lead to the idea that the assignment $x_{ii} = 1$ for all i will solve the problem, but, unfortunately, such an assignment does not define a tour. We must force x_{ii} to be 0, by taking the distance from P_i to P_i to be M, as shown in Table 1′ below.

Table 1′

	P_1	P_2	P_3	P_4
P_1	M	3	20	25
P_2	3	M	10	30
P_3	20	10	M	7
P_4	25	30	7	M

But even working with Table 1′ is not sufficient. *We must still rule out assignments which do not define a complete tour, that is, which contain subcycles involving fewer than four cities*, for example, $x_{12} = x_{21} = 1$ and $x_{34} = x_{43} = 1$, all other $x_{ij} = 0$.

The only way out of these difficulties is to realize that the traveling-salesman problem is not an assignment problem but an integer program. For the situation at hand, the appropriate program can be defined as follows: Let $x_{ij} = 1$ if the tour goes directly from P_i to P_j; otherwise $x_{ij} = 0$. Let d_{ij} be the distance from P_i to P_j, as shown in Table 1′. Then the objective is to

$$\text{Minimize} \quad Z = d_{11}x_{11} + \cdots + d_{44}x_{44}$$

$$\text{where} \quad x_{ij} = 1 \text{ or } 0$$

$$\text{and} \quad x_{ij} + x_{i2} + x_{i3} + x_{i4} = 1, \qquad \text{for } i = 1, 2, 3, 4 \qquad (1)$$

$$x_{ij} + x_{2j} + x_{3j} + x_{4j} = 1, \qquad \text{for } j = 1, 2, 3, 4 \qquad (2)$$

$$x_{ij} + x_{ji} \leq 1, \qquad \text{for all } i, j \qquad (3)$$

$$x_{it} + x_{tj} + x_{ji} \leq 2, \qquad \text{for all } i, t, j \qquad (4)$$

The constraints represented by (1) guarantee that there is only one path chosen emanating from each P_i; those represented by (2) guarantee that each P_j is visited only once; (3) rules out all assignments containing subcycles between two cities; and (4) rules out all assignments containing subcycles between three cities. [Constraint (4) is really redundant. If there are n cities it is only necessary to rule out subcycles of order $\frac{1}{2}n$ or lower, because higher-order subloops cannot exist if the lower-order ones have been prevented.] The number of constraints needed to define a traveling-salesman problem becomes quite large, even for networks of moderate size. Also, if we solve by the simplex algorithm, the x_{ij} may well turn out to be fractions.

As is so often the case, the important applications of the traveling-salesman model have nothing to do with travel. The real applications are to problems of sequencing of tasks. To illustrate, suppose the P_i represent products that must be made in some order on a continuous basis, and the setup cost for each product is a function of the product previously made. In other words, the "distances" d_{ij} represent the cost of changing over from P_i to P_j. The analog of a salesman's shortest tour is the determination of the sequence of production that will minimize total setup cost.

Exercises

1. The objective of the following assignment problem is to associate candidates P_i to jobs J_j, so as to minimize the total rating. The individual ratings are:

	J_1	J_2	J_3
P_1	8	4	7
P_2	5	3	4
P_3	4	1	5

(a) Verify, by means of linear programming, that the following assignment is optimal: $f_{12} = f_{23} = f_{31} = 1$, all other $f_{ij} = 0$.

(b) Verify that the above solution is optimal by examining all combinatorial assignments.

2. Consider the following rating table:

	T_1	T_2	T_3	T_4	T_5
S_1	15	19	20	12	13
S_2	20	18	10	13	17
S_3	18	21	15	21	20
S_4	20	21	13	21	17
S_5	18	21	16	17	21

The objective is to find that assignment of S_i's to T_j's that *maximizes* the total sum of ratings. Show that the following solution is optimal: $x_{13} = 1$, $x_{21} = 1$, $x_{32} = 1$, $x_{44} = 1$, $x_{55} = 1$, all other $x_{ij} = 0$.

3. A publisher's representative has five college towns to visit every month. He is based in one of the towns. How should he sequence his tour so as to minimize total distance traveled, when the distances are as shown below:

	1	*2*	*3*	*4*	*5*
1	0	150	190	120	250
2	150	0	180	170	300
3	190	180	0	250	400
4	120	170	250	0	300
5	250	300	400	300	0

10

EXCHANGE ALGEBRA

The material in this chapter will help the student understand different versions of the simplex algorithm, especially the $(C^j - Z^j)$ version. This form of the algorithm often appears in the literature on linear programming, and, on the surface, its steps are quite different from those we have previously presented. The chapter is devoted to five main topics. First, we develop the fundamental ideas of exchange algebra and apply them to describe the passage from one simplex tableau to another. Second the connection between exchange algebra and the so-called $(C^j - Z^j)$ version of the simplex algorithm is established. Third, we show how to handle certain problems of sensitivity analysis using the techniques developed in this chapter. Fourth, in Section 10.4, we reinterpret the material of Section 10.1 in a more concise fashion by stressing the exchange of column vectors. And fifth, in Section 10.5, some of the fundamental analytic results of linear programming are discussed, and the reader will see the ultimate justification for our insistence on working with dictionaries, or their tableau equivalent.

10.1 EXCHANGE ALGEBRA

We return to our favorite example.

PROBLEM 10.1. Maximize $P = 5x + 6y$

where $x \geq 0, \quad y \geq 0$

and $3x + y \leq 48$

$3x + 4y \leq 120$

As we saw in Chapter 1, this program corresponds to the product mix problem described in the following table:

	Products		
	X	*Y*	*Maximum time available*
Machine I	3	1	48 hr
Machine II	3	4	120 hr
Per unit contribution to net profit	\$5	\$6	

The table summarizes the following information:

(1) To create 1 unit of X, we must sacrifice 3 hours of the time available on machine I as well as 3 hours on machine II, but in return we *gain* 5 units of profit P.

(2) To create 1 unit of Y, we sacrifice 1 hour on machine I and 4 hours on machine II, thereby *gaining* 6 units of P.

These rules are *statements of exchange*, or *terms of trade*. Linear-programming problems usually take form in terms of trade. These terms are then translated into appropriate algebraic relations. Only by defining x as the number of units of product X can we write down the *mathematical* inequalities of Problem 10.1. The nontrivial inequalities correspond to information obtainable by reading across the rows of the above table. For example, "$3x + y$" describes the amount of time used on machine I when we make x units of product X and y units of product Y.

The phrase "create 1 unit of X" is the equivalent of "increase x by 1 unit," but we stress that production formulas are not algebraic equations or inequalities. Statements of exchange are like production formulas, or recipes. In this chapter we show how to read down the columns of any simplex tableau and associate a unique set of statements of exchange with each tableau.

The sequence of tableaux needed to solve Problem 10.1 is given below. For the present we use the P-downstairs format. Later we extend "exchange algebra" to the $-P$-upstairs format. In this chapter it is absolutely necessary to *use the left margin of each tableau to list the current basic variables*, one basic variable for each row in the body of a tableau.

Tableau 1

	x	y	s_1	s_2		
s_1	3	1	1	0	48	(48)
s_2	3	④	0	1	120	←(30)
	5	6	0	0	$P - 0$	
		↑				

The left margin of Tableau 1 shows that the basic variable in row 1 is s_1; and the right margin gives the current value of s_1. Similarly for s_2.

Tableau 2

	x	y	s_1	s_2		
s_1	$\textcircled{\frac{9}{4}}$	0	1	$-\frac{1}{4}$	18	←(8)
y	$\frac{3}{4}$	1	0	$\frac{1}{4}$	30	(40)
	$\frac{1}{2}$ ↑	0	0	$-\frac{3}{2}$	$P - 180$	

Tableau 3

	x	y	s_1	s_2	
x	1	0	$\frac{4}{9}$	$-\frac{1}{9}$	8
y	0	1	$-\frac{1}{3}$	$\frac{1}{3}$	24
	0	0	$-\frac{2}{9}$	$-\frac{13}{9}$	$P - 184$

Tableau 3 is terminal.

To discuss the ideas of exchange algebra, we pause and reexamine Tableau 2. The rows of this tableau read, in equation form,

$$\tfrac{9}{4}x \qquad + s_1 - \tfrac{1}{4}s_2 = 18 \tag{1}$$

$$\tfrac{3}{4}x + y \qquad + \tfrac{1}{4}s_2 = 30 \tag{2}$$

$$\tfrac{1}{2}x \qquad - \tfrac{3}{2}s_2 = P - 180 \tag{3}$$

The nonbasic variables are x and s_2. But Eq. (3) instructs us to increase x as much as possible. This can only be done by decreasing the current basic variables, that is, y and s_1. If s_2 is maintained at 0, the system of equations of Tableau 2 will reduce to

$$\tfrac{9}{4}x \quad + s_1 \quad = \quad 18 \tag{1'}$$

$$\tfrac{3}{4}x + y \quad = \quad 30 \tag{2'}$$

$$\tfrac{1}{2}x \quad - P = -180 \tag{3'}$$

Equations (1′), (2′), and (3′) each contain x and only one of the current basic variables. If we increase x by 1 unit, we must (a) decrease s_1 by $\frac{9}{4}$ units and (b) decrease y by $\frac{3}{4}$ units, and (c) P will increase by $\frac{1}{2}$.

The above three *statements of exchange* were deduced by considering the three rows of Tableau 2. However, they can be read off immediately, simply by focusing on the x column of Tableau 2 and formulating the correct terms of trade between x and the set of current basic variables in the left margin of this tableau, as the reader can see in the portion of Tableau 2 reproduced below:

	x	y	s_1	s_2
s_1	$\frac{9}{4}$	·	·	$-\frac{1}{4}$
		·	·	
y	$\frac{3}{4}$	·	·	$\frac{1}{4}$
	$\frac{1}{2}$	–	–	$-\frac{3}{2}$

Similarly, by focusing on the s_2 column above, we see that if we increase s_2 by 1 unit we must decrease s_1 by $-\frac{1}{4}$ unit and decrease y by $\frac{1}{4}$ unit, and that the net effect of increasing s_2 by 1 unit is to increase P by $-\frac{3}{2}$ units. Decreasing s_1 by $-\frac{1}{4}$ unit is equivalent to increasing s_1 by $\frac{1}{4}$ unit. These terms of trade for s_2 are the row-by-row consequences of setting $x = 0$ in the row equations of Tableau 2 and interpreting the resulting system:

$$\begin{aligned} s_1 + (-\tfrac{1}{4})s_2 &= 18 \\ y \quad + \quad (\tfrac{1}{4})s_2 &= 30 \\ + (-\tfrac{3}{2})s_2 &= P - 180 \end{aligned}$$

To illustrate, consider the first of these equations. When $s_2 = 0$, it follows that $s_1 = 18$. And if $s_2 = 1$, then s_1 will decrease by $-\frac{1}{4}$; that is, s_1 will become $18\frac{1}{4}$. Similarly, if $s_2 = 1$, then y decreases by $\frac{1}{4}$ and P decreases by $\frac{3}{2}$.

To facilitate the ensuing discussion we now introduce two symbols, $\oplus$ and $\ominus$, which we call the symbols of "exchange addition" and "exchange subtraction," respectively. The following examples serve to define these symbols and illustrate their use:

(1) By definition, the exchange equation

$$x = as_1 \oplus bs_2$$

is the symbolic equivalent of the English phrase: "To change x by 1 unit we must change s_1 by a units and s_2 by b units, *where the changes in s_1 and s_2 are in the opposite direction to the change in x.*" Thus, if we *increase* x by 1 unit we must *decrease* s_1 by a units and *decrease* s_2 by b units. But if we *decrease* x by 1 unit, we must *increase* s_1 by a units and *increase* s_2 by b units.

Intuitively, the reader can think of x as representing units of product X and s_1 and s_2 as representing available time on machines I and II. Then

$$x = as_1 \oplus bs_2$$

corresponds to the current production formula: "To make 1 unit of X we must use up a units of the time currently available on machine I and b units of the time available on machine II."

(2) If a is negative, "to increase s_1 by a units" must mean "to decrease

s_1 by $-a$ units." We are therefore forced to relate the symbols of exchange addition and exchange subtraction as follows:

$$\oplus(-k)y = \ominus ky$$

for any variable y. Thus, the equations

$$s_1 = cx \oplus -dy$$

and

$$s_1 = cx \ominus \quad dy$$

both represent the same phrase, namely, "To *increase* s_1 by 1 we must *decrease* x by c and *increase* y by d."

The reader can check that the s_2 column of Tableau 2 above informs us that

$$s_2 = \ominus\tfrac{1}{4}s_1 \oplus \tfrac{1}{4}y$$

(3) $x = as_1 \oplus bs_2$ and $kx = (ka)s_1 \oplus (kb)s_2$ are equivalent for all $k \neq 0$, provided we interpret "kx" to mean "to increase x by k units."

(4) It is left to the reader to show we can transpose terms from one side of an exchange equation to another, provided we change signs. Thus, from $x = 3s_1 \oplus 3s_2$ it follows that $3s_1 = x \ominus 3s_2$. And from (3) above, we see that

$$3s_1 = \quad x \ominus 3s_2$$

and

$$s_1 = \tfrac{1}{3}x \ominus s_2$$

are equivalent.

(5) *We can group terms and manipulate parentheses just as in ordinary algebra.* To illustrate, we explain why

$$3x_1 \ominus [x_2 \ominus 2x_1] = 3x_1 \ominus x_2 \oplus 2x_1 = 5x_1 \ominus x_2$$

To this end, let $v_1 = x_2 \ominus 2x_1$ and $w_1 = 3x_1 \ominus [x_2 \ominus 2x_1] = 3x_1 \ominus v_1$. Then, "by increasing w_1 by 1, we lose 3 units of x_1 and gain 1 unit of v_1." But by gaining 1 unit of v_1 we gain its components, namely, "a unit increase in x_2 and a decrease in x_1 of 2 units." Hence, by increasing w_1 by 1 we gain 1 unit of x_2 and lose 5 units of x_1, so that

$$w_1 = 5x_1 \ominus x_2$$

The validity of these rules will be established in a concise, mathematical form in Section 10.4. However, the reader should convince himself that they are true by reasoning meticulously as we have done above, considering in turn how the row equations of a simplex tableau determine the adjustments which must be made when we increase a nonbasic variable.

With the help of exchange notation, we can generalize somewhat, as follows. Let the following tableau stand for any tableau in our problem:

	x_1	x_2	x_3	x_4	
x_3	a_{11}	0	1	a_{14}	b_1
x_2	a_{21}	1	0	a_{24}	b_2
	a_{31}	0	0	a_{34}	$P - k$

Then the *body* of this tableau is synonymous with the following *exchange dictionary:*

$$x_1 = a_{11}x_3 \oplus a_{21}x_2$$
$$x_4 = a_{14}x_3 \oplus a_{24}x_2$$

Passing in the reverse direction is just as easy. For example, suppose we are given the exchange dictionary

$$s_1 = \tfrac{4}{9}x \oplus -\tfrac{1}{3}y$$
$$s_2 = -\tfrac{1}{9}x \oplus \tfrac{1}{3}y$$

Clearly, the body of the tableau which leads to this dictionary can only appear as

	x	y	s_1	s_2	
x	1	0	$\frac{4}{9}$	$-\frac{1}{9}$	
y	0	1	$-\frac{1}{3}$	$\frac{1}{3}$	

The three exchange dictionaries associated with the three tableaux in Problem 10.1 are

Exchange Dictionary 1.

$$x = 3s_1 \oplus 3s_2$$
$$y = s_1 \oplus 4s_2$$

Exchange Dictionary 2.

$$x = \tfrac{9}{4}s_1 \oplus \tfrac{3}{4}y$$
$$s_2 = -\tfrac{1}{4}s_1 \oplus \tfrac{1}{4}y$$

Exchange Dictionary 3.

$$s_1 = \tfrac{4}{9}x \oplus -\tfrac{1}{3}y$$
$$s_2 = -\tfrac{1}{9}x \oplus \tfrac{1}{3}y$$

Each of these dictionaries is *exchange equivalent* in the following sense: By the operations of exchange arithmetic it is possible to pass from one exchange dictionary to any other. Below, we show the equivalence of these dictionaries and, at the same time, build upon this equivalence to solve Problem 10.1 by the exchange algebra version of the simplex algorithm. This version proceeds as follows:

Start with Tableau 1 and determine the incoming and outgoing variables by the usual rules. The objective function of Tableau 1 instructs us to increase y as much as possible. It is the s_2 variable which limits the increase in y to $\frac{120}{4} = 30$. *Tableau 2* must therefore appear roughly as

	x	y	s_1	s_2	
s_1	?	0	1	?	?
y	?	1	0	?	30
	?	0	0	?	$P - 180$

Since we have already deduced that $x = 0$ and $y = 30$ in Tableau 2 and since we know that $P = 5x + 6y$, it follows that P must equal 180 in Tableau 2. We can replace the question marks with numerical values by thinking in exchange terms. Thus, from the pivot column of Tableau 1, we know that

$$y = s_1 \oplus 4s_2 \tag{1}$$

This means that as we pass from Tableau 1 to Tableau 2 by increasing y from 0 to 30, the value of s_1 must drop by 1 unit for every unit increase in y, and hence $s_1 = 48 - 30 = 18$ in Tableau 2. In order to replace the other question marks in the body of Tableau 2 all we need do is form the exchange dictionary needed to describe this tableau. In other words, we must express x and s_2 in terms of trade with s_1 and y. We begin with the "pivotal exchange equation of Tableau 1," that is, Eq. (1). By transposition we see immediately that

$$s_2 = -\tfrac{1}{4}s_1 \oplus \tfrac{1}{4}y \tag{2}$$

and now we can fill in the first two rows of the s_2 column.

To express x in terms of trade with s_1 and y, first note that

$$x = 3s_1 \oplus 3s_2 \tag{3}$$

from Tableau 1. To express x in terms of s_1 and y, we need only substitute Eq. (2) into Eq. (3). We obtain the new exchange equation

$$\begin{aligned} x &= 3s_1 \oplus 3(-\tfrac{1}{4}s_1 \oplus \tfrac{1}{4}y) \\ &= \tfrac{9}{4}s_1 \oplus \tfrac{3}{4}y \end{aligned} \tag{4}$$

This completes Exchange Dictionary 2 and allows us to fill in all the entries in Tableau 2, except for the coefficients of the nonbasic variables in the objective row, as shown below:

	x	y	s_1	s_2	
s_1	$\frac{9}{4}$	0	1	$-\frac{1}{4}$	18
y	$\frac{3}{4}$	1	0	$\frac{1}{4}$	30
	?	0	0	?	$P - 180$

We can calculate the missing coefficient of x by comparing *the direct effect on P* of increasing x by 1 unit with the loss in profit occasioned by the corresponding decreases in those variables which are needed to accommodate the increase in x. For example, if we increase x by 1, the *direct* increase in P is 5, as evident from the original objective function. However, when we increase x by 1, we must decrease s_1 and y, in accordance with the current exchange equation, that is, Eq. (4). It follows from this equation that C', the *cost* of increasing x by 1, is

$$\begin{aligned} C' &= \tfrac{9}{4}\text{ (direct contribution per unit of } s_1) + \tfrac{3}{4}\text{ (direct contribution per unit of } y) \\ &= \tfrac{9}{4}(0) + \tfrac{3}{4}(6) = \tfrac{18}{4} \end{aligned}$$

This *cost* is called the *indirect effect* on P of increasing x by 1 unit. Hence the *net gain* per unit increase in x, starting from the corner defined by Tableau 2, is the *difference* between the direct effect and the indirect effect, or

$$5 - \tfrac{18}{4} = \tfrac{2}{4}$$

This is the coefficient of x in row 3, Tableau 2.

Similarly, to find the coefficient of s_2, we must specify (1) the *direct effect* on P of increasing s_2 by 1, namely, 0, and (2) the *indirect effect* on P of increasing s_2 by 1.

The cost, or indirect effect on P of increasing s_2, follows from analyzing the current s_2 exchange equation. The terms of trade are $s_2 = -\tfrac{1}{4}s_1 \oplus \tfrac{1}{4}y$, and the cost is accordingly $-\tfrac{1}{4}(0) + \tfrac{1}{4}(6) = \tfrac{3}{2}$. Hence the net effect on P of increasing s_2 by 1 is $0 - \tfrac{3}{2} = -\tfrac{3}{2}$. This allows us to complete Tableau 2, as

Tableau 2

	x	y	s_1	s_2		
s_1	$\textcircled{\tfrac{9}{4}}$	0	1	$-\tfrac{1}{4}$	18	←(8)
y	$\tfrac{3}{4}$	1	0	$\tfrac{1}{4}$	30	(40)
	$\tfrac{1}{2}$ ↑	0	0	$-\tfrac{3}{2}$	$P - 180$	

This tableau is not optimal. We are instructed to increase x as much as possible. Now it is the basic variable s_1 which limits the increase in x to 8. The new pivotal exchange equation corresponds to the x column of Tableau 2, namely,

$$x = \tfrac{9}{4}s_1 \oplus \tfrac{3}{4}y$$

The rough form of Tableau 3 is

	x	y	s_1	s_2	
x	1	0	?	?	8
y	0	1	?	?	?
	0	0	?	?	?

To complete the stub of Tableau 3, note that the pivotal exchange equation of Tableau 2 shows that as we increase x from 0 to 8, we must decrease y by $(\tfrac{3}{4})(8) = 6$ units. The new y value is therefore $30 - 6 = 24$. Moreover, increasing x from 0 to 8 leads to an increase in P of $(\tfrac{1}{2})(8) = 4$ units. Hence the updated value of P is now 184. This new value of P can also be obtained by using the original profit function; when $x = 8$ and $y = 24$, then $P = 5x + 6y = 184$.

To complete Tableau 3 we must first build the exchange dictionary in terms of x and y. By transposition in the pivotal exchange equation of Tableau 2, we immediately obtain the correct equation for s_1, namely,

$$s_1 = \oplus\tfrac{4}{9}x \ominus \tfrac{1}{3}y = \tfrac{4}{9}x \oplus (-\tfrac{1}{3})y \tag{6}$$

As for s_2, if we substitute the new exchange equation for s_1, that is, Eq. (6), into the old exchange equation for s_2, that is, Eq. (2), we obtain the new form of s_2, namely,

$$\begin{aligned} s_2 &= -\tfrac{1}{4}(\tfrac{4}{9}x \ominus \tfrac{1}{3}y) \oplus \tfrac{1}{4}y \\ &= -\tfrac{1}{9}x \oplus \tfrac{1}{3}y \end{aligned} \tag{7}$$

At this point, we can fill in all the entries of Tableau 3, except for two entries in the objective row:

	x	y	s_1	s_2	
x	1	0	$\frac{4}{9}$	$-\frac{1}{9}$	8
y	0	1	$-\frac{1}{3}$	$\frac{1}{3}$	24
	0	0	?	?	$P - 184$

To find the coefficient of s_1 in the objective function we calculate the net effect on P of increasing s_1, as follows: From $s_1 = \frac{4}{9}(x) \oplus (-\frac{1}{3})y$, it follows that the indirect effect on P of increasing s_1 by 1 is equal to the sum of the monetary consequences of decreasing x by $\frac{4}{9}$ units and of decreasing y by $-\frac{1}{3}$. The amount of indirect effect is therefore $5(\frac{4}{9}) + 6(-\frac{1}{3}) = \frac{2}{9}$. Since the direct effect on P of increasing s_1 is 0, the net effect on P of increasing s_1 by 1 unit is $0 - \frac{2}{9} = -\frac{2}{9}$. Similarly, to find the net effect on P of increasing s_2 by 1 unit, note that the equation

$$s_2 = -\tfrac{1}{9}x \oplus \tfrac{1}{3}y$$

implies that the direct effect on P of increasing s_2 by 1 unit is 0, whereas the indirect effect on cost is $(-\frac{1}{9})(5) + (\frac{1}{3})(6) = \frac{13}{9}$. Hence the net effect is $0 - \frac{13}{9} = -\frac{13}{9}$.

It is instructive to verbalize Tableau 2 in terms of products and resources, just as we verbalized Tableau 1. These tableaux are equivalent, yet each one is the expression of the production formula corresponding to a different level of resources. To illustrate, if no resources have been utilized, then Tableau 1, or Exchange Dictionary 1, tells us how to produce *until we run into a bottleneck*. At the point of bottleneck, we must change the rules of production. Tableau 2 corresponds to the bottleneck "no more time on machine II," that is, $s_2 = 0$. But we have available for exchange 18 units of time on machine I and the 30 units of product Y already produced. Exchange Dictionary 2 gives the following current technical information:

(1) To make 1 unit of product X, we now have to sacrifice $\frac{9}{4}$ units of the remaining unutilized capacity of machine I, and, furthermore, we must reduce the output of product Y by $\frac{3}{4}$ units (thereby releasing time on machine II).

(2) To increase s_2 by 1, that is, to procure 1 unit of time on machine II, we must reduce the production of Y by $\frac{1}{4}$ units, and, furthermore, in so doing, we disembody or free $\frac{1}{4}$ units of time on machine I.

As an exercise, the reader should verbalize Tableau 3.

Generally speaking, each exchange dictionary can be interpreted as the statement of how to "combine chemically" the "currently basic, left-rim commodities" to form 1 unit of each "currently nonbasic commodity." This explains why the profit function changes from tableau to tableau. The reader may have wondered why profit, per unit increase in x, equals 5 in Tableau 1 but only $\frac{1}{2}$ in Tableau 2. This is because Tableau 1 views production and profit from the corner where $x = y = 0$, that is, where maximum resource capacity of $s_1 = 48$ and $s_2 = 120$ is available. Starting from this point of no prior production it is possible to increase x without decreasing y. But, if we start to produce from corner 2, where $x = s_2 = 0$ and $s_1 = 18$, $y = 30$, we can no longer increase x without decreasing y. In other words, we must now reckon with both the *direct effect and indirect effect* on profit of increasing x. The objective coefficient of x measures the *real*, or *net*, effect of increasing x by 1, that is, the difference between the direct and indirect effects.

The net effect is often referred to as the *reduced profit coefficient*. From the way we calculate the *reduced profit contribution* we see that if the direct effect of increasing x by 1 were to decrease by more than the *reduced coefficient*, that is, by more than $\frac{1}{2}$, then the objective row of Tableau 2 would no longer instruct us to increase x, and hence this tableau would be terminal.

Another example of the use of exchange algebra should help clarify the concept.

PROBLEM 10.2.

$$
\begin{aligned}
&\text{Maximize} \quad P = 3x + 6y \\
&\text{where} \quad x \geq 0, \quad y \geq 0 \\
&\text{and} \quad 3x + y \leq 48 \\
&\qquad\quad x + 3y \leq 48
\end{aligned}
$$

(This problem has already been studied in Chapter 5.)

Tableau 1

	x	y	s_1	s_2		
s_1	3	1	1	0	48	(48)
s_2	1	③	0	1	48	←(16)
	3	6	0	0	$P - 0$	
		↑				

The objective row of this tableau instructs us to increase y as much as possible. Since $s_1 = s_2 = 48$ and

$$y = s_1 \oplus 3s_2$$

s_2 will fall to 0 before s_1 does. When $y = 16$, the value of s_2 will have fallen from 48 to $48 - (3)(16) = 0$. Meanwhile, the value of s_1 decreases by 1 for every unit increase in y. Hence $s_1 = 48 - 16 = 32$ and $P = 3x + 6y = (3)(0) + (6)(16) = 96$ in the next tableau, whose general appearance must be

	x	y	s_1	s_2	
s_1	?	0	1	?	32
y	?	1	0	?	16
	?	0	0	0	$P - 96$

To complete this tableau we must first find the exchange equations for x and s_2 in terms of s_1 and y. From the pivot column of Tableau 1 we see that

$$y = s_1 \oplus 3s_2$$

It follows by transposition that

$$s_2 = \ominus\tfrac{1}{3}s_1 \oplus \tfrac{1}{3}y \tag{1}$$

which immediately gives us the entries for the body of the s_2 column of Tableau 2.

To obtain the x column, note that from Tableau 1 we have the exchange equation

$$x = 3s_1 \oplus s_2 \tag{2}$$

If we substitue Eq. (1) into Eq. (2) we obtain x in terms of s_1 and y:

$$\begin{aligned} x &= 3s_1 \oplus s_2 = 3s_1 \oplus (-\tfrac{1}{3}s_1 \oplus \tfrac{1}{3}y) \\ &= \tfrac{8}{3}s_1 \oplus \tfrac{1}{3}y \end{aligned} \tag{3}$$

To complete Tableau 2 we calculate the coefficients of x and s_2 in the objective row, as follows: The effect on P of increasing x by 1 unit follows from Eq. (3). From that equation, an increase in x of 1 unit must be compensated for by decreasing y by $\frac{1}{3}$ unit; and hence P will increase by the amount of the profit contribution of x, that is, 3 units, but also decrease because of the effect of decreasing y by an amount equal to $(\frac{1}{3})(6) = 2$ units. (Recall that 6 is the profit contribution of y.) Hence the net effect of increasing x by 1 unit is an increase in P of 1.

Likewise, the effect of increasing s_2 by 1 unit follows from the exchange equation

$$s_2 = -\tfrac{1}{3}s_1 \oplus \tfrac{1}{3}y$$

Increasing s_2 by 1 unit leads to a decrease in y of $\frac{1}{3}$ unit and hence an attendant net drop in P of $\frac{1}{3}(6) = 2$.

The complete form of Tableau 2 is therefore

Tableau 2

	x	y	s_1	s_2		
s_1	$\textcircled{\frac{8}{3}}$	0	1	$-\frac{1}{3}$	32	←(12)
y	$\frac{1}{3}$	1	0	$\frac{1}{3}$	16	(48)
	1 ↑	0	0	−2	$P - 96$	

To form Tableau 3 we must increase x as much as possible. When x increases, both s_1 and y will decrease. For every unit increase in x, s_1 decreases by $\frac{8}{3}$. The value of s_1 falls to 0 when x reaches $32/\frac{8}{3} = 12$. Moreover, as x increases to 12, y will decrease by $12(\frac{1}{3}) = 4$. As for profit, by increasing x to 12, we gain $(3)(12) = 36$, but we must forgo the profit on 4 units of y, namely, $(4)(6) = 24$. Net profit is therefore increased by $36 - 24 = 12$. Thus P will be $96 + 12 = 108$ in Tableau 3. As a check, if $x = 12$ and $y = 12$, then $P = 3x + 6y = 108$. We now have the rough form of the revised program:

Tableau 3

	x	y	s_1	s_2	
x	1	0	?	?	12
y	0	1	?	?	12
	0	0	?	?	$P - 108$

The key exchange equation needed to fill in Tableau 3 comes from the pivot column of Tableau 2, namely,

$$x = \tfrac{8}{3}s_1 \oplus \tfrac{1}{3}y \tag{4}$$

By transposition, it follows immediately that

$$s_1 = \tfrac{3}{8}x \oplus -\tfrac{1}{8}y \tag{5}$$

which gives us the new s_1 column.

To find the new s_2 column we substitute Eq. (5) into the old s_2 column, that is, Eq. (1). Thus

$$\begin{aligned} s_2 &= -\tfrac{1}{3}s_1 \oplus \tfrac{1}{3}y \\ &= -\tfrac{1}{3}(\oplus\tfrac{3}{8}x \ominus \tfrac{1}{8}y) \oplus \tfrac{1}{3}y \\ &= -\tfrac{1}{8}x \oplus \tfrac{3}{8}y \end{aligned} \tag{6}$$

The coefficients of s_1 and s_2 in the objective row of Tableau 3 can be found by examining Exchange Dictionary 3. From Eq. (5), the cost of increasing s_1

by 1 is

$$\tfrac{3}{8}(\text{direct profit contribution of 1 unit of } x) + (-\tfrac{1}{8})(\text{direct profit contribution of 1 unit of } y) = \tfrac{3}{8}(3) + (-\tfrac{1}{8})(6) = \tfrac{3}{8}$$

The net gain per unit increase in s_1 is therefore the direct effect on profit less the indirect effect, that is, $0 - \frac{3}{8} = -\frac{3}{8}$.

Similarly, from Eq. (6), the indirect effect of increasing s_2 by 1 is $(-\frac{1}{8})(3) + (\frac{3}{8})(6) = \frac{15}{8}$. The net effect of increasing s_2 by 1 is the direct effect minus the indirect effect, that is, $0 - \frac{15}{8} = -\frac{15}{8}$.

We leave it to the reader to complete Tableau 3, and, as an exercise, to think of Problem 10.2 as a product-mix program and verbalize Tableaux 2 and 3 as production formulas.

It is instructive to point up the similarities and differences between the earlier version of the simplex algorithm and the version using exchange equations. In both forms of the algorithm, we must first identify the incoming basic variable and then specify the maximum feasible increase in this variable.

In the version of the simplex algorithm presented in Chapters 3 and 5, the key to forming the updated tableau is the row of the outgoing variable. This row gives an ordinary algebraic equation that must be solved for the incoming variable in terms of the new corner variables. We substitute this new expression for the incoming variable into the remaining equations of the current dictionary and thereby obtain the revised dictionary, that is, the rows of the revised tableau.

On the other hand, when we use exchange algebra the key exchange equation is obtained from the column of the incoming variable. Once we have the correct equation, we must solve for the departing variable in terms of the new basic variables. Then we substitute this new expression for the departing variable into the current exchange dictionary and thereby obtain the revised exchange dictionary. The new exchange equations express the new corner variables in terms of the new basic variables and represent columns of the revised tableau.

Exercises

1. Do the exercises in Chapter 3 by exchange algebra, emphasizing both the *symbolism of exchange* and its *verbal equivalent.*

2. Explain why the accounting relations displayed below can be considered fundamental to linear programming:

(a)

Bill of Materials

	Product X	*Product Y*
Component a	6 per unit	2 per unit
Component b	10 per unit	12 per unit
Component c	15 per unit	10 per unit

(b)

		Per unit of X		*Per unit of Y*	
Component	*Unit cost*	*Number of components*	*Cost*	*Number of components*	*Cost*
a	\$0.12	6	\$0.72	2	\$0.24
b	\$0.20	10	\$2.00	12	\$2.40
c	\$0.50	15	\$7.50	10	\$5.00
		Total cost	\$10.22		\$7.64
		Selling price	\$15.00		\$10.00
		Profit	\$4.78		\$2.36

10.2 THE ($C^j - Z^j$) FORM OF THE SIMPLEX ALGORITHM

The ($C^j - Z^j$) method for performing simplex calculations is perhaps the most common of all the methods currently appearing in the literature of linear programming. It involves no new ideas, for it is really the exchange algebra method previously described, except for format. We begin by expressing Problem 10.2 in ($C^j - Z^j$) format. For reasons which will soon become apparent we designate all variables uniformly, and rename x, y, s_1, and s_2 as x_1, x_2, x_3, x_4, respectively. The initial ($C^j - Z^j$) tableau appears as

			Columns					
			1	2	3	4		
		C^j	3	6	0	0	P	
	C_i	*Basic solution*	x_1	x_2	x_3	x_4		
Rows 1	0	x_3	3	1	1	0	48	(48)
Rows 2	0	x_4	1	③	0	1	48	←(16)
		Z^j	0	0	0	0		
		$C^j - Z^j$	3	6 ↑	0	0		

First note that the rows and columns of the tableau are numbered. Starting with column 1 and reading across row 1, we see recorded the first nontrivial constraint of Problem 10.2 expressed in slack-variable form, namely,

$$3x_1 + x_2 + x_3 = 48$$

Likewise, starting with column 1 and reading across row 2, we see recorded the second nontrivial constraint. Obviously, the body of this tableau does not differ too much from the old simplex tableau.

In the column entitled "Basic solution" are listed the names of the current basic variables, each in its appropriate row. Notice that C_i heads a column of coefficients. (The subscript i stands for row number.) There is one entry in this column for each row corresponding to a constraint equation. The symbol C^j labels a row containing the objective coefficients of the basic and nonbasic variables. Recall that to form the simplex tableau the coefficients of these variables were detached and placed in columns.

Usually the symbol C denotes cost, but we shall use it in a more general sense to denote the objective function, whether it be cost or profit, to be minimized or maximized. In Tableau 1 the numbers listed across the C^j row (the superscript j stands for column number) are the coefficients of the variables in the initial objective function, taken in the order in which the variables are listed, that is,

$$P = 3x_1 + 6x_2 + 0(x_3) + 0(x_4)$$

The coefficients in the C^j row represent the *direct effect* of increasing each of the variables by one unit. The numbers listed under the C_i column are always a subset of the numbers listed across the C^j row. They represent the (direct) effect on the objective function of increasing each of the variables in the current basic solution by 1 unit—but what is more important, they will be used to calculate the indirect effect on the objective function of increasing a nonbasic variable. In Tableau 1, x_3 and x_4 are basic and hence $C_1 = C_2 = 0$.

Under each column there lies an entry in the row labeled Z^j. The Z^j entry under a given column represents the current *indirect effect* on the objective function of increasing by 1 unit the variable belonging to column j. In the preceding initial tableau, all Z^j are 0. Reading across the row labeled $C^j - Z^j$ we see repeated the current objective function. The significance of the entries under C_i and across the Z^j and $Z^j - C^j$ rows will become more apparent as we pass to the next tableau.

By studying the objective function and the stub of Tableau 1 we deduce, by familiar logic, that we should increase x_2 to 16, at which point x_4 will fall to 0. In the next tableau, x_2 will replace x_4 as a variable in the basic solution. The entries in the new tableau will reflect the changes caused by removing 48 units of x_4 from the solution and introducing 16 units of x_2.

The general aspect of Tableau 2 must be as indicated below.

			Columns				
			1	2	3	4	
		C^j	3	6	0	0	P
	C_i	*Basic solution*	x_1	x_2	x_3	x_4	
Rows 1	0	x_3	?	0	1	?	32
Rows 2	6	x_2	?	1	0	?	16
		Z^j	?	?	?	?	
		$C^j - Z^j$	?	?	?	?	

Note that columns 1, 2, 3, 4 are still the x_1, x_2, x_3, x_4 columns, respectively. The entries in the C^j row are also the same as in the initial tableau. They never change from one tableau to the next. On the other hand, the C_i column does change. When a new variable becomes part of the basic solution, the C_i coefficient appropriate to the row of this new variable must be changed to record its direct per unit contribution to profit. Thus, since x_2 is basic, the C_i entry in row 2 of Tableau 2 is 6.

The usual way of filling in the entries of Tableau 2 is to work on the columns first, just as we did in Section 10.1, by explicit or implicit use of exchange algebra. Let us assume that this has been accomplished and that the only task remaining is to fill in the C^j and Z^j rows. Then Tableau 2 will appear as

Tableau 2

			1	2	3	4	
		C^j	3	6	0	0	P
	C_i	*Basic solution*	x_1	x_2	x_3	x_4	
1	0	x_3	$\frac{8}{3}$	0	1	$-\frac{1}{3}$	32
2	6	x_2	$\frac{1}{3}$	1	0	$\frac{1}{3}$	16
		Z^j	?	?	?	?	
		$C^j - Z^j$	?	?	?	?	

It remains to fill in the Z^j and $C^j - Z^j$ rows. As previously emphasized, the $C^j - Z^j$ row will be nothing but our "new" profit function, that is, the profit function expressed as a function of the corner variables x_1 and x_4.

The C^j entries represent the *direct effect* on the objective function of increasing the variable associated with the *j*th column by 1 unit. The Z^j entries represent the *indirect effect* on the objective function of increasing this variable by one unit.

The missing Z^j entries can be found by referring to Exchange Dictionary 2. To illustrate, to find the Z^j entry in column 1, we start with the exchange equation for x_1, namely,

$$x_1 = \tfrac{8}{3}x_3 \oplus \tfrac{1}{3}x_2$$

From this it follows that the indirect effect on the objective function of increasing x_1 by 1 unit is $\frac{8}{3}$ of the profit contribution per unit of x_3 and $\frac{1}{3}$ of the profit contribution per unit of x_2, that is,

$$Z^1 = (\tfrac{8}{3})(0) + (\tfrac{1}{3})(6) = 2$$

The *net effect* on the objective function is obviously the direct effect less the indirect effect, that is,

$$C^1 - Z^1 = 3 - 2 = 1$$

To calculate any Z^j we need not write down an explicit exchange equation but instead simply compare the x_j and C_i columns, as our eye sees them in Tableau 2. Thus, when $j = 1$,

C_i	x_1
0	$\frac{8}{3}$
6	$\frac{1}{3}$

Z^1 is obviously $(0)(\frac{8}{3}) + (6)(\frac{1}{3}) = 2$, the inner product of the vectors in the C_i and x^1 columns above.

To calculate Z^4, we must focus on

C_i	x_4
0	$-\frac{1}{3}$
6	$\frac{1}{3}$

Hence, $Z^4 = (0)(-\frac{1}{3}) + (6)(\frac{1}{3}) = 2$.

In general, suppose the tableau under study has a C_i column and x_n column, each with k entries:

C_i	x_n
c_1	a_{1n}
c_2	a_{2n}
.	.
.	.
.	.
c_k	a_{kn}

Then, $Z^n = c_1a_{1n} + c_2a_{2n} + \cdots + c_ka_{kn}$. This is the product of the two column vectors appearing under the column headings C_i and x_n.

To summarize, the differences between the exchange-algebra method of Section 10.1 and the $(C^j - Z^j)$ method are not ones of substance but merely differences in format and in the degree of formalism used to calculate the new objective function.

Listed below is the complete sequence of $(C^j - Z^j)$ tableaux needed to treat Problem 10.2, beginning with the completed Tableau 2.

Tableau 2

			1	2	3	4		
		C^j	3	6	0	0	P	
	C_i	*Basic solution*	x_1	x_2	x_3	x_4		
1	0	s_1	$\left(\frac{8}{3}\right)$	0	-1	$-\frac{1}{3}$	32	←(12)
2	6	y	$\frac{1}{3}$	1	0	$\frac{1}{3}$	16	(48)
		Z^j	2	6	0	2		
		$C^j - Z^j$	1	0	0	-2		
			↑					

Tableau 3

			1	2	3	4	
		C^j	3	6	0	0	P
	C_i	*Basic solution*	x_1	x_2	x_3	x_4	
1	3	x_1	1	0	$\frac{3}{8}$	$-\frac{1}{8}$	12
2	6	x_2	0	1	$-\frac{1}{8}$	$\frac{3}{8}$	12
		Z^j	3	6	$\frac{3}{8}$	$\frac{15}{8}$	
		$C^j - Z^j$	0	0	$-\frac{3}{8}$	$-\frac{15}{8}$	

The $(C^j - Z^j)$ form of the simplex algorithm has an occasional advantage over the dictionary–tableau form. In earlier chapters we emphasized that the simplex algorithm presupposes that the current objective function is expressed solely in terms of the nonbasic variables; that is, all tableaux must be adjusted. When we use the $(C^j - Z^j)$ table we need never bother to check if a tableau is adjusted—the $C^j - Z^j$ row always contains the correct objective function. In a word, the $(C^j - Z^j)$ method is self-adjusting.

To illustrate, consider the following problem:

$$\text{Minimize} \quad C = 2x_1 + 5x_2$$
$$\text{where} \quad x_1 \geq 0, \quad x_2 \geq 0$$
$$\text{and} \quad 3x_1 + 3x_2 \geq 5$$
$$x_1 + 4x_2 \geq 6$$

This program was studied in Problem 6.1. The auxiliary program, with surplus and artificial variables, reads

$$\text{Minimize} \quad C = 2x_1 + 5x_2 + Mx_5 + Mx_6$$

where all variables are nonnegative

$$\begin{aligned} \text{and} \quad 3x_1 + 3x_2 - x_3 \quad\quad + x_5 \quad\quad &= 5 \\ x_1 + 4x_2 \quad\quad - x_4 \quad\quad + x_6 &= 6 \end{aligned}$$

Then the initial $(C^j - Z^j)$ table has the following form:

			1	2	3	4	5	6	
		C^j	2	5	0	0	M	M	C
	C_i	*Basic solution*	x_1	x_2	x_3	x_4	x_5	x_6	
1	M	x_5	3	3	−1	0	1	0	5
2	M	x_6	1	4	0	−1	0	1	6
		Z^j	?	?	?	?	?	?	
		$C^j - Z^j$	?	?	?	?	?	?	

The very way in which the $C^j - Z^j$ row must be filled in, automatically gives an adjusted objective function, as shown below.

			1	2	3	4	5	6	
		C^j	2	5	0	0	M	M	C
	C_i	*Basic solution*	x_1	x_2	x_3	x_4	x_5	x_6	
1	M	x_5	3	3	−1	0	1	0	5
2	M	x_6	1	4	0	−1	0	1	6
		Z^j	$4M$	$7M$	$-M$	$-M$	M	M	
		$C^j - Z^j$	$2 - 4M$	$5 - 7M$	M	M	0	0	

The $(C^j - Z^j)$ *viewpoint allows us to prove easily the following theorem*: Once all the artificial variables in a program have become nonbasic, their M objective coefficients become and remain what they were in the original auxiliary objective function, that is, the C^j row. This must be so, because if no artificial variable is in the basis then the current Z^j row cannot contain any entries involving M's. Hence the current entries in the $C^j - Z^j$ row equal those of the C^j row minus the ordinary constants appearing in the Z^j row.

To conclude this section, we point out that by using a tableau with $-P$ upstairs, among the variables, and $-P$ in the left-hand column, among the basic variables, we can easily extend exchange algebra to the objective row of a tableau.

To illustrate, consider the initial tableau of Problem 10.1 in $-P$-upstairs format:

Tableau 1

	x	y	s_1	s_2	$-P$		
s_1	3	1	1	0	0	48	
s_2	3	④	0	1	0	120	←(30)
$-P$	5	6 ↑	0	0	1	0	

The extended exchange dictionary associated with this tableau is, by definition,

$$x = 3s_1 \oplus 3s_2 \oplus 5(-P) \tag{1}$$

$$y = s_1 \oplus 4s_2 \oplus 6(-P) \tag{2}$$

The maximum increase in y is 30, at which point $s_2 = 0$. This means that in Tableau 2, y will be basic and s_2 nonbasic. Hence, to form Exchange Dictionary 2 we must express s_2 and x in terms of s_1 and y.

From Eq. (2) above, we obtain, after transposition,

$$s_2 = \tfrac{1}{4}y \oplus (-\tfrac{1}{4})s_1 \oplus (-\tfrac{6}{4})(-P) \tag{3}$$

Also, by substituting Eq. (3) into Eq. (1), we obtain

$$\begin{aligned} x &= 3s_1 \oplus 3[\tfrac{1}{4}y \oplus (-\tfrac{1}{4})s_1 \oplus (-\tfrac{6}{4})(-P)] \oplus 5(-P) \\ &= \tfrac{9}{4}s_1 \oplus \tfrac{3}{4}y \oplus \tfrac{2}{4}(-P) \end{aligned} \tag{4}$$

Equations (3) and (4) give us all the coefficients needed to fill in Tableau 2 by columns:

Tableau 2

	x	y	s_1	s_2	$-P$		
s_1	$\frac{9}{4}$	0	1	$\frac{1}{4}$	0	18	←(8)
y	$\frac{3}{4}$	1	0	$-\frac{1}{4}$	0	30	(40)
$-P$	$\frac{2}{4}$	0	0	$-\frac{6}{4}$	1	-180	

Exercises

1. Do the exercises in Chapter 6 by the $(C^j - Z^j)$ method.
2. Solve the illustrative problems of this chapter by treating the objective variable as a basic variable, that is, by extending exchange algebra to the objective rows.

3. (a) Complete the following tableau, assuming that it represents a maximization program:

	C^j	3	4	5	0	0	0	P
C_i	*Basic solution*	x_1	x_2	x_3	x_4	x_5	x_6	
	x_5	1	0	0	-1	1	2	4
	x_2	$\frac{1}{2}$	1	0	2	0	$-\frac{1}{2}$	10
	x_3	$-\frac{1}{2}$	0	1	3	0	1	6
	Z^j							
	$C^j - Z^j$							

(b) Find the optimum tableau.

(c) Assuming that the variables x_4, x_5, and x_6 are the slack variables, state the original problem.

4. The following incomplete tableau is Tableau 2 in the solution of a linear program by the simplex algorithm. All variables corresponding to zero coefficients in the objective function are slack variables. Complete the tableau. Then find Tableaux 1 and 3 and state the original problem (the objective is to maximize P).

	C^j				0	0	0	P
C_i	*Basic solution*	x_1	x_2	x_3	x_4	x_5	x_6	
	30	$\frac{3}{4}$	1	$\frac{1}{2}$	$\frac{1}{4}$	0	0	
	50	$\frac{5}{4}$	0	$\frac{3}{2}$	$-\frac{1}{4}$	1	0	
	70	$-\frac{5}{4}$	0	$\frac{1}{2}$	$-\frac{3}{4}$	0	1	
	Z^j		8		2	0	0	
	$C^j - Z^j$	-1	0	-2	-2	0	0	

5. Discuss: The $(C^j - Z^j)$ process, especially as applied to a program already in adjusted form, is inefficient both computationally and pedagogically as contrasted with the pivot procedure of Chapters 5 and 6.

10.3 POSTOPTIMALITY ANALYSIS IN $(C^j - Z^j)$ FORMAT

The $(C^j - Z^j)$ tableau lends itself well to certain problems in postoptimality analysis, especially problems concerned with analyzing the effect of modifying a coefficient in the objective function. Such a modification leaves the region of feasibility unchanged. This means that, if a solution is optimum under the old objective function, it surely is feasible under the new objective function. Accordingly, to solve the new problem, we can begin by evaluating the new objective function at the old optimum corner—simply replace the entries in the old optimal C^j row and C_i column with their new values and compute the new $C^j - Z^j$ row.

To illustrate, turn to Problem 10.1 and suppose that the objective function is changed to $P = 5x + 5y$. Then change the C^j row and C_i column of the old optimal tableau to reflect the new objective function:

			1	2	3	4	
		C^j	5	5	0	0	P
	C_i	Basic solution	x	y	s_1	s_2	
1	5	x	1	0	$\frac{4}{9}$	$-\frac{1}{9}$	8
2	5	y	0	1	$-\frac{1}{3}$	$\frac{1}{3}$	24
		Z^j	5	5	$\frac{5}{9}$	$\frac{10}{9}$	
		$C^j - Z^j$	0	0	$-\frac{5}{9}$	$-\frac{10}{9}$	

This tableau is terminal. The old optimal corner, characterized by $s_1 = s_2 = 0$, is still optimum, and the new optimum value of P is 160.

Now suppose we modify Problem 10.1 by taking $P = 5x + 7y$ as the new objective function. The appropriate new tableau reads

			1	2	3	4		
		C^j	5	7	0	0	P	
	C_i	Basic solution	x	y	s_1	s_2		
1	5	x	1	0	$\textcircled{\frac{4}{9}}$	$-\frac{1}{9}$	8	←(18)
2	7	y	0	1	$-\frac{1}{3}$	$\frac{1}{3}$	24	
		Z^j	5	7	$-\frac{1}{9}$	$\frac{16}{9}$		
		$C^j - Z^j$	0	0	$\frac{1}{9}$ ↑	$-\frac{16}{9}$		

This solution is not terminal. Continuing the algorithm, we obtain the following tableau

			1	2	3	4	
		C^j	5	7	0	0	P
	C_i	*Basic solution*	x	y	s_1	s_2	
1	0	s_1	$\frac{9}{4}$	0	1	$-\frac{1}{4}$	18
2	7	y	$\frac{3}{4}$	1	0	$\frac{1}{4}$	30
		Z^j	$\frac{21}{4}$	7	0	$\frac{7}{4}$	
		$C^j - Z^j$	$-\frac{1}{4}$	0	0	$-\frac{7}{4}$	

This tableau is optimal. The optimum value of P is 210; it occurs at the corner $x = 0, y = 30, s_1 = 18, s_2 = 0$.

Exercises

1. From the optimal tableau of Problem 10.1 deduce the optimal tableau of the following programs:

$$\text{Maximize} \quad P = \gamma x_1 + \delta x_2$$
$$\text{where} \quad x_1 \geq 0, \quad x_2 \geq 0$$
$$\text{and} \quad 3x_1 + x_2 \leq 48$$
$$3x_1 + 4x_2 \leq 120$$

and

(a) $y = 5, \delta = 5$
(b) $y = 2, \delta = 7$
(c) $\gamma = 15, \delta = 4$

2. Using exchange algebra, solve as many of the postoptimality problems of Chapter 8 as you can.

3. Clarify the following statement: In a diet problem, if any ingredient is nonbasic at optimality, the $(C^j - Z^j)$ value associated with this ingredient indicates how much the cost of the ingredient must be reduced before it should be used in the diet.

10.4 LINEAR COMBINATIONS OF VECTORS

In this section, another, more mathematical description of the *algebra of exchange* is presented. Whereas in Section 10.1 the emphasis was on exchange between variables, here the emphasis will be on exchange between vectors.

Consider the system of equations

$$\text{(I)} \qquad \begin{cases} x_1 + 4x_2 = 9 \\ 2x_1 + 3x_2 = 8 \end{cases}$$

We can write this system in vector form as

$$x_1\begin{pmatrix}1\\2\end{pmatrix} + x_2\begin{pmatrix}4\\3\end{pmatrix} = \begin{pmatrix}9\\8\end{pmatrix}$$

Let $\mathbf{v}_1 = (1, 2)'$, $\mathbf{v}_2 = (4, 3)'$, and $\mathbf{v}_3 = (9, 8)'$. The problem of solving system (I) is clearly equivalent to seeking a set of numbers x_1 and x_2 which enable us to form $\mathbf{v}_3$ by "blending" or combining the vectors $\mathbf{v}_1$ and $\mathbf{v}_2$. More precisely, we seek x_1 and x_2 such that

$$\mathbf{v}_3 = x_1\mathbf{v}_1 + x_2\mathbf{v}_2$$

The solution to system (I) is $x_1 = 1$, $x_2 = 2$, which means that we can represent $\mathbf{v}_3$ as a combination of $\mathbf{v}_1$ and $\mathbf{v}_2$, namely,

$$\mathbf{v}_3 = \begin{pmatrix}9\\8\end{pmatrix} = 1\begin{pmatrix}1\\2\end{pmatrix} + 2\begin{pmatrix}4\\3\end{pmatrix} = x_1\mathbf{v}_1 + x_2\mathbf{v}_2$$

Now consider the system

$$\text{(II)} \qquad \begin{cases} 3x_1 + x_2 = 1 \\ 6x_1 + 2x_2 = 8 \end{cases}$$

This system can be interpreted as a problem in the blending of column vectors:

$$x_1\begin{pmatrix}3\\6\end{pmatrix} + x_2\begin{pmatrix}1\\2\end{pmatrix} \stackrel{?}{=} \begin{pmatrix}1\\8\end{pmatrix}$$

Observe that the first column vector above is equal to 3 times the second column vector and hence

$$x_1\begin{pmatrix}3\\6\end{pmatrix} + x_2\begin{pmatrix}1\\2\end{pmatrix} = 3x_1\begin{pmatrix}1\\2\end{pmatrix} + x_2\begin{pmatrix}1\\2\end{pmatrix} = \lambda\begin{pmatrix}1\\2\end{pmatrix}$$

where $\lambda = 3x_1 + x_2$. Thus the given problem reduces to determining λ such that

$$\lambda\begin{pmatrix}1\\2\end{pmatrix} = \begin{pmatrix}1\\8\end{pmatrix}$$

This is impossible, so system (II) cannot have a solution.

These examples lead to the following definitions.

DEFINITION 1. Let $\mathbf{v}_1, \ldots, \mathbf{v}_n$ be a set of vectors, all of the same dimension m. Any vector $\mathbf{w}$ of the form

$$\mathbf{w} = c_1\mathbf{v}_1 + \cdots + c_n\mathbf{v}_n$$

is a *linear combination* of $\mathbf{v}_1, \ldots, \mathbf{v}_n$.

DEFINITION 2A. The set of all linear combinations of $\mathbf{v}_1, \ldots, \mathbf{v}_n$ is the (vector) *space spanned* by $\mathbf{v}_1, \ldots, \mathbf{v}_n$.

DEFINITION 2B. The set of all *nonnegative* linear combinations of $\mathbf{v}_1, \ldots, \mathbf{v}_n$ is the *positive cone* spanned by $\mathbf{v}_1, \ldots, \mathbf{v}_n$.

DEFINITION 2C. The *convex hull* determined by $\mathbf{v}_1, \ldots, \mathbf{v}_n$ is the set of all vectors $\mathbf{k}$, of the form

$$\mathbf{k} = \gamma_1\mathbf{v}_1 + \gamma_2\mathbf{v}_2 + \cdots + \gamma_n\mathbf{v}_n$$

where $\gamma_i \geq 0$, for $i = 1, \ldots, n$ and $\gamma_1 + \gamma_2 + \cdots + \gamma_n = 1$.

The vector $\mathbf{k}$ is called a *convex combination* of $\mathbf{v}_1, \ldots, \mathbf{v}_n$. The convex hull is often referred to as the set of all convex combinations of the $\mathbf{v}_i$, $i = 1, \ldots, n$.

Clearly, since the vectors under consideration are all of dimension m, we can think of them as *points* in m-space and speak of the positive cone or the convex hull generated by these points. (Throughout this text, *point* and *vector* mean *m-tuple*, or ordered array of m numbers.)

To illustrate Definitions 2A, 2B, and 2C, let $\mathbf{v}_1 = (1, 0)'$ and $\mathbf{v}_2 = (0, 1)'$. Then the space spanned by $\mathbf{v}_1$ and $\mathbf{v}_2$ is the set of all points with two components, or, equivalently, the set of all vectors emanating from the origin of the usual two-dimensional coordinate system. The positive cone spanned by $\mathbf{v}_1$ and $\mathbf{v}_2$ is the set of all vectors emanating from the origin and terminating in the first quadrant; finally, the convex hull determined by $\mathbf{v}_1$ and $\mathbf{v}_2$ corresponds to the set of vectors emanating from 0 and terminating on the segment joining $P_1(1, 0)$ and $P_2(0, 1)$, as shown in Figure 38.

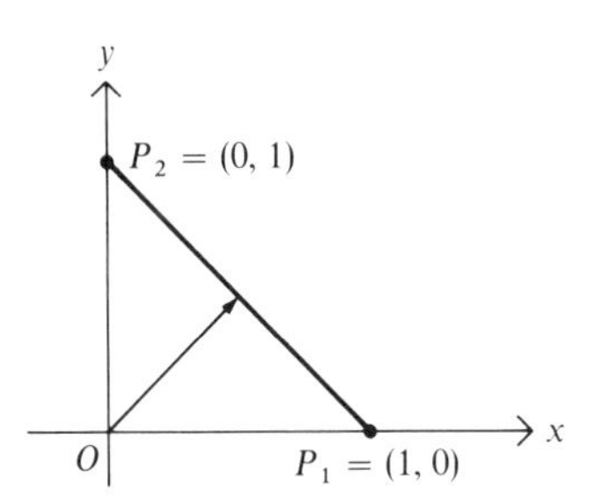

Figure 38

DEFINITION 3. Consider $\mathbf{v}_1, \ldots, \mathbf{v}_n$, all of the same dimension. Whenever the equation

$$c_1\mathbf{v}_1 + \cdots + c_n\mathbf{v}_n = \mathbf{0}$$

implies that

$$c_1 = \cdots = c_n = 0$$

the set of vectors $\mathbf{v}_1, \ldots, \mathbf{v}_n$ is *linearly independent*. If a set of vectors is not linearly independent, it is *linearly dependent*.

To illustrate, consider the following examples:

Example (a). The vectors $\mathbf{v}_1 = (1, 2)$ and $\mathbf{v}_2 = (3, 6)$ are linearly dependent, because it is possible to find numbers c_1 and c_2, not both 0, such that $c_1\mathbf{v}_1 + c_2\mathbf{v}_2 = 0$; for example, $c_1 = 3$ and $c_2 = -1$.

Example (*b*). The vectors $\mathbf{u}_1 = \begin{pmatrix}1\\0\end{pmatrix}$ and $\mathbf{u}_2 = \begin{pmatrix}0\\1\end{pmatrix}$ are linearly independent because

$$c_1\begin{pmatrix}1\\0\end{pmatrix} + c_2\begin{pmatrix}0\\1\end{pmatrix} = \begin{pmatrix}0\\0\end{pmatrix}$$

implies that $c_1 = 0$ and $c_2 = 0$.

The following theorems are fundamental.

THEOREM 10.1. If $\mathbf{v}_1, \ldots, \mathbf{v}_m$ are *linearly independent*, then

$$\mathbf{w} = c_1\mathbf{v}_1 + \cdots + c_m\mathbf{v}_m = d_1\mathbf{v}_1 + \cdots + d_n\mathbf{v}_m$$

implies $c_i = d_i$, for $i = 1, \ldots, m$.

In words, if a vector belongs to the space spanned by an independent set of vectors, there is only one way to express it as a combination of these vectors.

Proof. Suppose that

$$(c_1 - d_1)\mathbf{v}_1 + \cdots + (c_m - d_m)\mathbf{v}_m = \mathbf{0}$$

Since the $\mathbf{v}_i$'s are independent, it follows that $c_i - d_i = 0$, and hence $c_i = d_i$, for $i = 1, \ldots, m$.

THEOREM 10.2. If the vectors $\mathbf{v}_1, \ldots, \mathbf{v}_m$ are linearly dependent, then at least one of the vectors is a linear combination of the others.

Proof. Since $\mathbf{v}_1, \ldots, \mathbf{v}_m$ are linearly dependent, there exist numbers $c_1, \ldots, c_m$, not all 0, such that

$$c_1\mathbf{v}_1 + \cdots + c_m\mathbf{v}_m = \mathbf{0}$$

Suppose $c_k \neq 0$. Then

$$\mathbf{v}_k = -\frac{c_1}{c_k}\mathbf{v}_1 - \cdots - \frac{c_m}{c_k}\mathbf{v}_m \qquad \text{(Q.E.D.)}$$

To test the vectors $\mathbf{v}_1, \ldots, \mathbf{v}_m$ for independence, consider the relation

$$c_1\mathbf{v}_1 + c_2\mathbf{v}_2 + \cdots + c_m\mathbf{v}_m = \mathbf{0}$$

as a system of equations. Then: (1) If this system has no solution other than the trivial solution $c_1 = c_2 = \cdots = c_m = 0$, the vectors $\mathbf{v}_1, \ldots, \mathbf{v}_m$ are independent; (2) otherwise, they are dependent.

To illustrate, consider

Example (*c*). Let $\mathbf{v}_1 = (1, 4, 7)'$, $\mathbf{v}_2 = (2, 5, 8)'$, and $\mathbf{v}_3 = (3, 6, 10)'$.
To determine whether these vectors are dependent, solve the following system of equations by the methods of Chapter 4.

Tableau 1

c_1	c_2	c_3	
①	2	3	0
4	5	6	0
7	8	10	0

By pivoting as indicated, this tableau reduces to the following sequence of equivalent tableaux:

Tableau 2

c_1	c_2	c_3	
1	2	3	0
0	(−3)	−6	0
0	−6	−11	0

Tableau 3

c_1	c_2	c_3	
1	0	−1	0
0	1	2	0
0	0	(1)	0

Tableau 4

c_1	c_2	c_3	
1	0	0	0
0	1	0	0
0	0	1	0

Tableau 4 shows that the only solution to the original system is the trivial solution $c_1 = c_2 = c_3 = 0$, and hence the vectors $\mathbf{v}_1$, $\mathbf{v}_2$, and $\mathbf{v}_3$ are linearly independent.

We note in passing that a system of equations whose right-side constants are all 0 is said to be *homogeneous*. Such a system always possesses a solution, that is, the trivial solution where all the unknowns are 0.

Example (*d*). Let $\mathbf{v}_1 = \begin{pmatrix} 0 \\ 4 \\ 4 \end{pmatrix}$, $\mathbf{v}_2 = \begin{pmatrix} -1 \\ -1 \\ -2 \end{pmatrix}$, $\mathbf{v}_3 = \begin{pmatrix} 2 \\ -2 \\ 0 \end{pmatrix}$

To find all possible relations among these vectors, study the associated system of homogeneous equations:

Tableau 1

c_1	c_2	c_3	
0	−1	2	0
(4)	−1	−2	0
4	−2	0	0

Tableau 2

	c_1	c_2	c_3	
	0	(−1)	2	0
	1	$-\frac{1}{4}$	$-\frac{1}{2}$	0
	0	−1	2	0

Tableau 3

	c_1	c_2	c_3	
	0	1	−2	0
	1	0	−1	0
	0	0	0	0

Tableau 3 describes the general solution to the initial system, as $c_2 = 2c_3$ and $c_1 = c_3$, where c_3 is an arbitrary parameter. Thus, if $c_3 = 1$, then $c_1 = 1$ and $c_2 = 2$ and the vectors $\mathbf{v}_1$, $\mathbf{v}_2$, and $\mathbf{v}_3$ are dependent.

Example (*e*). Given the tableau

	3	2	−1	5	0
	5	1	4	−2	0
	1	−4	11	−19	0

Below we study simultaneously the relations among the columns of this tableau, viewed as vectors, and at the same time, the relations between the rows. To this end, consider Tableau 1:

Tableau 1

	c_1	c_2	c_3	c_4		R_1	R_2	R_3	
	(3)	2	−1	5	0	1	0	0	
	5	1	4	−2	0	0	1	0	
	1	−4	11	−19	0	0	0	1	

The first five columns of Tableau 1 contain the homogeneous system which determines the relations among the columns of the given tableau. The last three columns contain three distinct unit vectors that record the history of the passage from Tableau 1 to any subsequent tableau (as explained in Chapter 7).

By pivoting as indicated we obtain

Tableau 2

c_1	c_2	c_3	c_4		R_1	R_2	R_3
1	$\frac{2}{3}$	$-\frac{1}{3}$	$\frac{5}{3}$	0	$\frac{1}{3}$	0	0
0	**$-\frac{7}{3}$**	$\frac{17}{3}$	$-\frac{31}{3}$	0	$-\frac{5}{3}$	1	0
0	$-\frac{14}{3}$	$\frac{34}{3}$	$-\frac{62}{3}$	0	$-\frac{1}{3}$	0	1

Tableau 3

c_1	c_2	c_3	c_4		R_1	R_2	R_3
1	0	$\frac{9}{7}$	$-\frac{9}{7}$	0	$-\frac{1}{7}$	$\frac{2}{7}$	0
0	1	$-\frac{17}{7}$	$\frac{31}{7}$	0	$\frac{5}{7}$	$-\frac{3}{7}$	0
0	0	0	0	0	3	-2	1

The general solution of the original homogeneous system is

$$c_1 = -\tfrac{9}{7}c_3 + \tfrac{9}{7}c_4$$
$$c_2 = \tfrac{17}{7}c_3 - \tfrac{31}{7}c_4$$

Here both c_3 and c_4 are arbitrary parameters that can be assigned nonzero values; hence the four columns of the given tableau are dependent.

Now consider the "multiplier vector" of Tableau 3, that is, the vector appearing in row 3 and in the columns headed by the symbols R_1, R_2, and R_3, namely,

$$\mu_3 = (3, \quad -2, \quad 1)$$

As shown in Chapter 7, μ_3 records how row 3 of Tableau 3 was formed (disregarding the R_i columns):

$$\begin{aligned} \mathbf{0} &= \text{row 3 of Tbl 3} \\ &= 3\ (\text{row 1 of Tbl 1}) - 2\ (\text{row 2 of Tbl 1}) + 1(\text{ row 3 of Tbl 1}) \end{aligned}$$

This gives a nontrivial relation connecting the rows of Tableau 1.

Examples (c) to (e) illustrate the following theorem:

THEOREM 10.3. Any set of vectors which can be pivoted into a set of distinct unit columns must be linearly independent.

The proof is immediate, since these pivot operations reduce the associated homogeneous system to an equivalent system whose only solution is trivial.

The following results connect the theory of vectors to the simplex algorithm.

DEFINITION 4A. A vector with m components is called an *m vector*.

DEFINITION 4B. The set of all m vectors constitutes *m-space*.

DEFINITION 5. In m-space, any set of m linearly independent vectors forms a *basis* for the space.

For example, the distinct unit columns with m components form a basis of m-space, namely,

$$\mathbf{u}_1 = \begin{pmatrix} 1 \\ 0 \\ 0 \\ \cdot \\ \cdot \\ \cdot \\ 0 \end{pmatrix}, \quad \mathbf{u}_2 = \begin{pmatrix} 0 \\ 1 \\ 0 \\ \cdot \\ \cdot \\ \cdot \\ 0 \end{pmatrix}, \quad \ldots, \quad \mathbf{u}_m = \begin{pmatrix} 0 \\ 0 \\ 0 \\ \cdot \\ \cdot \\ \cdot \\ 1 \end{pmatrix}$$

This basis is called the *natural basis*, because, if $\mathbf{x} = (x_1, x_2, \ldots, x_m)$, then clearly

$$\mathbf{x} = x_1\mathbf{u}_1 + x_2\mathbf{u}_2 + \cdots + x_m\mathbf{u}_m$$

THEOREM 10.4A. In m-space, any $(m + 1)$ vectors are linearly dependent.

To illustrate, the four columns in Example (e) are dependent.

Theorem 10.4A plays a central role in the theory of vector spaces. By considering numerical examples, the reader should convince himself that Theorem 10.4A is equivalent to the following Theorem 10.4B.

THEOREM 10.4B. Any system of m homogeneous equations in n unknowns, where $m < n$, possesses a nontrivial solution.

THEOREM 10.5. If $\mathbf{v}_1, \ldots, \mathbf{v}_m$ is a basis of m-space, then every vector $\mathbf{w}$ in m-space can be written in exactly one way as a linear combination of $\mathbf{v}_1, \ldots, \mathbf{v}_m$.

Observe that Theorem 4A implies that the vectors

$$\mathbf{v}_1, \mathbf{v}_2, \ldots, \mathbf{v}_m, \mathbf{w}$$

are linearly dependent. Hence there exist c_i, for $i = 1, \ldots, m + 1$, such that

$$c_1\mathbf{v}_1 + c_2\mathbf{v}_2 + \cdots + c_m\mathbf{v}_m + c_{m+1}\mathbf{w} = \mathbf{0} \tag{1}$$

and not all $c_i = 0$.

We contend that $c_{m+1} \neq 0$, for if $c_{m+1} = 0$ then Eq. (1) would represent a nontrivial relation between the $\mathbf{v}_i$, $i = 1, \ldots, m$. Since $c_{m+1} \neq 0$, clearly

$$\mathbf{w} = -\frac{c_1}{c_{m+1}}\mathbf{v}_1 - \frac{c_2}{c_{m+1}}\mathbf{v}_2 - \cdots - \frac{c_m}{c_{m+1}}\mathbf{v}_m \tag{2}$$

Equation (2) represents $\mathbf{w}$ as a linear combination of $\mathbf{v}_1, \ldots, \mathbf{v}_m$. The uniqueness of this combination follows from Theorem 10.1.

Theorems 10.1 and 10.5 imply immediately that an $m \times m$ system of equations possesses a unique solution if and only if the columns of coefficients associated with the m unknowns are independent.

THEOREM 10.6 (THE EXCHANGE THEOREM FOR VECTORS). Let $\mathbf{v}_1, \mathbf{v}_2, \ldots, \mathbf{v}_m$ be a basis for m-space and let $\mathbf{w} \neq 0$ be an m vector. By Theorem 10.5, $\mathbf{w}$ can be expressed uniquely in terms of the given basis, as

$$\mathbf{w} = c_1\mathbf{v}_1 + \cdots + c_m\mathbf{v}_m \tag{1}$$

If any vector $\mathbf{v}_j$ for which $c_j \neq 0$ in Eq. (1) is removed from the set $\mathbf{v}_1, \ldots, \mathbf{v}_j, \ldots, \mathbf{v}_m$ and replaced by $\mathbf{w}$, then the new set of vectors is again a basis for m-space. But if $c_j = 0$, we do not obtain a new basis by exchanging $\mathbf{w}$ and $\mathbf{v}_j$.

The proof of this theorem is left as an exercise.

Let us now examine the simplex algorithm in terms of combinations of columns. For convenience, we repeat below the complete set of simplex tableaux needed to solve that perennial illustrative example:

$$\begin{aligned} &\text{Maximize} \quad P = 5x_1 + 6x_2 \\ &\text{subject to} \quad x_1 \geq 0, \quad x_2 \geq 0 \\ &\text{and} \quad 3x_1 + x_2 \leq 48 && (1) \\ &\qquad\quad 3x_1 + 4x_2 \leq 120 && (2) \end{aligned}$$

In the following tableaux x_3 and x_4 denote slack in Eqs. (1) and (2), respectively.

Tableau 1

	x_1	x_2	x_3	x_4	$-P$		
x_3	3	1	1	0	0	48	
x_4	3	(4)	0	1	0	120	←
$-P$	5	6	0	0	1	0	
		↑					

Tableau 2

	x_1	x_2	x_3	x_4	$-P$		
x_3	($\frac{9}{4}$)	0	1	$-\frac{1}{4}$	0	18	←
x_2	$\frac{3}{4}$	1	0	$\frac{1}{4}$	0	30	
$-P$	$\frac{2}{4}$	0	0	$-\frac{3}{2}$	1	-180	
	↑						

Tableau 3

	x_1	x_2	x_3	x_4	$-P$	
x_1	1	0	$\frac{4}{9}$	$-\frac{1}{9}$	0	8
x_2	0	1	$-\frac{1}{3}$	$\frac{1}{3}$	0	24
$-P$	0	0	$-\frac{2}{9}$	$-\frac{13}{9}$	1	-184

Each of these tableaux represents an extreme point:

(1) $x_1 = 0,\ x_2 = 0,\ x_3 = 8,\ x_4 = 120;\ -P = 0$
(2) $x_1 = 0,\ x_2 = 30,\ x_3 = 18,\ x_4 = 0;\ -P = -180$
(3) $x_1 = 8,\ x_2 = 24,\ x_3 = 0,\ x_4 = 0;\ -P = -184$

Since the tableaux are row equivalent, any solution to the system of equations displayed in one tableau must also be a solution to the other systems. We can, therefore, carry back each extreme solution to the initial tableau and interpret each solution as a combination of columns:

Solution 1.

$$0\begin{pmatrix}3\\3\\5\end{pmatrix} + 0\begin{pmatrix}1\\4\\6\end{pmatrix} + 48\begin{pmatrix}1\\0\\0\end{pmatrix} + 120\begin{pmatrix}0\\1\\0\end{pmatrix} + 0\begin{pmatrix}0\\0\\1\end{pmatrix} = \begin{pmatrix}48\\120\\0\end{pmatrix}$$

Solution 2.

$$0\begin{pmatrix}3\\3\\5\end{pmatrix} + 30\begin{pmatrix}1\\4\\6\end{pmatrix} + 18\begin{pmatrix}1\\0\\0\end{pmatrix} + 0\begin{pmatrix}0\\1\\0\end{pmatrix} - 180\begin{pmatrix}0\\0\\1\end{pmatrix} = \begin{pmatrix}48\\120\\0\end{pmatrix}$$

Solution 3.

$$8\begin{pmatrix}3\\3\\5\end{pmatrix} + 24\begin{pmatrix}1\\4\\6\end{pmatrix} + 0\begin{pmatrix}1\\0\\0\end{pmatrix} + 0\begin{pmatrix}0\\1\\0\end{pmatrix} - 184\begin{pmatrix}0\\0\\1\end{pmatrix} = \begin{pmatrix}48\\120\\0\end{pmatrix}$$

With each of the above vector equations we can associate an analogous contracted vector equation, obtained by omitting the $-P$ (or objective) column, as well as the basement (objective) component of each vector. This process is best explained by an example; thus the contraction of Solution 3 is

$$8\begin{pmatrix}3\\3\end{pmatrix} + 24\begin{pmatrix}1\\4\end{pmatrix} + 0\begin{pmatrix}1\\0\end{pmatrix} + 0\begin{pmatrix}0\\1\end{pmatrix} = \begin{pmatrix}48\\120\end{pmatrix}$$

The following conventions will be useful:

(1) If we disregard the basement, or objective, component of a column in a simplex tableau, we call the column a *contracted*, or *technical*, column vector. To emphasize that a column has not been contracted, we can call it *extended.* Observe that in discussing linear combinations of contracted columns we may disregard the $-P$ column, since its contraction is always the zero vector.

(2) To distinguish between columns of different tableaux we denote the jth column of the initial tableau by $\mathbf{x}_j$ and the jth column of any other tableau of interest by $\mathbf{k}_j$.

(3) The term *activity* will be reserved for the contracted columns of the initial tableau defining a linear program (whether or not the initial tableau is adjusted), the stub being excluded.

As an example, the activities in Problem 10.1 are

$$\mathbf{x}_1 = \begin{pmatrix}3\\3\end{pmatrix}, \quad \mathbf{x}_2 = \begin{pmatrix}1\\4\end{pmatrix}, \quad \mathbf{x}_3 = \begin{pmatrix}1\\0\end{pmatrix}, \quad \mathbf{x}_4 = \begin{pmatrix}0\\1\end{pmatrix}$$

Clearly, every simplex tableau describes a different way of expressing the technical stub as a linear combination of the technical columns of the original defining tableau. Think of the technical stub as describing the original set of limited resources. Then we can affirm, in both a loose and a precise sense, that *linear programming deals with the problem of allocating limited resources among competing activities in an optimal manner. Furthermore, every tableau in a simplex sequence specifies a different choice of linearly independent activities.* To be more precise, suppose that the basic variables in Tableau i are x_{i_1}, $x_{i_2}, \ldots, x_{i_m}$. Then the columns of Tableau i whose headings read $x_{i_1}, x_{i_2}, \ldots, x_{i_m}$ form a set of distinct unit columns, and hence, by Theorem 10.3, the initial activity vectors $\mathbf{x}_{i_1}, \mathbf{x}_{i_2}, \ldots, \mathbf{x}_{i_m}$ are linearly independent. To illustrate, in Tableau 2 the basic variables are x_2 and x_3, and, consequently, the initial activity vectors $\mathbf{x}_2$ and $\mathbf{x}_3$ must be linearly independent.

Furthermore, every tableau describes a different way of representing each initial technical column as a linear combination of the other initial technical columns. In fact, each exchange dictionary defines such a combination. To illustrate, the exchange dictionary for Tableau 2 above reads

$$x_1 = \tfrac{9}{4}x_3 \oplus \tfrac{3}{4}x_2$$
$$x_4 = -\tfrac{1}{4}x_3 \oplus \tfrac{1}{4}x_2$$

The reader should verify that the very same set of relations holds, *mutatis mutandis*, for the original activity vectors, namely,

$$\mathbf{x}_1 = \tfrac{9}{4}\mathbf{x}_3 + \tfrac{3}{4}\mathbf{x}_2$$
$$\mathbf{x}_4 = -\tfrac{1}{4}\mathbf{x}_3 + \tfrac{1}{4}\mathbf{x}_2$$

These results hold equally well for extended columns. In general, whenever *any* adjusted tableau shows the validity of the exchange equation

$$x_i = \alpha x_j \oplus \beta x_k \oplus \gamma x_q \oplus \delta(-P) \tag{1}$$

then the *initial* extended columns satisfy an equation of the same form, namely,

$$\mathbf{x}_i = \alpha\mathbf{x}_j + \beta\mathbf{x}_k + \gamma\mathbf{x}_q + \delta(-\mathbf{P}) \tag{2}$$

Note that Eq. (1) involves variables, that is, levels of activity, but that Eq. (2) involves vectors. For example, Tableau 3 of the problem at hand shows that

$$x_3 = \tfrac{4}{9}x_1 \oplus (-\tfrac{1}{3})(x_2) \oplus (-\tfrac{2}{9})(-P)$$

Hence, for extended columns, it follows that

$$\mathbf{x}_3 = \tfrac{4}{9}\mathbf{x}_1 + (-\tfrac{1}{3})(\mathbf{x}_2) + (-\tfrac{2}{9})(-\mathbf{P})$$

or

$$\begin{pmatrix}1\\0\\0\end{pmatrix} = \tfrac{4}{9}\begin{pmatrix}3\\3\\5\end{pmatrix} - \tfrac{1}{3}\begin{pmatrix}1\\4\\6\end{pmatrix} - \tfrac{2}{9}\begin{pmatrix}0\\0\\1\end{pmatrix}$$

It should now be clear to the reader that the simplex algorithm can be viewed as a method for comparing the relative profitability of exchanging each currently nonbasic activity with one of the current basic activities. Each activity entering the basis is chosen so as to increase the objective function, whereas the activity leaving the basis is chosen so that the stub **b** will remain in the convex hull of the new basis—which is another way of saying that **b** will be feasible.

For example, consider the passage from Tableau 1 to Tableau 2 above. The coefficients of the objective function instruct us to introduce the activity $\mathbf{x}_2 = (1, 4)'$ into the basis. This vector could have been exchanged for either $(1, 0)'$ or $(0, 1)'$ and a new basis formed, but observe that if it had replaced $(1, 0)'$, then **b** would not have been in the convex hull of the new basis, since

$$\mathbf{b} = \begin{pmatrix} 48 \\ 120 \end{pmatrix} = 48\begin{pmatrix} 1 \\ 4 \end{pmatrix} + (-72)\begin{pmatrix} 0 \\ 1 \end{pmatrix}$$

It is the θ_i rule which ensures that **b** *is a convex combination of the new basic vectors.*

PROBLEM 10.3. Maximize $P = x_1 + 3x_2 + 5x_3$

where all $x_i \geq 0$

and

$$\begin{aligned} 4x_1 + 6x_2 + 12x_3 + x_4 &= 22 \\ 6x_1 + 9x_2 + 18x_3 + x_5 &= 33 \\ 2x_1 + 3x_2 + 6x_3 + x_6 &= 11 \end{aligned}$$

If we set $x_4 = x_5 = x_6 = 0$, the 3×3 fundamental system that results possesses the solution

(a) $x_1 = x_2 = x_3 = 1;\ x_4 = x_5 = x_6 = 0$

However, this is not a basic solution, because the resulting system possesses other solutions, for example,

(b) $x_1 = 1,\ x_2 = 2,\ x_3 = \frac{1}{2};\ x_4 = x_5 = x_6 = 0$

(c) $x_1 = 1,\ x_2 = 0,\ x_3 = \frac{3}{2};\ x_4 = x_5 = x_6 = 0$

Moreover, since solution (a) is not unique, it may not be an extreme point of the region of feasibility. Observe that solution (a) is the average of solutions (b) and (c), and hence it is nonextreme, by definition.

As we have stated previously and as we shall soon demonstrate, every adjusted simplex tableau defines an extreme point of the region of feasibility. In other words, the nonextreme solution ($x_1 = x_2 = x_3 = 1;\ x_4 = x_5 = x_6 = 0$) could never occur as the solution corresponding to an adjusted simplex tableau.

The simplex solution of Problem 10.3 is

Tableau 1

	x_1	x_2	x_3	x_4	x_5	x_6	$-P$		
x_4	4	6	12	1	0	0	0	22	$(\frac{11}{6})$
x_5	6	9	18	0	1	0	0	33	$(\frac{11}{6})$
x_6	2	3	⑥	0	0	1	0	11	$\leftarrow(\frac{11}{6})$
$-P$	1	3	5 ↑	0	0	0	1	0	

(Note that we could just as well have pivoted in row 1 of column 3 or row 2 of column 3.)

Tableau 2

	x_1	x_2	x_3	x_4	x_5	x_6	$-P$	
x_4	0	0	0	1	0	-2	0	0
x_5	0	0	0	0	1	-3	0	0
x_3	$\frac{1}{3}$	($\frac{1}{2}$) circled	1	0	0	$\frac{1}{6}$	0	$\frac{11}{6}$
$-P$	$-\frac{2}{3}$	$\frac{1}{2}$ ↑	0	0	0	$-\frac{5}{6}$	1	$-\frac{55}{6}$

Tableau 3

	x_1	x_2	x_3	x_4	x_5	x_6	$-P$	
x_4	0	0	0	1	0	-2	0	0
x_5	0	0	0	0	1	-3	0	0
x_2	$\frac{2}{3}$	1	2	0	0	$\frac{1}{3}$	0	$\frac{11}{3}$
$-P$	-1	0	-1	0	0	-1	1	-11

Tableau 3 is optimal and degenerate.

This sequence of tableaux illustrates the following general results:

(1) If two activities, that is, initial contracted vectors, are proportional then the contracted vectors appearing under the same headings in any subsequent tableau will be in the same proportion as originally. Symbolically, if $\mathbf{x}_i = \lambda \mathbf{x}_j$, then $\mathbf{k}_i = \lambda \mathbf{k}_j$. Thus the first three contracted columns in the above tableaux are always in the proportion 2:3:6.

(2) Two or more linearly dependent activities cannot be basic at the same time. In particular, if an activity is a multiple of another, only one can be in solution (see Theorem 10.3).

(3) *The rows of an adjusted simplex tableau are linearly independent.* To understand why, consider the scheme:

$$
\begin{array}{ccc|ccccc||c}
a_{11} & \cdots & a_{1n} & 1 & 0 & \cdots & 0 & 0 & b_1 \\
\cdot & & \cdot & 0 & 1 & \cdots & 0 & 0 & \cdot \\
 & & & 0 & 0 & \cdots & 0 & 0 & \cdot \\
\cdot & & \cdot & \cdot & \cdot & & \cdot & \cdot & \\
 & & & \cdot & \cdot & & \cdot & \cdot & \\
\cdot & & \cdot & \cdot & \cdot & & \cdot & \cdot & \cdot \\
a_{m1} & \cdots & a_{mn} & 0 & 0 & \cdots & 1 & 0 & b_m \\
\hline
c_1 & \cdots & c_n & 0 & 0 & \cdots & 0 & 1 & 0
\end{array}
$$

Let $\mathbf{R}_i$ denote a typical row. Then

$$\mu_1 \mathbf{R}_1 + \cdots + \mu_m \mathbf{R}_m + \mu_{m+1} \mathbf{R}_{m+1} = \mathbf{0}$$

implies that $\mu_1 = \cdots = \mu_{m+1} = 0$, since each $\mu_i \mathbf{R}_i$ has μ_i as its $(n + i)$th component and 0 for all other components beyond the nth, for $i = 1, 2, \ldots, m + 1$.

Many linear programs contain rows or columns which are "partially dependent," as illustrated above. But the simplex algorithm is not affected by these partial dependencies. And if the rows of the defining tableau are "completely" dependent, Phase I of the simplex algorithm will discover and eliminate such redundancy (see Problem 6.6).

Exercises

1. Show that any set of vectors containing the zero vector is linearly dependent.

2. *True or false?*

(a) Any set of vectors containing a linearly independent set is itself linearly independent.

(b) Any subset of a linearly independent set of vectors is itself linearly independent.

(c) If each of the vectors $\mathbf{v}_1, \ldots, \mathbf{v}_n, \mathbf{v}_{n+1}$ belongs to the space spanned by $\mathbf{w}_1, \ldots, \mathbf{w}_n$, then the set of $\mathbf{v}_i$, $i = 1, \ldots, n + 1$ is linearly dependent.

(d) Any set of orthogonal vectors is linearly independent.

3. Given the three vectors $\mathbf{v}_1 = \begin{pmatrix} 1 \\ 2 \end{pmatrix}$, $\mathbf{v}_2 = \begin{pmatrix} 3 \\ 4 \end{pmatrix}$, and $\mathbf{v}_3 = \begin{pmatrix} 5 \\ 6 \end{pmatrix}$.

(a) Which theorem tells us that these vectors are dependent?

(b) Show that, for any constant k, $k\mathbf{v}_1 - 2k\mathbf{v}_2 + k\mathbf{v}_3 = \mathbf{0}$.

4. Redo Example (d) of the text to determine any relations among the rows of the given set of vectors (as well as those among columns).

5. *True or false?*

(a) If n vectors of dimension n have their last component equal to 0, they must be linearly dependent. [Example: $\mathbf{v}_1 = (1, 2, 0)$, $\mathbf{v}_2 = (3, 4, 0)$, $\mathbf{v}_3 = (5, 6, 0)$.]

(b) If $c_1\mathbf{v}_1 + c_2\mathbf{v}_2 + \cdots + c_n\mathbf{v}_n = 0$, then $c_i = 0$ implies that $\mathbf{v}_i$ *cannot* be expressed as a linear combination of the remaining vectors.

6. Consider the vectors $\mathbf{x}_1 = (1, 0, 1)'$, $\mathbf{x}_2 = (1, 1, 1)'$, and $\mathbf{x}_3 = (2, 1, 1)'$.

(a) Show that $\mathbf{x}_1$, $\mathbf{x}_2$, and $\mathbf{x}_3$ form a basis for 3-space.
(b) Express $\mathbf{u}_1 = (1, 0, 0)'$ as a linear combination of $\mathbf{x}_1$, $\mathbf{x}_2$, and $\mathbf{x}_3$.
(c) Express $\mathbf{x}_4 = (3, 2, 2)'$ as a linear combination of $\mathbf{x}_1$, $\mathbf{x}_2$, and $\mathbf{x}_3$.
(d) If we replace $\mathbf{x}_2$ by $\mathbf{x}_4$ will the remaining set be a basis?
(e) If we replace $\mathbf{x}_1$ by $\mathbf{x}_4$ will the remaining set be a basis?

7. The following tableau displays the coefficients of six column vectors $\mathbf{x}_1, \ldots, \mathbf{x}_6$ relative to a basis composed of $\mathbf{x}_4$, $\mathbf{x}_2$, and $\mathbf{x}_5$:

	x_1	x_2	x_3	x_4	x_5	x_6
x_4	2	0	2	1	0	−1
x_2	−2	1	4	0	0	0
x_5	3	0	5	0	1	2

(a) Exchange $\mathbf{x}_1$ for $\mathbf{x}_5$, then $\mathbf{x}_3$ for $\mathbf{x}_4$.
(b) Exchange $\mathbf{x}_3$ for $\mathbf{x}_4$, then $\mathbf{x}_1$ for $\mathbf{x}_5$.
(c) Compare the results in (a) and (b).
(d) What are the coordinates of $\mathbf{x}_6$ with respect to the basis of $\mathbf{x}_1$, $\mathbf{x}_3$, and $\mathbf{x}_5$?
(e) Can we exchange $\mathbf{x}_6$ for $\mathbf{x}_2$, or $\mathbf{x}_6$ for $\mathbf{x}_4$?

8. The following tableau displays a basic solution of a 3×7 system of equations:

	x_1	x_2	x_3	x_4	x_5	x_6	x_7	
x_3	−1	0	1	4	1	0	−4	2
x_2	−2	1	0	−1	1	0	3	1
x_6	−3	0	0	2	1	1	1	4

Successively insert $\mathbf{x}_5$ and $\mathbf{x}_4$ into the basis and determine the vector to be removed in each case so that the basic variables in each solution remain nonnegative.

9. Discuss:

(a) *Red* and *blue* are a basis for the space of all colors.
(b) The primary colors are a basis for the space of all colors.

10. Prove:

(a) Given a finite number of feasible solutions, any convex combination of them is again feasible.

(b) If the value of the objective function is P_0 at each of the above solutions, then its value is also P_0 at each point in the convex hull of these solutions.

(c) Whenever a linear program is optimal at r different corner points, then any point in the convex hull of these r points is also optimal.

(d) Given all the alternative optimal basic feasible solutions of a program, how can we write a general expression for all the nonbasic optimal solutions?

10.5 SIMPLEX DICTIONARIES AND EXTREME POINTS

In this section we demonstrate that, for the most general linear program, each dictionary, that is, (feasible) tableau with as many distinct unit columns as basic rows, can be associated with an extreme point of the region of feasibility. This fact is really not needed for a logical development of the theory of linear programming; the constructive approach employed in this text need never invoke the "geometry" of the region of feasibility, since the whole theory can be built around tableaux with the appropriate number of distinct unit columns, as follows:

(1) Given any program we can always form an auxiliary tableau containing enough artificial variables to furnish immediately an initial (basic feasibile) solution. This auxiliary tableau can be adjusted in a routine fashion.

(2) By associating M coefficients with these artificial variables, either all artificial variables will be driven out of solution as we perform the required simplex iterations, or else the optimum of the auxiliary problem will exhibit a nonzero artificial variable.

In the last case, the original problem is nonfeasible. Otherwise,

(3) The original problem is feasible, and, in solving the auxiliary problem, we solve the original problem.

(4) By construction, the simplex algorithm takes us from one adjusted tableau to another, until the algorithm necessarily terminates or we learn that the objective function is unbounded.

Below is sketched a somewhat different approach to the fundamental facts of linear programming. Our goal is to help the reader who may encounter a more analytical, less constructive approach than the one presented in this text.

The following equations describe the region of feasibility of the general linear program:

$$x_i \geq 0, \quad i = 1, 2, \ldots, n$$

and

$$a_{11}x_1 + a_{12}x_2 + \cdots + a_{1n}x_n = b_1 \qquad (1)$$

$$a_{21}x_1 + a_{22}x_2 + \cdots + a_{2n}x_n = b_2 \qquad (2)$$

$$\cdot \quad \cdot \quad \cdot \quad \cdot \quad \cdot \quad \cdot \quad \cdot \quad \cdot \qquad \vdots$$

$$a_{m1}x_1 + a_{m2}x_2 + \cdots + a_{mn}x_n = b_m \qquad (m)$$

Equations (1)–(m) are equivalent to the vector equation

$$x_1\mathbf{a}_1 + x_2\mathbf{a}_2 + \cdots + x_n\mathbf{a}_n = \mathbf{b}$$

where $\mathbf{a}_j$ denotes the jth column of coefficients.

Think of the solutions to the system formed by Eqs. (1)–(m) as a point in n-space, or as a vector with n components. In any case, the reader knows the definition of a convex set and of an extreme point in n-space—these concepts are immediate generalizations of those encountered in the discussion of 3-space given earlier. The set of feasible points is readily shown to be a convex set. Moreover, it is not difficult to show that the number of extreme points of the region of feasibility is finite, and that every point in the region is a convex combination of its extreme points. From this it follows immediately that if a linear function attains its optimum over the region, then this optimum is assumed at at least one extreme point. (If the feasible set is unbounded, the objective function can become arbitrarily large or small, but the simplex method will detect when this situation exists. Otherwise, we can always assume that the feasible set is bounded.*)

The next two theorems characterize the extreme points of the region of feasibility. They give an ultimate justification for our insistence on working with dictionaries, that is, tableaux having a unit column for each basic row.

THEOREM 10.7. Let $X = (x_1, x_2, \ldots, x_n)$ be an extreme point of the region of feasibility of the general linear program. Then the column vectors $\mathbf{a}_j$ for which $x_j > 0$ are linearly independent.

Without loss of generality, we may assume that the nonzero components of X are the first h components, and

$$\mathbf{b} = x_1\mathbf{a}_1 + x_2\mathbf{a}_2 + \cdots + x_h\mathbf{a}_h$$

Now suppose that the h column vectors $\mathbf{a}_1, \mathbf{a}_2, \ldots, \mathbf{a}_h$ are dependent. Then there exist constants $c_1, c_2, \ldots, c_h$, not all of which are 0, such that

$$c_1\mathbf{a}_1 + c_2\mathbf{a}_2 + \cdots + c_h\mathbf{a}_h = \mathbf{0}$$

Consider the n-dimensional point, or vector

$$P = (c_1, c_2, \ldots, c_h, 0, 0, \ldots, 0)$$

Observe that the coordinates of P may be negative, but they must necessarily satisfy Eqs. (1) to (m) above. To obtain a nonnegative solution to these equations let us form

$$U = X + \epsilon P$$
$$V = X - \epsilon P$$

where $\epsilon > 0$ is a constant to be determined. The coordinates of U and V are, respectively,

$$U = (x_1 + \epsilon c_1, x_2 + \epsilon c_2, \ldots, x_h + \epsilon c_h, 0, \ldots, 0)$$
$$V = (x_1 - \epsilon c_1, x_2 - \epsilon c_2, \ldots, x_h - \epsilon c_h, 0, \ldots, 0)$$

* For further details, we refer the reader to F. A. Ficken, *The Simplex Method of Linear Programming* (Holt, Rinehart and Winston, New York, 1961).

We stress that, for ϵ small enough, the points U and V must both be nonnegative—simply take ϵ so small that the absolute value of each coordinate of ϵP is less than the corresponding coordinate of X. Note further that U and V are feasible points of the region of feasibility—for example,

$$
\begin{aligned}
(x_1 + \epsilon c_1)\mathbf{a}_1 + (x_2 + \epsilon c_2)\mathbf{a}_2 + \cdots + (x_h + \epsilon c_h)\mathbf{a}_h + (0)\mathbf{a}_{h+1} + \cdots + (0)\mathbf{a}_n \\
= [x_1\mathbf{a}_1 + x_2\mathbf{a}_2 + \cdots + x_h\mathbf{a}_h] + \epsilon[c_1\mathbf{a}_1 + \cdots + c_h\mathbf{a}_h] \\
= \mathbf{b} + \mathbf{0} = \mathbf{b}
\end{aligned}
$$

Finally, observe that $X = \frac{1}{2}(U + V)$.

Hence the assumption that $\mathbf{a}_1, \mathbf{a}_2, \ldots, \mathbf{a}_h$ are dependent contradicts the fact that X is an extreme point.

COROLLARY 10.1. At most m of the x_i are strictly positive.

Since every set of $(m + 1)$ vectors in m-space is linearly dependent, it follows from Theorem 10.7 that an extreme point of the above general program cannot have more than m positive components x_i.

THEOREM 10.8. Given the general program defined above, suppose a set of k independent $\mathbf{a}_j$ can be found such that $\mathbf{b}$ is a strictly positive combination of these k columns. Assume that these columns are $\mathbf{a}_1, \mathbf{a}_2, \ldots, \mathbf{a}_k$, the first k columns, and that

$$x_1\mathbf{a}_1 + x_2\mathbf{a}_2 + \cdots + x_k\mathbf{a}_k = \mathbf{b}$$

$x_i > 0$, $i = 1, \ldots, k$. Then the point X in n-space, defined by

$$X = (x_1, x_2, \ldots, x_k, 0, \ldots, 0)$$

is an extreme point of the region of feasibility.

Proof. To establish this contention, suppose that X is not extreme. From this it follows that X can be expressed as the average of two distinct feasible points, say,

$$X = \tfrac{1}{2}(U + V)$$

where U and V are nonnegative solutions of Eqs. (1)–(m).

We make the important observation that, like X, both U and V must have their last $(n - k)$ components equal to 0—otherwise the relation $2X = U + V$ would be impossible. Hence we can write

$$
\begin{aligned}
U &= (u_1, u_2, \ldots, u_k, 0, \ldots, 0) \\
V &= (v_1, v_2, \ldots, v_k, 0, \ldots, 0)
\end{aligned}
$$

Thus it has been shown that

$$
\begin{aligned}
x_1\mathbf{a}_1 + x_2\mathbf{a}_2 + \cdots + x_k\mathbf{a}_k &= \mathbf{b} \\
u_1\mathbf{a}_1 + u_2\mathbf{a}_2 + \cdots + u_k\mathbf{a}_k &= \mathbf{b} \\
v_1\mathbf{a}_1 + v_2\mathbf{a}_2 + \cdots + v_k\mathbf{a}_k &= \mathbf{b}
\end{aligned}
$$

But $\mathbf{a}_1, \ldots, \mathbf{a}_k$ are linearly independent, and hence $u_i = v_i = x_i$ for $i = 1, 2, \ldots, k$. In other words, X cannot be the average of two distinct feasible points, which shows that X is extreme.

Exercises

1. Consider the following set of constraints:

$$\begin{aligned} x_1 + 3x_2 + x_3 + x_4 \qquad\qquad &= 4 \\ x_1 + 3x_2 + \tfrac{1}{4}x_3 \qquad + x_5 \qquad &= 4 \\ x_1 + x_2 + x_3 \qquad\qquad + x_6 &= 7 \end{aligned}$$

(a) Is the solution $(x_1 = 1, x_2 = 1, x_6 = 5; x_3 = x_4 = x_5 = 0)$ an extreme point?
(b) Is the solution $(x_1 = x_2 = x_3 = 0; x_4 = 4, x_5 = 4, x_6 = 7)$ an extreme point?

2. Criticize the following statement: Given a system of m equations in n unknowns, where $m < n$, a *basic solution* can be generated as follows:

(a) Give $n - m$ of the variables values of 0 (thereby providing a "partial solution" and, at the same time, eliminating these variables from the system of equations).

(b) Solve the remaining m equations in m unknowns to obtain values for the remaining m variables.

Together, steps (a) and (b) yield solution values for all n variables, and the solution obtained is a basic solution.

3. *True or false?* Given m equations in n unknowns, there will be $n!/m!\,(n - m)!$ basic solutions. [Recall that $k!$ equals the product of the first k integers, for any k; that is, $k! = (1)(2)(3)\cdots(k - 1)(k)$.]

4. Consider the following initial tableau,

	x_1	x_2	x_3	x_4	x_5	$-P$	
	1	0	1	1	2	0	7
	0	1	0	1	1	0	4
	-2	-3	-3	-11	-12	1	$-k$

(a) Is this tableau adjusted?
(b) If the objective is to maximize P, is this tableau optimal?
(c) Is either of the points $(x_1 = 7, x_2 = 4; x_3 = x_4 = x_5 = 0)$, $(x_1 = 0, x_2 = 4, x_3 = 7, x_4 = 0, x_5 = 0)$ an extreme point?
(d) Can the activities $\mathbf{x}_1$ and $\mathbf{x}_3$ both be in solution in the same adjusted tableau?

11

DUALITY

Several times previously the reader may have sensed the possibility of reversing the roles of the rows and columns in a linear program. This is accomplished in the present chapter, which is subdivided into three sections: an introduction to some dual problems which arise "naturally," a discussion of the fundamental theorems on duality, and a comparison of the optimal tableaux of a pair of dual problems.

11.1 INTRODUCTION TO DUALITY

Consider once again the following example.

PROBLEM 11.1. Maximize $P = 5x_1 + 6x_2$

where $x_1 \geq 0, \quad x_2 \geq 0$

and
$$3x_1 + x_2 \leq 48 \tag{1}$$
$$3x_1 + 4x_2 \leq 120 \tag{2}$$

Except for a change in the names of the variables, this is the problem with which we began this text and to which we have often returned to illustrate new ideas or techniques. Originally we imagined a manufacturer who produced two products, A and B, processed on two machines, I and II. The variables x_1 and x_2 represent the amounts of A and B manufactured, respectively. Equations (1) and (2) describe how machine time is utilized, according to the following production formula: To produce 1 unit of A we need 3 units of time on machine I and 3 units on machine II; to produce 1 unit of B we need 1 unit of time on machine I and 4 units on machine II. We have available, at most, 48 units of time on machine I and 120 units on machine II.

The solution to Problem 11.1 is shown in the following terminal tableau, where x_3 and x_4 measure slack against the available time on machines I and II, respectively.

	x_1	x_2	x_3	x_4	$-P$	
x_1	1	0	$\frac{4}{9}$	$-\frac{1}{9}$	0	8
x_2	0	1	$-\frac{1}{3}$	$\frac{1}{3}$	0	24
$-P$	0	0	$-\frac{2}{9}$	$-\frac{13}{9}$	1	-184

Now consider another problem that our manufacturer might have to face. Suppose he must decide on the minimum price he would take in exchange for his bundle of resources, namely, 48 units of time on machine I and 120 units of time on machine II.

Before analyzing this "seller's problem," we must make the following assumptions:

(a) We are willing to determine the minimum selling price in a *mathematical and linear fashion.* To be more precise, let y_1 denote the selling price of 1 unit of time on machine I, and y_2 the price of 1 unit on machine II. The function to be minimized is then

$$W = 48y_1 + 120y_2$$

(b) *The seller is never willing to sell units of resource at a price less than what he can earn by combining these units of resources into products.*

Assumptions (a) and (b) imply that y_1 and y_2 must satisfy the constraints

$$y_1 \geq 0, \quad y_2 \geq 0$$

and

$$3y_1 + 3y_2 \geq 5 \tag{3}$$

$$y_1 + 4y_2 \geq 6 \tag{4}$$

It is not difficult to show that inequality (3) can be obtained by analyzing the initial production formula for x_1. We know that

$$x_1 = 3x_3 \oplus 3x_4$$

where x_3 measures units of time on machine I and x_4 measures units of time on machine II. Hence, by assumption (a),

$$\begin{aligned}\text{Selling price of } x_1 &= 3(\text{price } x_3) + 3(\text{price } x_4)\\ &= 3y_1 + 3y_2\end{aligned}$$

And, by assumption (b), the selling price of a unit of x_1 must be at least 5.

Likewise, inequality (4) comes from the exchange equation for x_2, namely, $x_2 = x_3 \oplus 4x_4$.

We have shown that the solution of the seller's problem can be found by solving the following linear program:

PROBLEM 11.2. Minimize $W = 48y_1 + 120y_2$

where y_1 and y_2 are nonnegative

and
$$3y_1 + 3y_2 \geq 5 \tag{3}$$
$$y_1 + 4y_2 \geq 6 \tag{4}$$

Problem 11.2 can be solved as if it were completely unrelated to Problem 11.1, but to do so would be tantamount to throwing away information already in our possession. We have already solved Problem 11.1 and learned that the "implicit prices" associated with constraints (1) and (2) are $\frac{2}{9}$ and $\frac{13}{9}$, respectively. By definition, these implicit prices measure the marginal contributions to profit, and hence, because of assumption (b), they should be good guesses for the optimal values of y_1 and y_2 in Problem 11.2. Furthermore, if $y_1 = \frac{2}{9}$, $y_2 = \frac{13}{9}$ does happen to be the optimal solution to Problem 11.2, then the minimum value of W will be 184. In other words, the minimum selling price for the bundle of resources in question will be equal to the maximum contribution to profit which these resources can generate.

It is a fact that $y_1 = \frac{2}{9}$, $y_2 = \frac{13}{9}$ is the optimal solution to Problem 11.2. To demonstrate this, note that we can relate W to Problem 11.1 and its constraints by multiplying inequality (1) by y_1, inequality (2) by y_2, and then adding the products obtained. Thus,

$$(3x_1 + x_2)(y_1) + (3x_1 + 4x_2)(y_2) \leq 48y_1 + 120y_2 = W \tag{5}$$

The left side of inequality (5) can be rewritten as

$$(3y_1 + 3y_2)(x_1) + (y_1 + 4y_2)(x_2)$$

Furthermore, it follows from inequalities (3) and (4) of Problem 11.2 that

$$5x_1 + 6x_2 \leq (3y_1 + 3y_2)(x_1) + (y_1 + 4y_2)(x_2)$$

Hence

$$P = 5x_1 + 6x_2 \leq 48y_1 + 120y_2 = W \tag{6}$$

Inequality (6) implies that the minimum selling price for the complete set of resources should be at least as large as the maximum profit obtainable by combining all the resources.

We know the maximum value of P is 184. Hence $184 \leq W$. But $y_1 = \frac{2}{9}$ and $y_2 = \frac{13}{9}$ is a feasible solution to Problem 11.2. Furthermore, as we know from the multiplier theorem, at the point in question,

$$W = 48y_1 + 120y_2 = (48)(\tfrac{2}{9}) + (120)(\tfrac{13}{9}) = 184$$

This verifies that 184 is the minimum of W.

Problems 11.1 and 11.2 are said to be a pair of *dual problems*. In general, every linear program formulated as a problem of maximization with nontrivial inequalities of the "less-than-or-equal-to" type has associated with it a companion program which can be formulated as a minimization problem with nontrivial constraints of the "greater-than-or-equal-to" type. Each of these problems is said to be dual to the other. We shall see later that whenever one of these problems has a feasible solution so does the other, and the optimum values of their objective functions are equal.

In Section 11.2 we pursue more rigorously the study of duality. Now, we show by means of an example how to "verbalize" the passage from a simple diet problem to its dual: A housewife must buy baby food to satisfy minimum daily requirements (MDR) of 9, 5, and 6 units of vitamins 1, 2, and 3, respectively. These vitamins are only available in foods F_1 and F_2. Food F_1 contains 1 unit of vitamin 1, 1 unit of vitamin 2, and 2 units of vitamin 3; food F_2 contains 1 unit of each of these vitamins. The prices of these foods are \$5 for 1 unit of F_1 and \$3 for 1 unit of F_2.

Let y_1 denote the number of units of F_1 to be purchased and y_2 the number of units of F_2. Then the housewife's problem reduces to the following linear program.

PROBLEM 11.3.

$$\begin{aligned} &\text{Minimize} \quad C = 5y_1 + 3y_2 \\ &\text{where} \quad y_1 \geq 0, \quad y_2 \geq 0 \\ &\text{and} \quad y_1 + y_2 \geq 9 \\ &\qquad\quad y_1 + y_2 \geq 5 \\ &\qquad\quad 2y_1 + y_2 \geq 6 \end{aligned}$$

Now consider a vitamin salesman. He sells pure vitamins and wishes to price them just low enough so that the housewife will be motivated to procure the MDR directly from him. He is aware of how the housewife can obtain the MDR by purchasing foods F_1 and F_2; that is, he knows all the relevant facts of Problem 11.3. Let x_1, x_2, and x_3 denote the prices that the salesman should assign to a unit of vitamins 1, 2, and 3, respectively. At these prices, the (minimum) amount the salesman will receive for selling the MDR is

$$P = 9x_1 + 5x_2 + 6x_3$$

The salesman's objective is to determine x_1, x_2, and x_3 so as to *maximize P*, subject to certain restrictions on the prices. To formulate those restrictions, we assume, as in Problem 11.3, that the housewife purchases y_1 units of F_1 and y_2 units of F_2. Let us focus our attention on F_1. The cost of y_1 units of F_1 is $5y_1$ dollars. For this price the housewife receives 1 unit of vitamin 1, 1 unit of vitamin 2, and 2 units of vitamin 3—a group of vitamins which the salesman will sell for

$$[y_1x_1 + y_1x_2 + y_1(2x_3)] \quad \text{dollars}$$

We assume that the salesman's prices are less than the cost of obtaining vitamins by purchasing food, that is,

$$y_1x_1 + y_1x_2 + y_1(2x_3) \leq 5y_1$$

or

$$x_1 + x_2 + 2x_3 \leq 5$$

This is the first nontrivial constraint in the pricing problem.

By focusing attention on food F_2, we obtain the second constraint:

$$x_1 + x_2 + x_3 \leq 3$$

To sum up, the salesman's pricing problem is:

PROBLEM 11.4. Maximize $P = 9x_1 + 5x_2 + 6x_3$

where $x_1 \geq 0, \quad x_2 \geq 0, \quad x_3 \geq 0$

and $x_1 + x_2 + 2x_3 \leq 5$

$x_1 + x_2 + x_3 \leq 3$

Exercises

1. Form the dual of:

Maximize $P = 2x_1 + 5x_2$

where $x_1 \geq 0, \quad x_2 \geq 0$

and $x_1 \leq 4$

$x_2 \leq 3$

$x_1 + x_2 \leq 8$

2. Form the dual of:

Minimize $C = 4w_1 + 3w_2 + 8w_3$

where $w_1 \geq 0, \quad w_2 \geq 0, \quad w_3 \geq 0$

and $w_1 + w_3 \geq 2$

$w_2 + 2w_3 \geq 5$

3. Check that Exercises 1 and 2 above are duals.

11.2 THE FUNDAMENTAL THEOREMS ON DUALITY

This section contains a systematic survey of the main theorems on duality. Most of these theorems are proven, but our primary concern is to clarify their meaning, by example and counterexample.

By definition, the following two problems are said to form *a pair of dual problems.*

PROBLEM MAX. Maximize $P = c_1x_1 + c_2x_2 + \cdots + c_nx_n$

where all variables are nonnegative

$$\begin{aligned} \text{and}\quad a_{11}x_1 + a_{12}x_2 + \cdots + a_{1n}x_n &\le b_1 \\ a_{21}x_1 + a_{22}x_2 + \cdots + a_{2n}x_n &\le b_2 \\ &\cdots \\ a_{m1}x_1 + a_{m2}x_2 + \cdots + a_{mn}x_n &\le b_m \end{aligned}$$

PROBLEM MIN. Minimize $C = b_1y_1 + b_2y_2 + \cdots + b_my_m$

where all variables are nonnegative

$$\begin{aligned} \text{and}\quad a_{11}y_1 + a_{21}y_2 + \cdots + a_{m1}y_m &\ge c_1 \\ a_{12}y_1 + a_{22}y_2 + \cdots + a_{m2}y_m &\ge c_2 \\ &\cdots \\ a_{1n}y_1 + a_{2n}y_2 + \cdots + a_{mn}y_m &\ge c_n \end{aligned}$$

The vector of coefficients in row k of Problem Max is the transpose of the vector of coefficients in column k of Problem Min. The nontrivial inequalities in the maximization problem are all of one type, namely, "less than or equal to." The nontrivial inequalities in the minimization problem are also all of one type, namely, "greater than or equal to." Moreover, the coefficients of the objective function in one problem are the limiting numbers in the other problem.

We point out that in the definition of a pair of dual problems no assumptions have been made as to the nature of the a_{ij}'s, the b_i's, or the c_j's. These numbers may be positive, negative, or zero. We reserve the term *primal* for the first of a pair of dual problems brought to our attention. No ambiguity should arise.

We must exercise care when formulating the dual to a given problem. To illustrate, consider forming the dual to the following program:

Example (*a*). Minimize $Z = 5x_1 + 6x_2$

where $x_1 \ge 0, \quad x_2 \ge 0$

$$\begin{aligned} \text{and}\quad 3x_1 + x_2 &\le 48 \\ 3x_1 + 4x_2 &\le 120 \end{aligned}$$

This program does not quite fit into the format for dualization, because it is a minimization problem with constraints of the "less than" type. However, if we write it as a maximization problem with objective function equal to $(-5)(x_1) + (-6)(x_2)$, then its dual is easily seen to be

Minimize $W = 48y_1 + 120y_2$

where $y_1 \ge 0, \quad y_2 \ge 0$

$$\begin{aligned} \text{and}\quad 3y_1 + 3y_2 &\ge -5 \\ y_1 + 4y_2 &\ge -6 \end{aligned}$$

Another way to dualize the primal is to change the direction of its nontrivial inequalities and write it as

$$\begin{aligned}
&\text{Minimize} \quad Z = 5x_1 + 6x_2 \\
&\text{where} \quad x_1 \geq 0, \quad x_2 \geq 0 \\
&\text{and} \quad -3x_1 - x_2 \geq -48 \\
&\qquad\quad -3x_1 - 4x_2 \geq -120
\end{aligned}$$

The dual of this problem can be written immediately as

$$\begin{aligned}
&\text{Maximize} \quad R = -48y_1 - 120y_2 \\
&\text{where} \quad y_1 \geq 0, \quad y_2 \geq 0 \\
&\text{and} \quad -3y_1 - 3y_2 \leq 5 \\
&\qquad\quad -y_2 - 4y_2 \leq 6
\end{aligned}$$

The two duals are equivalent; that is, the maximum of R occurs at the same point as the minimum of W. (Note that $R = -W$ and that both these objectives are evaluated over the *same* region of feasibility.)

This last example points up a possible discrepancy in our definition of a pair of dual problems. After all, whether a problem is one of maximization or minimization depends on how we choose to represent it. Suppose, for example, that we write the objective function of the primal as

$$\text{Minimize} \quad -P = -c_1x_1 - c_2x_2 - \cdots - c_nx_n$$

Then, in order to dualize, we must describe the region of feasibility by nontrivial inequalities of the "$\geq$" type. This region can be described as follows:

$$\begin{gathered}
-a_{11}x_1 - a_{12}x_2 - \cdots - a_{1n}x_n \geq -b_1 \\
\cdots \\
-a_{m1}x_1 - a_{m2}x_2 - \cdots - a_{mn}x_n \geq -b_m
\end{gathered}$$

The dual of this "minimization problem" is therefore

$$\begin{aligned}
&\text{Maximize} \quad (-b_1y_1 - b_2y_2 - \cdots - b_my_m) \\
&\text{where} \quad y_1 \geq 0, \ldots, y_m \geq 0 \\
&\text{and} \quad -a_{11}y_1 - a_{21}y_2 - \cdots - a_{m1}y_m \leq -c_1 \\
&\qquad\qquad\qquad \cdots \\
&\qquad\quad -a_{1n}y_1 - a_{2n}y_2 - \cdots - a_{mn}y_m \leq c_n
\end{aligned}$$

But this is Problem Min, merely expressed in maximization form. It should now be perfectly clear to the reader that there is no ambiguity in the definition of duality and that "the dual of the dual is the primal."

It is often useful to represent both the primal and dual simultaneously in one table as:

	x_1	x_2	$\cdots$	x_n	$\leq$
y_1	a_{11}	a_{12}	$\cdots$	a_{1n}	b_1
y_2	a_{21}	a_{22}	$\cdots$	a_{2n}	b_2
.					.
.	....	....	....	....	.
.					.
y_m	a_{m1}	a_{m2}	$\cdots$	a_{mn}	b_m
$\geq$	c_1	c_2	$\cdots$	c_n	

The maximization problem follows from reading this table across rows, the minimization problem from reading down columns. We must, of course, pay attention to the nature of the nontrivial inequalities. Note that there are no slack variables listed in the above scheme. In this chapter we consistently denote these slack variables as follows:

PROBLEM MAX.

Maximize $P = c_1x_1 + c_2x_2 + \cdots + c_nx_n + 0(x_{n+1}) + 0(x_{n+2}) + \cdots + 0(x_{n+m})$

where all variables are nonnegative

and
$$
\begin{aligned}
a_{11}x_1 + a_{12}x_2 + \cdots + a_{1n}x_n + x_{n+1} \qquad\qquad &= b_1 \\
a_{21}x_1 + a_{22}x_2 + \cdots + a_{2n}x_n \qquad + x_{n+2} \qquad &= b_2 \\
\cdots \qquad\qquad & \\
a_{m1}x_1 + a_{m2}x_2 + \cdots + a_{mn}x_n \qquad\qquad + x_{n+m} &= b_m
\end{aligned}
$$

PROBLEM MIN.

Minimize $C = b_1y_1 + b_2y_2 + \cdots + b_my_m + 0(y_{m+1}) + 0(y_{m+2}) + \cdots + 0(y_{m+n})$

where all variables are nonnegative

and
$$
\begin{aligned}
a_{11}y_1 + a_{21}y_2 + a_{31}y_3 + \cdots + a_{m1}y_m - y_{m+1} \qquad\qquad &= c_1 \\
a_{12}y_1 + a_{22}y_2 + a_{32}y_3 + \cdots + a_{m2}y_m \qquad - y_{m+2} \qquad &= c_2 \\
\cdots \qquad\qquad & \\
a_{1n}y_1 + a_{2n}y_2 + a_{3n}y_3 + \cdots + a_{mn}y_m \qquad\qquad - y_{m+n} &= c_n
\end{aligned}
$$

In formulating the fundamental theorems on duality it is helpful to match up primal and dual variables, according to the following *law of correspondence:* The *slack* variables of the primal correspond to the *structural* variables of the dual; in order, the slack variable of the first constraint of the primal corresponds to the first structural variable of the dual, the slack variable of the second constraint of the primal corresponds to the second structural variable of the dual, and so on.

The exactly analogous relationship holds between the slack variables of the dual and the structural variables of the primal. Thus x_{n+r} corresponds to y_r and y_{m+i} corresponds to x_i. The reason for this choice of correspondence will become clear as we examine the following major theorems of duality.

THEOREM 11.1. Let $x_1, x_2, \ldots, x_n$ be *any* feasible solution to the maximization problem, and let $y_1, y_2, \ldots, y_m$ be *any* feasible solution to the minimization problem. Then

(1) $c_1x_1 + c_2x_2 + \cdots + c_nx_n \leq b_1y_1 + b_2y_2 + \cdots + b_my_m$.

(2) $P^* \leq C^*$, where P^* denotes the optimum value of the objective function in the maximization problem and C^* denotes the optimum in the minimization problem.

Theorem 11.1 is true irrespective of the signs of the quantities a_{ij}, b_i, c_j. For the sake of notational simplicity, we sketch the proof of Theorem 11.1 in a low-dimensional case. However, the method used is general. Consider the following pair of dual problems:

Maximize $P = c_1x_1 + c_2x_2$	Minimize $C = b_1y_1 + b_2y_2 + b_3y_3$
where $x_1 \geq 0, \quad x_2 \geq 0$	where $y_1 \geq 0, \quad y_2 \geq 0, \quad y_3 \geq 0$
and $a_{11}x_1 + a_{12}x_2 \leq b_1$ (1)	and $a_{11}y_1 + a_{21}y_2 + a_{31}y_3 \geq c_1$ (I)
$a_{21}x_1 + a_{22}x_2 \leq b_2$ (2)	$a_{12}y_1 + a_{22}y_2 + a_{32}y_3 \geq c_2$ (II)
$a_{31}x_1 + a_{32}x_2 \leq b_3$ (3)	

Multiply inequalities (1), (2), and (3) by y_1, y_2, and y_3, respectively, and then add them together, obtaining

$$(a_{11}y_1 + a_{21}y_2 + a_{31}y_3)x_1 + (a_{12}y_1 + a_{22}y_2 + a_{32}y_3)x_2 \leq b_1y_1 + b_2y_2 + b_3y_3 \tag{4}$$

Now multiply inequalities (I) and (II) by x_1 and x_2, respectively, and add these products, obtaining

$$(a_{11}x_1 + a_{12}x_2)y_1 + (a_{21}x_1 + a_{22}x_2)y_2 + (a_{31}x_1 + a_{32}x_2)y_3 \geq c_1x_1 + c_2x_2 \tag{5}$$

If we expand the expressions on the left side of inequalities (4) and (5), we see that these expressions are identical. Hence,

$$c_1x_1 + c_2x_2 \leq a_{11}x_1y_1 + a_{12}x_2y_1 + a_{21}x_1y_2 + a_{22}x_2y_2 + a_{31}x_1y_3 + a_{32}x_2y_3 \leq b_1y_1 + b_2y_2 + b_3y_3$$

This inequality is called the *fundamental inequality of duality theory*. It holds for any two vectors **x** and **y**, provided only that **x** is a feasible solution to Problem Max and y is a feasible solution to Problem Min. Part (1) of Theorem 11.1 is an immediate consequence of this fundamental inequality. It is left as an exercise to prove that part (2) of Theorem 11.1 follows quite easily from part (1).

A mnemonic for holding the fundamental inequality in mind is the "inequality"

$$\text{Max} \leq \text{Mix} \leq \text{Min}$$

where Mix denotes the expression in both x's and y's which appears as the inner term of the inequality. (Observe that adjacent expressions in this mnemonic differ in only one letter of the alphabet.) We can also hold the fundamental inequality in mind through the tabular representation of a pair of dual problems. Consider the scheme

	x_1	$\cdots$	x_n	
y_1	a_{11}	$\cdots$	a_{1n}	b_1
$\cdot$	$\cdot$		$\cdot$	$\cdot$
$\cdot$	$\cdot$		$\cdot$	$\cdot$
$\cdot$	$\cdot$		$\cdot$	$\cdot$
y_m	a_{m1}	$\cdots$	a_{mn}	b_m
	c_1	$\cdots$	c_n	

The smallest expression in the fundamental inequality is the inner product of the top and bottom row vectors in this table. The largest expression is the inner product of the left and right column vectors. And the center, or mixed, expression in the fundamental inequality can be found by taking each a_{ij}, multiplying it by its row and column headings, and then summing over all i and j.

We can use Theorem 11.1 to put bounds on the optimal value of a linear program. Consider the following example:

Example (*b*).

$$\begin{aligned} &\text{Maximize} \quad P = 5x_1 + 6x_2 \\ &\text{where} \quad x_1 \geq 0, \quad x_2 \geq 0 \\ &\text{and} \quad 3x_1 + x_2 \leq 48 \\ &\qquad\;\; 3x_1 + 4x_2 \leq 120 \end{aligned}$$

The dual problem is

$$\begin{aligned} &\text{Minimize} \quad C = 48y_1 + 120y_2 \\ &\text{where} \quad y_1 \geq 0, \quad y_2 \geq 0 \\ &\text{and} \quad 3y_1 + 3y_2 \geq 5 \\ &\qquad\quad y_1 + 4y_2 \geq 6 \end{aligned}$$

Feasible solutions to the dual problem can readily be found, for example, $y_1 = 0$, $y_2 = \frac{5}{3}$. For this solution we get $C = 200$. Hence, prior to actually solving the maximization problem, we can affirm that the maximum value of P is at most 200.

Theorem 11.1 is only true if both primal and dual possess feasible solutions. To illustrate, consider

Example (*c*).

$$\text{Maximize} \quad P = x_1 + 2x_2$$
$$\text{where} \quad x_1 \geq 0, \quad x_2 \geq 0$$
$$\text{and} \quad x_1 + 0(x_2) \leq 10$$

The value of P can be made as large as desired because the region of feasibility is not bounded; x_2 can be made arbitrarily large. If Theorem 11.1 were to hold in this case, the minimum value of the dual objective function would have to be arbitrarily large at every possible point in the dual region of feasibility. This is impossible, and hence the region of feasibility of the dual must be empty. The reader can check this contention by direct examination of the dual, namely,

$$\text{Minimize} \quad 10y_1$$
$$\text{where} \quad y_1 \geq 0$$
$$\text{and} \quad 1(y_1) \geq 1$$
$$0(y_1) \geq 2$$

To construct an example of a maximization problem with an empty region of feasibility, all we need do is represent the above minimization problem as a maximization problem. For a more substantial example, consider the following:

Example (*d*).

$$\text{Maximize} \quad P = x_1 + 2x_2$$
$$\text{where} \quad x_1 \geq 0, \quad x_2 \geq 0$$
$$\text{and} \quad x_1 + 2x_2 \leq 10$$
$$-x_1 - x_2 \leq -30$$

The reader can check that these constraints are contradictory. The dual is

$$\text{Minimize} \quad C = 10y_1 - 30y_2$$
$$\text{where} \quad y_1 \geq 0, \quad y_2 \geq 0$$
$$\text{and} \quad y_1 - y_2 \geq 1$$
$$2y_1 - y_2 \geq 2$$

These constraints put no restriction on y_2, and hence C can be made arbitrarily small.

It is possible for both the primal and dual to have empty regions of feasibility. The experience of Examples (c) and (d) shows that to construct such a pair is equivalent to constructing an example where both primal and dual objective functions are unbounded. Accordingly, let us return to Example (d) and ask how we may make its objective function "look unbounded" and still keep its region of feasibility empty. One idea is to add an extra variable to the objective function but omit it from the contradictory constraints:

Example (e).

$$\text{Maximize} \quad P = x_1 + 2x_2 + x_3$$
$$\text{where} \quad x_1 \geq 0, \quad x_2 \geq 0, \quad x_3 \geq 0$$
$$\text{and} \quad x_1 + 2x_2 + 0(x_3) \leq 10$$
$$-x_1 - x_2 + 0(x_3) \leq -30$$

The dual problem is

$$\text{Minimize} \quad C = 10y_1 - 30y_2$$
$$\text{where} \quad y_1 \geq 0, \quad y_2 \geq 0$$
$$\text{and} \quad y_1 - y_2 \geq 1$$
$$2y_1 - y_2 \geq 2$$
$$0y_1 + 0y_2 \geq 1$$

Both regions of feasibility are now empty. To sum up, we state the following corollaries to Theorem 11.1.

COROLLARY 11.1. If the primal objective function is unbounded, then the region of feasibility of the dual is empty.

COROLLARY 11.2. If feasible solutions exist for both primal and dual, then each problem possesses a finite optimum.

Later we sketch a proof of Corollary 11.2 which depends upon the simplex algorithm. However, note that the value of C at any feasible point is an upper bound to P. In such a case, mathematical analysis shows that P^* must exist. By a similar argument, C^* is also finite.

THEOREM 11.2 (THE FUNDAMENTAL DUALITY THEOREM). Let $x_1, x_2, \ldots, x_n$ be a feasible solution for the maximum problem and $y_1, y_2, \ldots, y_m$ be a feasible solution for the dual minimum problem. Furthermore, suppose these solutions render their respective objective functions equal, that is,

$$P = c_1x_1 + c_2x_2 + \cdots + c_nx_n = b_1y_1 + b_2y_2 + \cdots + b_my_m = C$$

Then both these solutions are, in fact, optimal solutions of their respective problems.

Proof. To show that $x_1, x_2, \ldots, x_n$ is an optimal solution to the maximum problem, suppose that $x_1', x_2', \ldots, x_n'$ is any other solution. Then, by Theorem 11.1,

$$c_1x_1' + c_2x_2' + \cdots + c_nx_n' \leq b_1y_1 + b_2y_2 + \cdots + b_my_m$$

But, by hypothesis,

$$b_1y_1 + b_2y_2 + \cdots + b_my_m = c_1x_1 + c_2x_2 + \cdots + c_nx_n$$

and hence

$$c_1x_1' + c_2x_2' + \cdots + c_nx_n' \leq c_1x_1 + c_2x_2 + \cdots + c_nx_n$$

proving our contention. A similar proof shows that $y_1, \ldots, y_m$ is optimal.

The following theorem is another rather simple consequence of the fundamental inequality.

THEOREM 11.3A (THE COMPLEMENTARY-SLACKNESS THEOREM). If the feasible solutions $(x_1, \ldots, x_n)$ and $(y_1, \ldots, y_m)$ render their objective functions equal, then

(a) $x_{n+j} \neq 0$ implies $y_j = 0$
(b) $y_j \neq 0$ implies $x_{n+j} = 0$
(c) $x_i \neq 0$ implies $y_{m+i} = 0$
(d) $y_{m+i} \neq 0$ implies $x_i = 0$

Note that x_{n+j} represents the slack in the jth nontrivial constraint of Problem Max, that is,

$$x_{n+j} = b_j - (a_{j1}x_1 + a_{j2}x_2 + \cdots + a_{jn}x_n)$$

Hence the condition $x_{n+j} \neq 0$ is equivalent to

$$a_{j1}x_1 + a_{j2}x_2 + \cdots + a_{jn}x_n < b_j$$

Likewise, to affirm that $y_{m+i} \neq 0$ is equivalent to stating that

$$a_{1i}y_1 + a_{2i}y_2 + \cdots + a_{mi}y_m > c_i$$

The hypothesis of Theorem 11.3A forces all three expressions in the fundamental inequality to be equal. Thus,

$$\begin{aligned} c_1x_1 + \cdots + c_nx_n & \\ &= a_{11}x_1y_1 + \cdots + a_{1n}x_ny_1 + \cdots + a_{m1}x_1y_m + \cdots + a_{mn}x_ny_m \\ &= b_1y_1 + \cdots + b_my_m \end{aligned}$$

and

$$(c_1x_1 + \cdots + c_nx_n) - (a_{11}x_1y_1 + \cdots + a_{mn}x_ny_m) = 0$$

or, equivalently,

$$x_1(c_1 - a_{11}y_1 - \cdots - a_{m1}y_m) + \cdots + x_n(c_n - a_{1n}y_1 - \cdots - a_{mn}y_m) = 0$$

or finally,

$$x_1y_{m+1} + \cdots + x_ny_{m+n} = 0 \tag{1}$$

In this last equation every summand is the product of the two "corresponding dual variables," each of which is constrained to be nonnegative. The total sum can therefore be 0 only if at least one of the two factors in a summand is 0. This proves contentions (c) and (d); (a) and (b) follow by similar reasoning.

Complementary slackness does not imply that if one of a pair of corresponding variables is 0, the other must be nonzero. Both x_i and y_{m+i} can be 0, as the following example will show:

Example (f).

Primal. Maximize $P = x_1 + 2x_2$

where $x_1 \geq 0, \quad x_2 \geq 0$

and $x_1 + 2x_2 \leq 10$

One optimum solution is $x_1 = 10$, $x_2 = 0$, and $x_3 = 0$, where x_3 denotes slack in the above nontrivial constraint.

Dual. Minimize $10y_1$

where $y_1 \geq 0$

and $y_1 \geq 1$

$2y_1 \geq 2$

The optimum occurs when $y_1 = 1$, at which point both y_2 and y_3 are 0, where y_2 and y_3 are the surplus variables in the nontrivial constraints.

Corresponding variables and their values at optimality are

Primal	$x_1 = 10$	$x_2 = 0$	$x_3 = 0$
Dual	$y_2 = 0$	$y_3 = 0$	$y_1 = 1$

x_2 and y_3 correspond to each other and both are 0 at optimality.

To illustrate the theorems, consider the following examples.

Example (g).

Maximize $P = 5x_1 + 6x_2$

where $x_1 \geq 0, \quad x_2 \geq 0$

and
$$3x_1 + x_2 \leq 48 \quad (1)$$
$$3x_1 + 4x_2 \leq 120 \quad (2)$$
$$4x_1 + x_2 \leq 60 \quad (3)$$

Minimize $C = 48y_1 + 120y_2 + 60y_3$

where $y_1 \geq 0, \quad y_2 \geq 0, \quad y_3 \geq 0$

and
$$3y_1 + 3y_2 + 4y_3 \geq 5 \quad (\mathrm{I})$$
$$y_1 + 4y_2 + y_3 \geq 6 \quad (\mathrm{II})$$

Suppose we know that both problems have solutions and that the optimum to the primal is such that all inequalities, except the third, are satisfied as equalities.

Thus the primal optimum is at the intersection of the limiting lines

$$3x_1 + x_2 = 48$$
$$3x_1 + 4x_2 = 120$$

that is, $x_1 = 8$, $x_2 = 24$, and therefore $P = 184$. Moreover, since inequality (3) does have slack, the corresponding variable in the dual must equal zero at optimality, that is, $y_3 = 0$. In addition, inequalities (I) and (II) must be equalities, for otherwise x_1 and x_2 could not be nonzero. The optimum to the minimization problem is therefore found by solving

$$3y_1 + 3y_2 = 5$$
$$y_1 + 4y_2 = 6$$

giving $y_1 = \frac{2}{9}$, $y_2 = \frac{13}{9}$, and $C = 184$.

Sometimes the structure of a program is so special that, by using ad hoc procedures, we can find feasible solutions to both primal and dual for which the values of the objective functions are equal. Consider the following *scheduling problem:*

PROBLEM 11.3. The following table shows the minimum number of interns required at a hospital during the four periods into which the day is divided:

Period	*Time*	*Required number of interns*
(1)	Midnight to 6 AM	9
(2)	6 AM to Noon	12
(3)	Noon to 6 PM	10
(4)	6 PM to Midnight	6

Interns report for duty at midnight, 6 AM, noon, and 6 PM, and each works a 12-hour shift. The scheduling problem is to determine how many interns should report for duty at the start of each time period to minimize the total number of interns required for a day's operation.

Let x_i denote the number of interns who report for work at the start of period i, for $i = 1, 2, 3, 4$. The given problem is equivalent to the following linear program (provided the optimum is integral):

$$\text{Minimize} \quad (x_1 + x_2 + x_3 + x_4)$$

$$\text{subject to} \quad x_i \geq 0, \quad i = 1, 2, 3, 4$$

$$\text{and} \quad x_1 + x_2 \geq 12 \tag{1}$$

$$x_2 + x_3 \geq 10 \tag{2}$$

$$x_3 + x_4 \geq 6 \tag{3}$$

$$x_1 + x_4 \geq 9 \tag{4}$$

The dual program is:

$$\text{Maximize} \quad (12y_1 + 10y_2 + 6y_3 + 9y_4)$$

$$\text{subject to} \quad y_i \geq 0, \quad i = 1, 2, 3, 4$$

$$\text{and} \quad y_1 + y_4 \leq 1 \tag{1'}$$

$$y_1 + y_2 \leq 1 \tag{2'}$$

$$y_2 + y_3 \leq 1 \tag{3'}$$

$$y_3 + y_4 \leq 1 \tag{4'}$$

Note the special structure of this dual pair of programs. The coefficients in each constraint are 1 or 0, and each constraint contains exactly two variables with a nonzero coefficient. Obviously, no y_i can be larger than 1.

An intelligent way of solving the dual is to choose the greatest number of y_i's such that no two are in the same constraint and then set each such y_i equal to 1. There are only two possible solutions that can be obtained by this procedure, namely,

(a) $y_1 = 1, y_3 = 1; y_2 = y_4 = 0.$
(b) $y_2 = 1, y_4 = 1; y_1 = y_3 = 0.$

Solution (a) gives an objective value of 18; solution (b) gives a value of 19. To test solution (b) we seek a feasible solution to the primal for which the objective equals 19. Furthermore, by complementary slackness, if $y_2 = y_4 = 1$, then Eqs. (2) and (4) must be equalities. Accordingly, we seek a solution to the following system which is feasible for the minimization program and for which the objective function equals 19:

$$\begin{aligned} x_2 + x_3 \qquad &= 10 \\ x_1 \qquad\qquad + x_4 &= \ \ 9 \end{aligned}$$

Consider the solution to the above system of equations for which $x_1 = 9, x_2 = 3, x_3 = 7, x_4 = 0$. This is a primal solution for which the objective function $x_1 + x_2 + x_3 + x_4$ does take on the value of 19. Hence this solution is optimal. (Find alternate optima.)

The converse of Theorem 11.3A is true; that is,

THEOREM 11.3B. A sufficient condition for any pair of feasible solutions to the primal and dual to be optimal is the condition of complementary slackness, that is,

(1) $x_j y_{m+j} = 0$, for $j = 1, \ldots, n$, where y_{m+j} denotes the jth dual surplus variable.

(2) $y_i x_{n+i} = 0$, for $i = 1, \ldots, m$, where x_{n+i} is the ith slack variable in the primal.

This theorem can be established by showing that condition (1) above implies that Max = Mix and that condition (2) implies that Mix = Min. To illustrate, suppose that $m = 3, n = 4$ and that the pair of problems are as shown below:

	x_1	x_2	x_3	x_4	
y_1	a_{11}	a_{12}	a_{13}	a_{14}	b_1
y_2	a_{21}	a_{22}	a_{23}	a_{24}	b_2
y_3	a_{31}	a_{32}	a_{33}	a_{34}	b_3
	c_1	c_2	c_3	c_4	

Then

$$\begin{aligned} \text{Mix} - \text{Max} &= (a_{11}x_1y_1 + a_{21}x_1y_1 + a_{31}x_1y_3) + \cdots \\ &\quad + (a_{14}x_4y_1 + a_{24}x_4y_2 + a_{34}x_4y_3) - [c_1x_1 + \cdots + c_4x_4] \\ &= -x_1(c_1 - a_{11}y_1 - a_{21}y_2 - a_{31}y_3) - \cdots \\ &\quad - x_4(c_4 - a_{14}y_1 - a_{24}y_2 - a_{34}y_3) \\ &= x_1y_4 + x_2y_5 + x_3y_6 + x_4y_7 \end{aligned}$$

where y_4 denotes surplus in the first dual column, and so on. But, by hypothesis, $x_1y_4 = 0$, $x_2y_5 = 0$, $x_3y_6 = 0$, $x_4y_7 = 0$. Hence Max = Mix.

By a similar argument, Mix = Min. Thus Max = Min, and the solution under consideration must be optimal.

We remark in passing that complementary slackness plays an important role in many of the recent algorithms of linear programming.

THEOREM 11.4. Suppose that a finite optimum for a primal problem has been obtained by the simplex algorithm. Then the set of multipliers appearing in the objective row constitutes, except for a possible change in sign, an optimal solution to the dual problem.

Before examining the general case, consider the following simple pair of dual problems:

Example (*h*).

Maximize $P = 4x_1 + 5x_2$	Minimize $C = 16y_1 + 5y_2$
where $x_1 \geq 0,\quad x_2 \geq 0$	where $y_1 \geq 0,\quad y_2 \geq 0$
and $x_1 + 3x_2 \leq 16$	and $y_1 + 2y_2 \geq 4$
$2x_1 + 2x_2 \leq 5$	$3y_1 + 2y_2 \geq 5$

We solve the maximization problem below.

Tableau 1

	x_1	x_2	x_3	x_4	$-P$		
x_3	1	3	1	0	0	16	$(\frac{16}{3})$
x_4	2	②	0	1	0	5	$\leftarrow(\frac{5}{2})$
$-P$	4	5	0	0	1	0	
		↑					

Tableau 2

	x_1	x_2	x_3	x_4	$-P$	
x_3	-2	0	1	$-\frac{3}{2}$	0	$\frac{17}{2}$
x_2	1	1	0	$\frac{1}{2}$	0	$\frac{5}{2}$
$-P$	-1	0	0	$-\frac{5}{2}$	1	$-\frac{25}{2}$

Tableau 2 is terminal.

The absolute values of the objective-row multipliers appearing with x_3 and x_4 are the values of the y correspondents of x_3 and x_4 needed to give an optimum solution to the minimization problem. Thus x_3 corresponds to y_1 and x_4 to y_2, and we claim that $y_1 = 0$ and $y_2 = \frac{5}{2}$ is an optimum solution to the minimization problem. To verify, note that the point in question is feasible, because

$$y_1 + 2y_2 = 0 + (2)(\tfrac{5}{2}) = 5 \geq 4$$

and

$$3y_1 + 2y_2 = 0 + (2)(\tfrac{5}{2}) = 5 \geq 5$$

Furthermore, the value of C is

$$C = 16y_1 + 5y_2 = (16)(0) + (5)(\tfrac{5}{2}) = \tfrac{25}{2}$$

Since the value of C at this point equals the maximum of P, the point $y_1 = 0$, $y_2 = \frac{5}{2}$ gives us the minimum of C. This is a general result.

To see why Theorem 11.4 is true in all cases, consider a more general example:

$$\begin{aligned}
&\text{Maximize} \quad P = c_1x_1 + c_2x_2 + c_3x_3 \\
&\text{where} \quad x_1 \geq 0, \quad x_2 \geq 0, \quad x_3 \geq 0 \\
&\text{and} \quad a_{11}x_1 + a_{12}x_2 + a_{13}x_3 \leq b_1 \\
&\qquad\quad a_{21}x_1 + a_{22}x_2 + a_{23}x_3 \leq b_2 \\
&\qquad\quad a_{31}x_1 + a_{32}x_2 + a_{33}x_3 \leq b_3 \\
&\qquad\quad a_{41}x_1 + a_{42}x_2 + a_{43}x_3 \leq b_4
\end{aligned}$$

The initial tableau for this problem reads

	x_1	x_2	x_3	x_4	x_5	x_6	x_7	$-P$	
x_4	a_{11}	a_{12}	a_{13}	1	0	0	0	0	b_1
x_5	a_{21}	a_{22}	a_{23}	0	1	0	0	0	b_2
x_6	a_{31}	a_{32}	a_{33}	0	0	1	0	0	b_3
x_7	a_{41}	a_{42}	a_{43}	0	0	0	1	0	b_4
	c_1	c_2	c_3	0	0	0	0	1	0

Suppose the objective row of the terminal tableau reads

$$m_1,\ m_2,\ m_3,\ m_4,\ m_5,\ m_6,\ m_7,\ 1 \ \|\ -k$$

According to the conventions we have been following, all $m_i \leq 0$. Now the dual problem is

$$\begin{aligned}
&\text{Minimize} \quad C = b_1y_1 + b_2y_2 + b_3y_3 + b_4y_4 \\
&\text{where} \quad y_1 \geq 0, \quad y_2 \geq 0, \quad y_3 \geq 0, \quad y_4 \geq 0 \\
&\text{and} \quad a_{11}y_1 + a_{21}y_2 + a_{31}y_3 + a_{41}y_4 \geq c_1 \\
&\qquad\quad a_{12}y_1 + a_{22}y_2 + a_{32}y_3 + a_{42}y_4 \geq c_2 \\
&\qquad\quad a_{13}y_1 + a_{23}y_2 + a_{33}y_3 + a_{43}y_4 \geq c_3
\end{aligned}$$

The correspondence between x's, m's, and y's is:

m's	m_1	m_2	m_3	m_4	m_5	m_6	m_7
x's	x_1	x_2	x_3	x_4	x_5	x_6	x_7
y's	y_5	y_6	y_7	y_1	y_2	y_3	y_4

Note that corresponding x's and m's bear the same subscript because we link x_i and m_i only if m_i is the coefficient of x_i in the terminal objective function. The correspondence between x's and y's is the primal–dual correspondence. These two correspondences induce a natural correspondence between m's and y's, namely, m_i and y_j correspond to each other if they correspond to the same x_i. Thus m_4 corresponds to y_1, m_5 to y_2, and so on.

To prove that the set of simplex multipliers in the terminal primal tableau furnish the optimum of the dual problem, we first show that $y_1 = -m_4$, $y_2 = -m_5$, $y_3 = -m_6$, and $y_4 = -m_7$ is a feasible solution. Consider column 1 of the initial tableau of the primal. By the theory of multipliers we know that

$$(m_4, m_5, m_6, m_7, 1) \begin{pmatrix} a_{11} \\ a_{21} \\ a_{31} \\ a_{41} \\ c_1 \end{pmatrix} = m_1$$

That is,

$$a_{11}m_4 + a_{21}m_5 + a_{31}m_6 + a_{41}m_7 + c_1 = m_1$$

Hence

$$a_{11}(-m_4) + a_{21}(-m_5) + a_{31}(-m_6) + a_{41}(-m_7) - (-m_1) = c_1 \qquad (1)$$

or

$$a_{11}(-m_4) + a_{21}(-m_5) + a_{31}(-m_6) + a_{41}(-m_7) \geq c_1$$

This shows that $y_1 = -m_4, \ldots, y_4 = -m_7$ satisfies the first nontrivial dual constraint. By similar reasoning it is easy to see that this point satisfies each of the dual constraints.

Again, by the theory of multipliers,

$$(m_4, m_5, m_6, m_7, 1) \begin{pmatrix} b_1 \\ b_2 \\ b_3 \\ b_4 \\ 0 \end{pmatrix} = -k$$

This shows that the value of the dual objective function at the point under study is equal to the optimum value of the primal objective function, and hence the point in question provides an optimum solution of the dual problem.

It should be emphasized that m_1, m_2, and m_3 are not simplex multipliers, but they do represent the slack at optimality in the three nontrivial dual equations, as the reader can see by examining Eq. (1). In other words, the optimal values of y_5, y_6, and y_7 are, respectively, $-m_1$, $-m_2$, and $-m_3$. Furthermore, if x_1 is a basic variable in the optimal primal tableau, then $m_1 = 0$, and the first nontrivial dual constraint is satisfied exactly, that is, with no slack, at optimality in the dual problem.

Thus we have demonstrated the somewhat sharper statement than Theorem 11.4:

THEOREM 11.5. The multiplier (or implicit value) of the ith nonbasic variable in the optimum primal solution is equal to the numerical value of the corresponding dual basic variable, except for a possible change in sign. Similarly, the numerical value of the jth basic variable at the optimum primal solution equals the multiplier (or implicit value) of the corresponding dual nonbasic variable (except for a possible change in sign).

To illustrate, let us return to Example (h). The optimal tableau of the primal problem provides us with the following information:

Max problem	*Basic variables*		*Nonbasic variables*
$P^* = \frac{25}{2}$	$x_2 = \frac{5}{2}$	$x_3 = \frac{17}{2}$	x_1, with objective coefficient -1 x_4, with objective coefficient $-\frac{5}{2}$

By Theorem 11.5 we deduce that for the dual problem, at optimality the basic variables are y_2 and y_3, with $y_2 = \frac{5}{2}$ and $y_3 = 1$, and the nonbasic variables are y_1 and y_4, with objective coefficients of $\frac{17}{2}$ and $\frac{5}{2}$, respectively. In addition, $C^* = P^* = \frac{25}{2}$.

The following set of simple numerical examples is intended to clarify the meaning of Theorems 11.3, 11.4, and 11.5. We give the optimum solutions to both primal and dual and ask the reader to determine corresponding variables and verify Theorems 11.3, 11.4, and 11.5.

Example (i).

Primal.

Maximize $P = 5x_1 + 6x_2 + 7x_3$

subject to $x_i \geq 0$, for $i = 1, 2, 3$

and $3x_1 + x_2 + 2x_3 \leq 10$

The optimum solution is $x_2 = 10$, and all other $x_i = 0$. The slack variable x_4 associated with the nontrivial constraint above is 0, and $P^* = 60$.

Dual. Minimize $C = 10y_1$

subject to $y_1 \geq 0$

and $3y_1 \geq 5$

$y_1 \geq 6$

$2y_1 \geq 7$

The optimum occurs when $y_1 = 6$. If y_2, y_3, and y_4 denote slack in the nontrivial constraints, then at optimality, $y_2 = 13$, $y_3 = 0$, $y_4 = 5$. The minimum of C is 60.

Example (j).

Primal. Maximize $P = 5x_1 - 6x_2$
where $x_1 \geq 0, \quad x_2 \geq 0$
and $x_1 \leq 7$
$x_2 \leq 10$

The optimum solution is $x_1 = 7$, $x_2 = 0$, $x_3 = 0$, $x_4 = 10$, where x_3 and x_4 denote slack in the nontrivial constraints. $P^* = 35$.

Dual. Minimize $C = 7y_1 + 10y_2$
where $y_1 \geq 0, \quad y_2 \geq 0$
and $y_1 \geq 5$
$y_2 \geq -6$

The optimum occurs at $y_1 = 5$. $C^* = 35$ and $y_2 = 0$, $y_3 = 0$, $y_4 = 6$, where y_3 and y_4 denote slack in the nontrivial constraints.

Example (k).

Primal. Maximize $P = 5x_1 + 6x_2$
where $x_1 \geq 0, \quad x_2 \geq 0$
and $x_1 \leq 4$
$x_1 \leq 7$
$x_2 \leq 10$
$x_2 \leq 20$

The optimum value of P is 80; it occurs at the point $x_1 = 4$, $x_2 = 10$. The values of the slack variables are $x_3 = 0$, $x_4 = 3$, $x_5 = 0$, $x_6 = 10$.

Dual.
Minimize $C = 4y_1 + 7y_2 + 10y_3 + 20y_4$
where all $y_i \geq 0$
and $y_1 + y_2 \geq 5$
$y_3 + y_4 \geq 6$

The optimum occurs when $y_1 = 5$, $y_2 = 0$, $y_3 = 6$, $y_4 = 0$. $C^* = 80$ and the values of the slack variables are $y_5 = 0$ and $y_6 = 0$.

The fundamental theorems of duality establish a precise connection between the optimal multipliers of the primal and the optimal values of the dual variables. Furthermore, the optimal multipliers of the primal are equal to the shadow prices which measure the change in the optimum objective as a function of the individual b_i, that is, the maximum capacity of resource i. This means that it is also possible to analyze the effect on the optimum value of the objective of varying the b_i solely in terms of dual variables rather than primal multipliers. Indeed, many authors treat postoptimality analysis by means of duality theory (see Exercise 18 below). However, the reader should realize that theorems are like cards in a deck. They can be ordered in many ways. It is usually false to assert that Theorem A must necessarily precede Theorem B. *In fact, it is false to assert that duality theory is a necessary prerequisite to the study of shadow prices and postoptimality analysis.*

There is one other version of dual correspondence worth memorizing, the so-called *nonsymmetric case*. Suppose the primal reads

$$\begin{aligned}
&\text{Maximize} \quad c_1x_1 + c_2x_2 + \cdots + c_nx_n \\
&\text{where} \quad x_i \geq 0, \quad i = 1, 2, \ldots, n \\
&\text{and} \quad a_{11}x_1 + a_{12}x_2 + \cdots + a_{in} = b_1 && (1) \\
&\qquad \cdots \\
&\qquad a_{i1}x_1 + a_{i2}x_2 + \cdots + a_{in} = b_i && (i) \\
&\qquad \cdots \\
&\qquad a_{m1}x_1 + a_{m2}x_2 + \cdots + a_{mn} = b_m && (m)
\end{aligned}$$

In other words, suppose that the nontrivial primal constraints are all *equalities*. In this case the dual variables are *unrestricted in sign*. To prove this, convert the primal into standard form by replacing each equation (i) with the two inequalities

$$a_{i1}x_1 + a_{i2}x_2 + \cdots + a_{in} \leq \ \ b_i$$
$$-a_{i1}x_1 - a_{i2}x_2 - \cdots - a_{in} \leq -b_i$$

The primal–dual table will then appear as

	x_1	x_2	$\cdots$	x_n	
y_1	a_{11}	a_{12}	$\cdots$	a_{1n}	$\leq b_1$
$\vdots$	$\cdots$	$\cdots$	$\cdots$	$\cdots$	$\vdots$
y_m	a_{m1}	a_{m2}	$\cdots$	a_{mn}	$\leq b_m$
y_{m+1}	$-a_{11}$	$-a_{12}$	$\cdots$	$-a_{1n}$	$\leq -b_1$
$\vdots$	$\cdots$	$\cdots$	$\cdots$	$\cdots$	$\vdots$
y_{2m}	$-a_{m1}$	$-a_{m2}$	$\cdots$	$-a_{mn}$	$\leq -b_m$
	c_1	c_2	$\cdots$	c_n	

By the usual rules, the dual is

Minimize $b_1(y_1 - y_{m+1}) + b_2(y_2 - y_{m+2}) + \cdots + b_m(y_m - y_{2m})$

where all $y_i \geq 0$, $i = 1, \ldots, m, m+1, \ldots, 2m$

and $a_{11}(y_1 - y_{m+1}) + a_{21}(y_2 - y_{m+2}) + \cdots + a_{m1}(y_m - y_{2m}) \geq c_1$

$$\cdots$$

$$a_{1n}(y_1 - y_{m+1}) + a_{2n}(y_2 - y_{m+2}) + \cdots + a_{mn}(y_m - y_{2m}) \geq c_n$$

Since each y_i is nonnegative, if we let $w_i = y_i - y_{m+i}$, then clearly w_i is a free variable, thus completing the proof that we can consider the dual variables w_i to be unrestricted.

To illustrate, if the primal is

$$\text{Maximize} \quad (x_1 + x_2 + x_3)$$
$$\text{where} \quad x_i \geq 0, \ i = 1, 2, 3$$
$$\text{and} \quad x_1 + 2x_2 + 4x_3 = b_1$$
$$-x_2 + 3x_3 = b_2$$

then the dual program is

$$\text{Minimize} \quad (b_1y_1 + b_2y_2)$$
where y_1 and y_2 are both unrestricted
$$\text{and} \quad y_1 \geq 1$$
$$2y_1 - y_2 \geq 1$$
$$4y_1 + 3y_2 \geq 1$$

We stress that $y_1 = 1$, $y_2 = -1$ is dual feasible, and hence the primal objective function is bounded above by $b_1 - b_2$.

As a first exercise, the reader should show that in the nonsymmetric case "the dual of the dual is the primal." The reader should also demonstrate that whenever the primal maximization program involves nonnegative variables subject to both inequality and equality constraints, then the dual variables corresponding to the primal inequalities are constrained to be nonnegative, but the dual variables corresponding to primal equalities are unrestricted. For example, suppose the primal reads

$$\begin{aligned} &\text{Maximize} \quad (x_1 + x_2) \\ &\text{where} \quad x_1 \geq 0, \quad x_2 \geq 0 \\ &\text{and} \quad 3x_1 + x_2 \leq 48 \\ &\phantom{\text{and} \quad} 3x_1 + 5x_2 = 100 \end{aligned}$$

Then the dual must be

$$\begin{aligned} &\text{Minimize} \quad (48y_1 + 100y_2) \\ &\text{where} \quad y_1 \geq 0, \ y_2 \text{ free} \\ &\text{and} \quad 3y_1 + 3y_2 \geq 1 \\ &\phantom{\text{and} \quad 3} y_1 + 5y_2 \geq 1 \end{aligned}$$

As the reader should expect, no contradiction can arise if we dualize starting from any possible form of the primal. To illustrate, suppose the original Problem 11.1 is rewritten as follows:

$$\begin{aligned} &\text{Maximize} \quad P = 5x_1 + 6x_2 + 0x_3 + 0x_4 \\ &\text{where} \quad x_1 \geq 0, \quad x_2 \geq 0, \quad x_3 \geq 0, \quad x_4 \geq 0 \\ &\text{and} \quad 3x_1 + x_2 + x_3 = 48 \\ &\phantom{\text{and} \quad} 3x_1 + 4x_2 + x_4 = 120 \end{aligned}$$

Suppose we dualize by the nonsymmetric rules. Then the dual obtained must read

$$\begin{aligned} &\text{Minimize} \quad 48y_1 + 120y_2 \\ &\text{where both } y_1 \text{ and } y_2 \text{ are unrestricted} \\ &\text{and} \quad 3y_1 + 3y_2 \geq 5 && (1) \\ &\phantom{\text{and} \quad 3} y_1 + 4y_2 \geq 6 && (2) \\ &\phantom{\text{and} \quad 3} y_1 \geq 0 && (3) \\ &\phantom{\text{and} \quad 3y_1 +} y_2 \geq 0 && (4) \end{aligned}$$

Equations (3) and (4) above show that y_1 and y_2 are really not free, a result which would have been immediate if we had started from the inequality form of Problem 11.1 and had dualized by the symmetric rules.

We conclude with an example of how duality can be used to verify that a certain solution is optimal. The reader may recall, from Chapter 9, the problem of finding the line of best Tchebycheff fit to the points $P_1(1, 2)$, $P_2(3, 4)$, and $P_3(4, 7)$. If b_0 and b_1 are assumed to be free variables then the Tchebycheff line is $y_p = b_0 + b_1 x$, where $b_0 = -\frac{1}{3}$, and $b_1 = \frac{5}{3}$, and the maximum residual is $z = \frac{2}{3}$. To test the validity of this contention, consider the original primal row problem and its dual:

	b_0	b_1	z	
y_1	1	1	−1	2
y_2	−1	−1	−1	−2
y_3	1	3	−1	4
y_4	−1	−3	−1	−4
y_5	1	4	−1	7
y_6	−1	−4	−1	−7
	0	0	−1	P

If we substitute the optimal values of b_0, b_1, and z into the row problem, we obtain the following information about the slackness of its constraints:

$$s_1 \neq 0, \quad s_2 = 0, \quad s_3 = 0, \quad s_4 \neq 0, \quad s_5 \neq 0, \quad s_6 = 0$$

By the theorem on complementary slackness, at the optimum of the dual, the variables y_1, y_4, and y_5 must all be 0. Furthermore, since the primal variables b_0, b_1, and z are nonzero at optimality, all the dual constraints must be satisfied with no slack at optimality. Accordingly, we can determine the remaining optimum y_i by solving the system

$$\begin{aligned} -y_2 + y_3 - y_6 &= 0 \\ -y_2 + 3y_3 - 4y_6 &= 0 \\ -y_2 - y_3 - y_6 &= -1 \end{aligned}$$

This system has the unique solution $y_2 = \frac{1}{6}$, $y_3 = \frac{1}{2}$, and $y_6 = \frac{1}{3}$. The dual objective function takes on the value

$$-(2)(\tfrac{1}{6}) + (4)(\tfrac{1}{2}) - (7)(\tfrac{1}{3}) = -\tfrac{2}{3}$$

which agrees with the primal optimum.

Exercises

1. Form the duals of the following programs:

(a) Minimize $Z = u_1 - u_2$

where $u_1 \geq 0, \quad u_2 \geq 0$

and $3u_1 + u_2 \leq 10$

$2u_1 + u_2 \geq 5$

(b) Minimize $Q = 10v_1 - 5v_2$

where $v_1 \geq 0, \quad v_2 \geq 0$

and $3v_1 - 2v_2 \geq -1$

$v_1 - v_2 \geq 1$

2. Verify that (a) and (b) of Exercise 1 are dual to each other.

3. Form the duals of:

(a) Maximize $P = 5x_1$

where $x_1 \geq 0$

and $x_1 \leq 5$

$x_1 \leq 20$

(b) Minimize $I = 2y_1 + y_2$

where $y_1 \geq 0, \quad y_2 \geq 0$

and $3y_1 + 2y_2 = 12$

4. (a) Form the dual of:

$$\begin{aligned} &\text{Minimize} \quad C = 3x_1 + 2x_2 \\ &\text{where} \quad x_1 \geq 0, \quad x_2 \geq 0 \\ &\text{and} \quad x_1 + x_2 \leq 8 \\ &\qquad 6x_1 + 4x_2 \geq 12 \\ &\qquad 5x_1 + 8x_2 = 20 \end{aligned}$$

(b) Verify that the dual of the dual of (a) is the given program.

5. *True or false?*

(a) If a program is unbounded, its dual is infeasible.
(b) If a program is infeasible, its dual is unbounded.
(c) If one of a pair of dual problems is infeasible, the other must fail to have an optimal vector.
(d) Given the pair of corresponding variables, x_i and y_{n+i}; at optimality, if $x_i = 0$ then $y_{n+i} \neq 0$.

6. Solve the following problem and its dual, then verify that the solutions obtained satisfy the complementary-slackness condition:

$$\begin{aligned} &\text{Maximize} \quad P = x_1 + x_2 \\ &\text{where} \quad x_1 \geq 0, \quad x_2 \geq 0 \\ &\text{and} \quad x_1 + x_2 \leq 10 \\ &\qquad 2x_1 + 3x_2 \leq 40 \end{aligned}$$

7. Given the primal program:

$$\begin{aligned} &\text{Maximize} \quad x \\ &\text{where} \quad x \geq 0 \\ &\text{and} \quad 0(x) \leq 1 \end{aligned}$$

(a) Is this program feasible?
(b) Is this program bounded?
(c) What can we say about the dual?

8. Solve the following program and its dual, then verify that the solutions obtained satisfy the complementary slackness condition:

$$\begin{aligned} &\text{Maximize} \quad P = 7x \\ &\text{where} \quad x \geq 0 \\ &\text{and} \quad x \leq 5 \\ &\qquad x \geq 3 \\ &\qquad x \leq 7 \end{aligned}$$

9. Find the dual of each of the following programs:

(a)

$$\begin{aligned} &\text{Maximize} \quad 3x_1 + 2x_2 + 9x_3 \\ &\text{subject to } x_1 \geq 0, \quad x_2 \geq 0, \quad x_3 \geq 0 \\ &\text{and} \qquad x_2 + x_3 \leq 11 \\ &\qquad 2x_1 + 9x_2 + 7x_3 \leq 12 \end{aligned}$$

(b)

$$\begin{aligned} &\text{Maximize} \quad 3x_1 + 4x_3 \\ &\text{where} \quad x_1 \textit{ unrestricted}, x_2 \geq 0, x_3 \textit{ unrestricted} \\ &\text{and} \qquad x_2 + 6x_3 = 14 \\ &2x_1 + 11x_2 + 3x_3 \geq 10 \end{aligned}$$

10. *Give the dual of the following linear-programming problem in such a form that the dual variables are nonnegative and no strict equalities appear in the dual constraints:*

$$\begin{aligned} &\text{Maximize} \quad P = 10x_1 - 3x_2 + 4x_4 \\ &\text{where} \quad 13x_1 + 2x_2 + 4x_3 + x_4 \leq 27 \\ &\qquad\quad 12x_1 + 5x_2 - 6x_3 + 9x_4 = 6 \\ &\qquad\quad 11x_1 - x_2 + 2x_3 + 4x_4 \geq 8 \\ &\text{and} \quad x_1 \geq 0, \quad x_2 \geq 0; \quad x_3 \text{ and } x_4 \text{ are both free} \end{aligned}$$

11. Given the program:

$$\begin{aligned} &\text{Maximize} \quad P = x_1 + x_2 + x_3 \\ &\text{where} \quad x_i \geq 0, \quad i = 1, 2, 3 \\ &\text{and} \qquad x_1 + x_2 - x_3 \leq 4 \\ &\qquad\qquad x_2 \leq 3 \\ &\qquad\quad 3x_1 + x_2 \leq 5 \end{aligned}$$

(a) Show that P can be increased beyond any limit simply by increasing x_3.

(b) Before forming the dual to this program, what can we say about its region of feasibility.

(c) Form the dual.

(d) In general, how can we estimate a range for the optimum value of the objective before carrying out the simplex algorithm?

12. The following program is Problem (P) [(P) for primal]:

$$\begin{aligned} &\text{Minimize} \quad (c_1x_1 + c_2x_2 + c_3x_3 + c_4x_4) \\ &\text{where} \quad x_1 \geq 0, \quad x_2 \geq 0, \quad x_3 \text{ and } x_4 \text{ are both free} \\ &\text{and} \quad a_{11}x_1 + a_{12}x_2 + a_{13}x_3 + a_{14}x_4 \leq b_1 \\ &\qquad\quad a_{21}x_1 + a_{22}x_2 + a_{23}x_3 + a_{24}x_4 = b_2 \end{aligned}$$

(a) State a dual to Problem (P).

(b) Suppose that $(c_1x_1 + c_2x_2 + c_3x_3 + c_4x_4)$ can be made arbitrarily small for feasible (x_1, x_2, x_3, x_4). What can be said about the dual to Problem (P)?

(c) Suppose there is an extreme optimal solution to (P) in which the slack variable of the first inequality is positive. What can we say about any optimal solution to the dual?

13. Given the set of observations

x	0	1	2
y	0	0	1

It is claimed that the line of best fit, in the minimax sense, to the above observed points is $y_p = -\frac{1}{4} + \frac{1}{2}x$. Check this contention by considering the problem of determining the best predicting line as a linear program. Do not solve this program but employ duality theory to guide the check.

14. Given the observations

x	0	1	2	6	7
y	0	0	1	2	3

Check the contention that $y_p = -\frac{1}{10} + \frac{2}{5}x$ is the line of best fit, in the minimax sense, to these observations.

15. One unit of products X, Y, or Z sells for 6, 4, and 5, respectively. The following table shows how much of resources I and II are required for 1 unit of X, Y, or Z, and also how much of each resource is available:

	X	*Y*	*Z*	*Available*
I	4	2	4	15
II	5	1	3	18

Impute costs to the resources so that the total cost of all resources equals the total selling price for the finished products, and so that no product contains embodied resources whose total imputed cost is less than the selling price of that product. .

16. Clarify the following statement: The simplex algorithm can be viewed as a search for a set of factor prices, $p_j, j = 1, \ldots, m$, such that:

(a) The total change in value of the product is *exhausted*, that is,

$$\Delta P_0 = p_1 b_1 + \cdots + p_m b_m$$

(b) All activities used at a positive level, that is, basic variables, have a zero profit.

(c) All activities not in use, that is, nonbasic variables, are nonprofitable.

17. Express the following linear-programming results mathematically:

(a) If some other use can be found for the unused resources it will not require any decrease in the returns from the present use (as long as the excess remains).

(b) If an activity is used at optimality, then its value in terms of the implicit values of its resources equals the unit profit of that activity.

18. Show that at any iteration of the simplex algorithm, by substituting the current simplex multipliers for the corresponding variables in the original dual constraints and then taking the differences between the left- and right-hand sides of these dual constraints, we obtain, up to sign, the $(C^j - Z^j)$ of the primal.

19. Clarify the following statement: An important economic interpretation of the dual variables is in terms of the contribution of the scarce resources to the objective.

11.3 DUAL OPTIMAL TABLEAUX

In this section we show how to pass from the optimal tableau of a given problem to the optimal tableau of its dual.

Repeated below are the optimal tableaux of the dual problems, 11.1 and 11.2, rewritten in general notation.

Optimal Tableau (Problem 11.1)

	x_1	x_2	x_3	x_4	$-P$	
x_1	1	0	$\frac{4}{9}$	$-\frac{1}{9}$	0	8
x_2	0	1	$-\frac{1}{3}$	$\frac{1}{3}$	0	24
$-P$	0	0	$-\frac{2}{9}$	$-\frac{13}{9}$	1	-184

Optimal Tableau (Problem 11.2)

	y_1	y_2	y_3	y_4	$-C$	
y_1	1	0	$-\frac{4}{9}$	$\frac{1}{3}$	0	$\frac{2}{9}$
y_2	0	1	$\frac{1}{9}$	$-\frac{1}{3}$	0	$\frac{13}{9}$
	0	0	8	24	1	-184

Suppose we have solved only Problem 11.1 and wish to write the optimal tableau for Problem 11.2. How would we proceed? The methods of Section 11.2 are of some help. The exact correspondence among variables can be obtained from the initial tableau of Problem 11.1:

	x_1	x_2	x_3	x_4	
y_1	3	1	1	0	48
y_2	3	4	0	1	120
y_3	-1	0	0	0	0
y_4	0	-1	0	0	0
	5	6	0	0	0

The correspondence between variables is

$$x_1 \leftrightarrow y_3$$

$$x_2 \leftrightarrow y_4$$

$$x_3 \leftrightarrow y_1$$

$$x_4 \leftrightarrow y_2$$

Since x_3 and x_4 are nonbasic at optimality, the basic variables in the dual problem are y_1 and y_2. Furthermore, by Theorem 11.5, the optimal tableau of Problem 11.2 must appear as

	y_1	y_2	y_3	y_4	$-C$	
y_1	1	0	?	?	0	$\frac{2}{9}$
y_2	0	1	?	?	0	$\frac{13}{9}$
$-C$	0	0	8	24	1	-184

To fill in the question marks we quantify the optimal exchange dictionary of Problem 11.1. This optimal exchange dictionary reads

$$x_3 = \tfrac{4}{9}x_1 \oplus -\tfrac{1}{3}(x_2)$$
$$x_4 = -\tfrac{1}{9}x_1 \oplus \tfrac{1}{3}x_2$$

To quantify these equations we must pass to the set of arithmetical relations

$$v(x_3) = \tfrac{4}{9}v(x_1) + (-\tfrac{1}{3})v(x_2)$$
$$v(x_4) = -\tfrac{1}{9}v(x_1) + (\tfrac{1}{3})v(x_2)$$

where, by $v(x)$, we mean the *value of x*. If we define $v(x_i)$ to be the dual correspondent y_j of x_i, then we obtain the equations linking the y_i needed to complete the optimal tableau of Problem 11.2. Thus $v(x_1) = y_3$, $v(x_2) = y_4$, $v(x_3) = y_1$, $v(x_4) = y_2$, and

$$y_1 = \tfrac{4}{9}y_3 - \tfrac{1}{3}y_4$$
$$y_2 = -\tfrac{1}{9}y_3 + \tfrac{1}{3}y_4$$

To fill in values across row 1 of the optimal tableau, we must first transpose these equations to read

$$y_1 - \tfrac{4}{9}y_3 + \tfrac{1}{3}y_4 = \text{the current value of } y_1 = \tfrac{2}{9}$$
$$y_2 + \tfrac{1}{9}y_3 - \tfrac{1}{3}y_4 = \text{the current value of } y_2 = \tfrac{13}{9}$$

For another example, suppose we wish to

$$\begin{aligned} &\text{Minimize} \quad C = 3y_1 + 5y_2 \\ &\text{where} \quad y_1 \geq 0, \quad y_2 \geq 0 \\ &\text{and} \quad y_1 + 3y_2 \geq 9 \\ &\qquad\quad y_1 + y_2 \geq 5 \\ &\qquad\quad 2y_1 + y_2 \geq 6 \end{aligned}$$

The dual problem is to

$$\begin{aligned} &\text{Maximize} \quad P = 9x_1 + 5x_2 + 6x_3 \\ &\text{where} \quad x_1 \geq 0, \quad x_2 \geq 0, \quad x_3 \geq 0 \\ &\text{and} \quad x_1 + x_2 + 2x_3 \leq 3 \\ &\qquad\quad 3x_1 + x_2 + x_3 \leq 5 \end{aligned}$$

The sequence of tableaux needed to solve this problem is

Tableau 1

	x_1	x_2	x_3	x_4	x_5	$-P$		
x_4	1	1	2	1	0	0	3	(3)
x_5	③	1	1	0	1	0	5	←($\frac{5}{3}$)
$-P$	9 ↑	5	6	0	0	1	0	

Tableau 2

	x_1	x_2	x_3	x_4	x_5	$-P$		
x_4	0	$\frac{2}{3}$	($\frac{5}{3}$)	1	$-\frac{1}{3}$	0	$\frac{4}{3}$	←($\frac{4}{5}$)
x_1	1	$\frac{1}{3}$	$\frac{1}{3}$	0	$\frac{1}{3}$	0	$\frac{5}{3}$	(5)
$-P$	0	2	3 ↑	0	-3	1	-15	

Tableau 3

	x_1	x_2	x_3	x_4	x_5	$-P$		
x_3	0	($\frac{2}{5}$)	1	$\frac{3}{5}$	$-\frac{1}{5}$	0	$\frac{4}{5}$	←(2)
x_1	1	$\frac{1}{5}$	0	$-\frac{1}{5}$	$\frac{2}{5}$	0	$\frac{7}{5}$	(7)
$-P$	0	$\frac{4}{5}$ ↑	0	$-\frac{9}{5}$	$-\frac{12}{5}$	1	$-\frac{87}{5}$	

Tableau 4

	x_1	x_2	x_3	x_4	x_5	$-P$	
x_2	0	1	$\frac{5}{2}$	$\frac{3}{2}$	$-\frac{1}{2}$	0	2
x_1	1	0	$-\frac{1}{2}$	$-\frac{1}{2}$	$\frac{1}{2}$	0	1
$-P$	0	0	-2	-3	-2	1	-19

This tableau is optimal.

The correspondence between primal and dual variables is $x_1 \leftrightarrow y_3$, $x_2 \leftrightarrow y_4$, $x_3 \leftrightarrow y_5$, $x_4 \leftrightarrow y_1$, $x_5 \leftrightarrow y_2$. Hence, the rough form of the optimal tableau for the minimization problem is

	y_1	y_2	y_3	y_4	y_5	$-C$	
y_1	1	0	?	?	0	0	3
y_2	0	1	?	?	0	0	2
y_5	0	0	?	?	1	0	2
$-C$	0	0	1	2	0	1	-19

To complete this tableau, we turn to the optimal exchange dictionary of the maximization problem. Thus,

$$x_3 = \tfrac{5}{2}x_2 \oplus -\tfrac{1}{2}x_1$$
$$x_4 = \tfrac{3}{2}x_2 \oplus -\tfrac{1}{2}x_1$$
$$x_5 = -\tfrac{1}{2}x_2 \oplus \tfrac{1}{2}x_1$$

After quantification, we obtain the following equations between y_i's:

$$y_5 = \tfrac{5}{2}y_4 - \tfrac{1}{2}y_3$$
$$y_1 = \tfrac{3}{2}y_4 - \tfrac{1}{2}y_3$$
$$y_2 = -\tfrac{1}{2}y_4 + \tfrac{1}{2}y_3$$

The optimal tableau for the minimization problem must therefore read

	y_1	y_2	y_3	y_4	y_5	$-C$	
y_1	1	0	$\frac{1}{2}$	$-\frac{3}{2}$	0	0	3
y_2	0	1	$-\frac{1}{2}$	$\frac{1}{2}$	0	0	2
y_5	0	0	$\frac{1}{2}$	$-\frac{5}{2}$	1	0	2
$-C$	0	0	1	2	0	1	-19

Exercises

Obtain the optimal tableau for each of the following problems by working with its dual:

1.

$$\begin{aligned} &\text{Minimize} \quad C = y_1 + y_2 + y_3 \\ &\text{where} \quad y_1 \geq 0, \quad y_2 \geq 0, \quad y_3 \geq 0 \\ &\text{and} \quad 3y_1 + y_2 + 2y_3 \geq 18 \\ &\qquad 3y_1 + 4y_2 + 3y_3 \geq 21 \end{aligned}$$

2.

$$\begin{aligned} &\text{Minimize} \quad C = 2y_1 + 5y_2 \\ &\text{where} \quad y_1 \geq 0, \quad y_2 \geq 0 \\ &\text{and} \quad 3y_1 + 3y_2 \geq 5 \\ &\qquad y_1 + 4y_2 \geq 6 \end{aligned}$$

(See Problem 6.1.)

3.

$$\begin{aligned} &\text{Minimize} \quad C = 3y_1 + 5y_2 \\ &\text{where} \quad y_1 \geq 0, \quad y_2 \geq 0 \\ &\text{and} \quad y_1 + 3y_2 \geq 9 \\ &\qquad y_1 + y_2 \geq 5 \\ &\qquad 2y_1 + y_2 \geq 6 \end{aligned}$$

(See Problem 1.4.)

4.

$$\begin{aligned} &\text{Minimize} \quad I = y_1 + y_2 \\ &\text{where} \quad y_1 \geq 0, \quad y_2 \geq 0 \\ &\text{and} \quad 3y_1 + 10y_2 \geq 20 \\ &\qquad\quad 5y_1 + 2y_2 \geq 20 \\ &\qquad\quad 5y_1 + 6y_2 \geq 40 \end{aligned}$$

(See Problem 2.5.)

5. Formulate and solve the dual to Exercise 6, p. 204.

12

CONDENSED TABLEAUX AND DUALITY; INTRODUCTION TO GAME THEORY

In this chapter the simplex algorithm is streamlined by *condensing* the simplex tableau. We accomplish this by omitting all unit-vector columns—after all, it is sufficient to designate the current basic variables along the left rim of a tableau. This condensation will force us to reformulate the old pivot rules. By the use of condensed tableaux and new pivot rules both primal and dual can be solved simultaneously. We can then pass from the optimal tableau of one problem to the optimal tableau of its dual in a much simpler way than explained in Section 11.3, simply by "reading down columns."

12.1 INTRODUCTION TO CONDENSED TABLEAUX AND CONDENSED PIVOTING

Repeated below is the sequence of tableaux needed to solve Problem 11.1. To the left of the page is the usual extended tableau; to the right, the equivalent condensed tableau.

Tableau 1

	x_1	x_2	x_3	x_4	$-P$		
x_3	3	1	1	0	0	48	
x_4	3	④	0	1	0	120	←
$-P$	5	6	0	0	1	0	
		↑					

Tableau 1′

	x_1	x_2		
x_3	3	1	48	
x_4	3	④	120	←
	5	6	0	
		↑		

Tableau 2

	x_1	x_2	x_3	x_4	$-P$		
x_3	$\textcircled{\frac{9}{4}}$	0	1	$-\frac{1}{4}$	0	18	←
x_2	$\frac{3}{4}$	1	0	$\frac{1}{4}$	0	30	
$-P$	$\frac{1}{2}$ ↑	0	0	$-\frac{3}{2}$	1	-180	

Tableau 2′

	x_1	x_4		
x_3	$\textcircled{\frac{9}{4}}$	$-\frac{1}{4}$	18	←
x_2	$\frac{3}{4}$	$\frac{1}{4}$	30	
	$\frac{1}{2}$ ↑	$-\frac{3}{2}$	-180	

Tableau 3

	x_1	x_2	x_3	x_4	$-P$	
x_1	1	0	$\frac{4}{9}$	$-\frac{1}{9}$	0	8
x_2	0	1	$-\frac{1}{3}$	$\frac{1}{3}$	0	24
$-P$	0	0	$-\frac{2}{9}$	$-\frac{13}{9}$	1	-184

Tableau 3′

	x_3	x_4	
x_1	$\frac{4}{9}$	$-\frac{1}{9}$	8
x_2	$-\frac{1}{3}$	$\frac{1}{3}$	24
	$-\frac{2}{9}$	$-\frac{13}{9}$	-184

The top row of a condensed tableau carries the names of the current non-basic variables, and the left-most column contains the names of the current basic variables. A condensed tableau should be read like the equivalent extended tableau. Thus, Tableau 2′ is only an abbreviation for the system of equations

$$\begin{aligned} \tfrac{9}{4}x_1 - \tfrac{1}{4}x_4 + x_3 \qquad &= 18 \\ \tfrac{3}{4}x_1 + \tfrac{1}{4}x_4 \qquad + x_2 &= 30 \\ \tfrac{1}{2}x_1 - \tfrac{3}{2}x_4 \qquad &= P - 180 \end{aligned}$$

The following set of rules show how to pass from one condensed simplex tableau to the next:

(1) The logic for choosing the pivot is the same for condensed tableaux as for extended tableaux.

(2) In passing from the current condensed tableau to the updated condensed tableau we interchange the headings of the current pivot column and current pivot row. (The heading of the current pivot column represents the incoming basic variable; the heading of the current pivot row represents the outgoing basic variable.)

(3) The entries in the updated tableau can be deduced from the entries in the current tableau by observing the following rules:

(a) The pivot entry is changed into its reciprocal.

(b) All other entries in the pivot row, except for the pivot itself, are divided by the value of the pivot.

(c) All other entries in the pivot column, except for the pivot itself, are divided by the *negative* of the pivot.

(d) Any entry not lying in the pivot row or pivot column is transformed by the usual extended "rectangle rule."

(4) The test for optimality in a condensed tableau is the same as for an extended tableau.

Rules (2) and (3) are depicted schematically below, where we pass from Tableau n to Tableau (n + 1) by condensed pivoting.

Tableau n

	x_k	x_j	
x_r	ⓟ	q	b_1 ←
x_i	r	s	b_2
	c_1 ↑	c_2	k

Tableau (n + 1)

	x_r	x_j	
x_k	$1/p$	q/p	b_1/p
x_i	$-r/p$	$s - rq/p$	$b_2 - rb_1/p$
	$-c_1/p$	$c_2 - c_1q/p$	$k - c_1b_1/p$

It is easy to justify the rules for condensed pivoting. Simply write Condensed Tableau n in extended form, pivot on the extended tableau by the old rules, and then condense the result:

Tableau n (Extended Form)

	x_k	x_j	x_r	x_i		
x_r	ⓟ	q	1	0	b_1	←
x_i	r	s	0	1	b_2	
	c_1 ↑	c_2	0	0	k	

Tableau (n + 1) (Extended Form)

	x_k	x_j	x_r	x_i	
x_k	1	q/p	$1/p$	0	b_1/p
x_i	0	$s - rq/p$	$-r/p$	1	$b_2 - rb_1/p$
	0	$c_2 - c_1q/p$	$-c_1/p$	0	$k - c_1b_1/p$

Tableau (n + 1) (Condensed Form)

	x_j	x_r	
x_k	q/p	$1/p$	b_1/p
x_i	$s - rq/p$	$-r/p$	$b_2 - rb_1/p$
	$c_2 - c_1q/p$	$-c_1/p$	$k - c_1b_1/p$

To fix ideas we now give some numerical examples of condensed pivoting.

Example (a). Pivot as indicated:

	x_1	x_2		
x_3	①	2	3	←
x_4	3	7	14	
	9	4	0	
	↑			

Solution.

	x_3	x_2	
x_1	1	2	3
x_4	−3	1	5
	−9	−14	−27

Example (b). Pivot as indicated:

	x_1	x_2		
x_3	1	2	3	
x_4	③	−7	4	←
x_5	−2	5	9	
	9	4	0	
	↑			

Solution.

	x_4	x_2	
x_3	$-\frac{1}{3}$	$\frac{13}{3}$	$\frac{5}{3}$
x_1	$\frac{1}{3}$	$-\frac{7}{3}$	$\frac{4}{3}$
x_5	$\frac{2}{3}$	$\frac{1}{3}$	$\frac{35}{3}$
	−3	25	−12

Exercises

1. Solve the exercises proposed in Chapter 5 using condensed tableaux. Check your answers in as many ways as possible.

2. Use condensed pivoting to do Exercises 7 and 8, p. 261.

12.2 CONDENSED TABLEAUX AND DUALITY

It can be argued that the high point of elegance in linear programming is achieved by using condensed tableaux and treating simultaneously both primal and dual. To illustrate, consider the following problem.

PROBLEM 12.1. Maximize $P = 9x_1 + 5x_2 + 6x_3$

$$\text{where } x_1 \geq 0, \quad x_2 \geq 0, \quad x_3 \geq 0$$

$$\begin{aligned} \text{and} \quad x_1 + x_2 + 2x_3 &\leq 3 \\ 3x_1 + x_2 + x_3 &\leq 5 \end{aligned}$$

This author likes to place both primal and dual into the following condensed framework:

	x_1y_3	x_2y_4	x_3y_5	
x_4y_1	1	1	2	3
x_5y_2	3	1	1	5
	9	5	6	0

The headings of the columns and rows contain both primal and dual variables linked together according to the law of correspondence. *It is up to the user to interpret the tableau according to the purpose at hand.* Thus we can read off the primal by focusing on rows and equations in x to obtain, for example,

$$\begin{aligned} x_1 + x_2 + 2x_3 + x_4 \qquad &= 3 \\ 3x_1 + x_2 + x_3 \qquad + x_5 &= 5 \end{aligned}$$

$$\text{Maximize } 9x_1 + 5x_2 + 6x_3$$

To read the dual we must, of course, focus on columns, on equations in y, and on *surplus* variables, to obtain the constraints

$$\begin{aligned} y_1 + 3y_2 - y_3 \qquad\qquad &= 9 \\ y_1 + y_2 \qquad - y_4 \qquad &= 5 \\ 2y_1 + y_2 \qquad\qquad - y_5 &= 6 \end{aligned}$$

and the objective function: Minimize $3y_1 + 5y_2$.

If we maintain these conventions, each pivot operation performed can be interpreted as a transformation in either the primal or the dual. Below the given problem is solved by choosing the pivot according to the primal simplex rules.

Tableau 1

	x_1y_3	x_2y_4	x_3y_5		
x_4y_1	1	1	2	3	(3)
x_5y_2	③	1	1	5	$\leftarrow(\frac{5}{3})$
	9	5	6	0	
	↑				

Tableau 2

	x_5y_2	x_2y_4	x_3y_5		
x_4y_1	$-\frac{1}{3}$	$\frac{2}{3}$	$\textcircled{\frac{5}{3}}$	$\frac{4}{3}$	$\leftarrow(\frac{4}{5})$
x_1y_3	$\frac{1}{3}$	$\frac{1}{3}$	$\frac{1}{3}$	$\frac{5}{3}$	(5)
	-3	2	3 $\uparrow$	-15	

Tableau 3

	x_5y_2	x_2y_4	x_4y_1		
x_3y_5	$-\frac{1}{5}$	$\textcircled{\frac{2}{5}}$	$\frac{3}{5}$	$\frac{4}{5}$	$\leftarrow(2)$
x_1y_3	$\frac{2}{5}$	$\frac{1}{5}$	$-\frac{1}{5}$	$\frac{7}{5}$	(7)
	$-\frac{12}{5}$	$\frac{4}{5}$ $\uparrow$	$-\frac{9}{5}$	$-\frac{87}{5}$	

Tableau 4

	x_5y_2	x_3y_5	x_4y_1	
x_2y_4	$-\frac{1}{2}$	$\frac{5}{2}$	$\frac{3}{2}$	2
x_1y_3	$\frac{1}{2}$	$-\frac{1}{2}$	$-\frac{1}{2}$	1
	-2	-2	-3	-19

This tableau is terminal. We can read off the optimal primal dictionary by reading across rows, and the optimal dual dictionary by reading down columns. Thus the optimal dual dictionary is

$$\begin{aligned}
-\tfrac{1}{2}y_4 + \tfrac{1}{2}y_3 - y_2 \qquad\qquad &= -2\\
\tfrac{5}{2}y_4 - \tfrac{1}{2}y_3 \qquad - y_5 \qquad &= -2\\
\tfrac{3}{2}y_4 - \tfrac{1}{2}y_3 \qquad\qquad - y_1 &= -3\\
2y_4 + \ y_3 \qquad\qquad\qquad &= \ C - 19
\end{aligned}$$

A moment's reflection on the matter of appropriate algebraic signs shows that $y_4 = y_3 = 0$ defines the dual optimum corner.

Observe that we can also read off equations of exchange, between variables or vectors, at a glance from any of the above tableaux. For example, Tableau 4 informs us that

$$\begin{aligned}
\mathbf{x}_5 &= -\tfrac{1}{2}\mathbf{x}_2 + \tfrac{1}{2}\mathbf{x}_1\\
\mathbf{x}_3 &= \ \ \tfrac{5}{2}\mathbf{x}_2 - \tfrac{1}{2}\mathbf{x}_1\\
\mathbf{x}_4 &= \ \ \tfrac{3}{2}\mathbf{x}_2 - \tfrac{1}{2}\mathbf{x}_1
\end{aligned}$$

These equations describe relations among the contracted columns of Tableau 1:

$$\mathbf{x}_1 = \begin{pmatrix} 1 \\ 3 \end{pmatrix} \quad \mathbf{x}_2 = \begin{pmatrix} 1 \\ 1 \end{pmatrix} \quad \mathbf{x}_3 = \begin{pmatrix} 2 \\ 1 \end{pmatrix} \quad \mathbf{x}_4 = \begin{pmatrix} 1 \\ 0 \end{pmatrix} \quad \mathbf{x}_5 = \begin{pmatrix} 0 \\ 1 \end{pmatrix}$$

and the relations can be easily verified.

The reader should now formulate the correct vector equations between extended columns.

As for contracted rows, Tableau 4 can be read as follows:

$$\begin{aligned} -\mathbf{y}_4 &= -\tfrac{1}{2}\mathbf{y}_2 + \tfrac{5}{2}\mathbf{y}_5 + \tfrac{3}{2}\mathbf{y}_1 \\ -\mathbf{y}_3 &= \quad \tfrac{1}{2}\mathbf{y}_2 - \tfrac{1}{2}\mathbf{y}_5 - \tfrac{1}{2}\mathbf{y}_1 \end{aligned}$$

Here $\mathbf{y}_1$ represents row 1 of Tableau 1, that is, $\mathbf{y}_1 = (1, 1, 2)$. Similarly, $\mathbf{y}_2 = (3, 1, 1)$, $\mathbf{y}_3 = (-1, 0, 0)$, $\mathbf{y}_4 = (0, -1, 0)$, and $\mathbf{y}_5 = (0, 0, -1)$. Then

$$\begin{aligned} -\mathbf{y}_4 = (0, 1, 0) &= -\tfrac{1}{2}\mathbf{y}_2 + \tfrac{5}{2}\mathbf{y}_5 + \tfrac{3}{2}\mathbf{y}_1 \\ &= -\tfrac{1}{2}(3, 1, 1) + \tfrac{5}{2}(0, 0, -1) + \tfrac{3}{2}(1, 1, 2) \end{aligned}$$

as contended.

Again we leave it to the reader to formulate the correct exchange equations between extended row vectors.*

Exercises

1. Redo Exercises (1) to (4), pp. 296–297, using condensed tableaux.
2. Clarify the following statements:
   (a) As the primal strives for optimality, the dual strives for feasibility.
   (b) The simplex algorithm preserves feasibility in the primal and strives for feasibility in the dual.

12.3 TWO-PERSON, ZERO-SUM GAMES

No discussion of duality should omit reference to matrix games. A *matrix game* is a mathematical model of a conflict situation. Our interest here is not with how well such games serve as models of reality, but rather with matrix games as mathematical objects.

To fix ideas, consider the 3 × 3 "payoff matrix,"

$$M = \begin{pmatrix} -1 & 1 & 1 \\ 2 & -2 & 2 \\ 3 & 3 & -3 \end{pmatrix}$$

* For a slightly different schematization of the primal-dual tableau than the one shown above, see the work of A. Tucker and M. Balinski, as presented in A. Spivey and R. Thrall, *Linear Optimization* (Holt, Rinehart and Winston, New York, 1970), Chapter VIII; or in M. Hall, *Combinatorial Theory* (Blaisdell, New York, 1967), Chapter VIII.

Suppose we have a row player R and a column player K. We assume that each player is aware of the entries in the payoff matrix and that, independently of each other, the player R chooses a row and the player K a column. Then K must pay R the amount at the intersection of the chosen row and column. Thus, if R chooses row 2 and K column 3, then R receives the amount $+2$; if R chooses row 2 and K column 2, then R will receive -2—of course, to say that R receives -2 is the same as saying that K receives $+2$. But, by definition, all payments are equal to the entries in the payoff matrix and flow from the column player to the row player. The sum of the row player's net gain and the column player's net loss is 0, which is why matrix games are often called *two-person zero-sum games.*

We assume the game is played over and over again and that R and K behave *rationally*, in such a way that their play exhibits the following properties:

(1) Neither player will play the same row or column all the time, because the other player would then be able to make unnecessarily large gains.

(2) Each player must vary the sequence in which he plays the rows or columns, because if one player can predict the other's moves, he will always be able to determine an optimal counterplay.

By definition, a *strategy* for R is a decision to play the various rows in a thoroughly mixed or random fashion, but according to some fixed probability distribution (r_1, r_2, r_3). In other words, R will choose rows at random, choosing row i with probability, or relative frequency r_i, for $i = 1, 2, 3$, where $r_1 + r_2 + r_3 = 1$. Similarly, a strategy for K is a decision to play the various columns at random according to a probability distribution (k_1, k_2, k_3), where k_i denotes the probability with which K will play column i.

A strategy, one of whose components equals 1, is said to be a *pure strategy;* otherwise it is called a *mixed strategy*. Thus $(r_1 = 0, r_2 = 1, r_3 = 0)$ represents the pure strategy where R plays only row 2. The reader should convince himself that a weighted average of pure strategies is a mixed strategy.

Now let us return to the matrix M and record the minimum entry in each row and the maximum in each column:

	β_1	β_2	β_3	α_i
α_1	-1	1	1	-1
α_2	2	-2	2	-2
α_3	3	3	-3	-3
β_j	3	3	2	

The entries in the first three rows and first three columns of this table constitute the matrix M itself. The constants in the α_i column are the minimum entries in the rows. Similarly, the entries in the β_j row are the maximum elements in each column.

The table shows clearly that if R were to adopt the strategy ($r_1 = 1$, $r_2 = 0$, $r_3 = 0$) then, irrespective of K's strategy, the worst R could do would be to receive the amount $\alpha_1 = -1$, that is, pay out 1 unit each time the game is played. If his objective is to maximize his minimum return then the pure strategy "play row 1 always" is better than any other pure strategy. Similarly, if K's objective is to minimize his maximum payout, and if he can only choose among pure strategies, the best he can do is to choose to play column 3 with relative frequency 1.

The maximum of the α_i is called the *lower value* of the game and is denoted by α_*. The minimum of the β_j is called the *upper value* of the game and is denoted by β^*. Note that α_* and β^* are equal to entries of M and that

$$\alpha_* \leq \beta^*$$

By definition, the lower value of the game is the maximum return that R can obtain by adhering to a pure strategy, whereas the upper value is the minimum payoff that K must make if he adheres to a pure strategy. If $\alpha_* = \beta^*$, the entry of M which simultaneously equals α_* and β^* is called a *saddle point* of the game.

Now, suppose that R is permitted to adopt mixed strategies but that K is allowed to use only pure strategies. What mixed strategy (r_1, r_2, r_3) should R adopt so as to maximize his minimum expected return V? A moment's thought shows that R must solve the following program:

PROBLEM 12.2. Maximize V

subject to $r_1 \geq 0, \quad r_2 \geq 0, \quad r_3 \geq 0$

and

$$-r_1 + 2r_2 + 3r_3 \geq V \tag{1}$$

$$r_1 - 2r_2 + 3r_3 \geq V \tag{2}$$

$$r_1 + 2r_2 - 3r_3 \geq V \tag{3}$$

$$r_1 + r_2 + r_3 \quad = 1 \tag{4}$$

The left-hand side of each constraint (i) is, by definition, R's *expected return* against K's ith pure strategy, for $i = 1, 2, 3$. Clearly, V represents a lower bound to each of these expected returns. The constraints can be put into more established form simply by transposing the variable V to the left and rewriting constraints (1) to (3) as

$$-r_1 + 2r_2 + 3r_3 - V \geq 0 \tag{1'}$$

$$r_1 - 2r_2 + 3r_3 - V \geq 0 \tag{2'}$$

$$r_1 + 2r_2 - 3r_3 - V \geq 0 \tag{3'}$$

We point out that $r_1 = \frac{6}{11}$, $r_2 = \frac{3}{11}$, $r_3 = \frac{2}{11}$, $V = \frac{6}{11}$ is a feasible solution to R's problem. By adopting this mixed strategy R can obtain an expected return of at least $\frac{6}{11}$, which is clearly superior to $\alpha_* = -1$, the lower value of the game.

On the other hand, suppose that K is allowed to adopt any mixed strategy but that R can employ only pure strategies. If he wishes to minimize the maximum expected payout, K must solve the program:

PROBLEM 12.3. Minimize W

$$\text{subject to} \quad k_1 \geq 0, \quad k_2 \geq 0, \quad k_3 \geq 0$$

$$\text{and} \quad -k_1 + k_2 + k_3 \leq W \quad (1)$$

$$2k_1 - 2k_2 + 2k_3 \leq W \quad (2)$$

$$3k_1 + 3k_2 - 3k_3 \leq W \quad (3)$$

$$k_1 + k_2 + k_3 = 1 \quad (4)$$

In this problem the left side of each inequality (i) represents K's expected payout against R's ith pure strategy, for $i = 1, 2, 3$. The variable W is an upper bound to each of these expected payouts. And clearly, the established form of inequalities (1) to (3) can be obtained by transposing W. Thus constraint (1) is usually written

$$-k_1 + k_2 + k_3 - W \leq 0$$

Let us now form the dual of Problem 12.3. We treat the objective as the maximization of $-W$, as shown in the following scheme:

	k_1	k_2	k_3	W		
u_1	-1	1	1	-1	$\leq$	0
u_2	2	-2	2	-1	$\leq$	0
u_3	3	3	-3	-1	$\leq$	0
u_4	1	1	1	0	$=$	1
	0	0	0	-1		Max $(-W)$

By the usual rules, the dual can be enunciated as follows:

PROBLEM 12.4. Minimize u_4

$$\text{where} \quad u_1 \geq 0, \quad u_2 \geq 0, \quad u_3 \geq 0, \quad u_4 \text{ is free}$$

$$\text{and} \quad -u_1 + 2u_2 + 3u_3 + u_4 \geq 0$$

$$u_1 - 2u_2 + 3u_3 + u_4 \geq 0$$

$$u_1 + 2u_2 - 3u_3 + u_4 \geq 0$$

$$-u_1 - u_2 - u_3 = -1$$

Note that if we let $u_1 = r_1$, $u_2 = r_2$, $u_3 = r_3$, and $u_4 = -V$, it becomes obvious that Problem 12.4 is the same as Problem 12.2 Hence the row player's problem and the column player's problem are dual to each other. If the reader solves Problem 12.3 directly he will see that the minimum of W is $W^* = \frac{6}{11}$, when $k_1 = \frac{5}{22}$, $k_2 = \frac{8}{22}$, $k_3 = \frac{9}{22}$. By duality, the optimal value of V is $V^* = \frac{6}{11}$.

The common optimal value of each player's objective is called the *optimal value* of the game and is denoted by V^*. The optimal value of the game lies between the lower and upper values of the game, that is, $\alpha_* \leq V^* \leq \beta^*$.

It can happen that $\alpha_* = \beta^*$, and hence $V^* = \alpha_* = \beta^*$. (Such a game is said to possess a saddle point.) Games with a saddle point are *strictly determined* because in this case the optimal strategies are manifest—R should play the row determined by the lower value of the game and K the column determined by the upper value. The entry in the matrix M at the intersection of R's optimal row and K's optimal column is V^*. To illustrate, consider the game defined by

$$M = \begin{pmatrix} 6 & 6 & 5 \\ 2 & 7 & 3 \end{pmatrix}$$

Here $\alpha_1 = 5$, $\alpha_2 = 2$, $\alpha_* = \alpha_1 = 5$; $\beta_1 = 6$, $\beta_2 = 7$, $\beta_3 = 5$, $\beta^* = 5$. Thus R's optimal strategy is always to play row 1 and K's optimal strategy is always to play column 3.

One important point must still be clarified. In defining each player's problem, we sought the best *mixed* strategy against any of the adversary's *pure* strategies. The question arises as to whether this "best" mixed strategy is still optimal if we allow the adversary to employ mixed strategies. The answer is yes. To see why, consider the following typical game:

	k_1	k_2	k_3	k_4	
r_1	a_{11}	a_{12}	a_{13}	a_{14}	
r_2	a_{21}	a_{22}	a_{23}	a_{24}	
r_3	a_{31}	a_{32}	a_{33}	a_{34}	

If V is the value of this game, then, by definition,

$$r_1a_{11} + r_2a_{21} + r_3a_{31} \geq V \tag{1}$$

$$r_1a_{12} + r_2a_{22} + r_3a_{32} \geq V \tag{2}$$

$$r_1a_{13} + r_2a_{23} + r_3a_{33} \geq V \tag{3}$$

$$r_1a_{14} + r_2a_{24} + r_3a_{34} \geq V \tag{4}$$

Equations (1) to (4) determine the row player's optimal strategies against the column player's pure strategies. Now multiply Eq. (1) by k_1, Eq. (2) by k_2, Eq. (3) by k_3, Eq. (4) by k_4, and add these products to obtain

$$k_1[\text{Eq. (1)}] + k_2[\text{Eq. (2)}] + k_3[\text{Eq. (3)}] + k_4[\text{Eq. (4)}] \geq V \tag{5}$$

But the left side of Eq. (5) is R's expected payoff against K's mixed strategy (again, by definition).

Conversely, suppose that Eq. (5) is valid for all possible k_i, where $i = 1, 2, 3, 4$ such that $k_1 + k_2 + k_3 + k_4 = 1$. Now take $k_1 = 1$, $k_2 = 0$, $k_3 = 0$,

$k_4 = 0$. This implies the validity of Eq. (1) above. Similarly, if we take $k_1 = 0$, $k_2 = 1$, $k_3 = 0$, $k_4 = 0$, then Eq. (2) is established; and so on for k_3 and k_4, each equal to 1 in turn.

Exercises

1. Given the matrix

$$M = \begin{pmatrix} 1 & -1 \\ -1 & 1 \end{pmatrix}$$

(a) Solve the associated game.

(b) Solve the game determined by the matrix obtained by adding 2 to all the entries in M.

(c) Compare the optimal strategies in (a) and (b).

2. Given

$$M = \begin{pmatrix} 6 & -1 & 5 & -3 \\ 19 & 9 & -5 & 17 \end{pmatrix}$$

(a) Show that if R consistently plays row 1, then K can guarantee himself a return of 3.

(b) Solve completely the game determined by M.

3. Each of the players R and K simultaneously shows one or two fingers. If the sum of fingers shown is even, R wins from K an amount equal to the sum of the fingers shown; if the sum is odd, K wins this sum from R.

(a) Construct the payoff matrix for this game.

(b) Show that the game is favorable to K.

4. *True or false?*

(a) In a strictly determined game each player has an optimal strategy which is pure.

(b) If each entry in a matrix game is increased by the same constant, then the optimal strategies do not change.

(c) If each entry in a matrix game is multiplied by the same constant, the optimal strategies do not change.

5. Solve the following game: Each player is dealt at random one of the cards K, Q, or A. After the row player sees his card he announces his guess as to the card still in the deck. The column player not only sees his own card but also hears the row player's guess before he announces his guess as to the third card. If either player guesses correctly he wins $\$x$ from his opponent ($x \neq 0$).

6. Find the value of the following game:

$$\begin{matrix} 2 & 8 & 4 & 5 \\ 6 & 7 & 5 & 6 \\ 8 & 3 & 1 & 3 \end{matrix}$$

7. Prove that a game can never have more than one saddle point.

8. Scissors, stone, and paper: Two players simultaneously expose either (i) two open fingers (scissors), (ii) a clenched fist (stone), or (iii) an open hand (paper). If both show the same hand, there is no payoff; otherwise the winner is determined according to the rule *scissors cut paper, stone breaks scissors, and paper covers stone.* More precisely, the payoff matrix reads

		(K)		
		Scissors	*Stone*	*Paper*
	Scissors	0	1	−1
(R)	Stone	−1	0	1
	Paper	1	−1	0

(a) Solve this game by the simplex algorithm.

(b) Could the optimal strategies and value of the game have been predicted in advance?

9. Labor and management are involved in collective bargaining over alternatives A, B, and C. The following payoff matrix for a two-person zero-sum game has been established, where labor is viewed as the maximizing player:

		(*Management*)		
		A	B	C
	A	$\frac{1}{4}$	-1	1
(Labor)	B	-1	$\frac{1}{3}$	3
	C	1	1	1

(a) Find the value of the game and an optimal strategy for each player.

(b) From past records of similar bargaining situations management has found that typically labor has chosen alternatives A, B, and C at random with probabilities $\frac{1}{2}$, $\frac{1}{4}$, and $\frac{1}{4}$, respectively. Compute the value of the game under the assumption that history will probably repeat itself, and compute the optimal strategy for management.

13
THE TRANSPORTATION FRAMEWORK

In previous chapters we have shown how the so-called *transportation problem* can be reduced to a linear program. In this chapter the problem is solved by placing it in a framework adapted to the special features of transportation. The connections between the transportation framework and simplex theory are explored in Section 13.1. In Section 13.2 the stepping-stone algorithm is developed for perfectly balanced transportation problems. We show how to handle unbalanced problems in Section 13.3 and exhibit some of the versatility of the transportation model by applying it to several situations not involving the physical transportation of goods.

13.1 BASIC CONCEPTS

As usual, we introduce our subject by a set of examples.

PROBLEM 13.1. In Problem 2.4 a transportation problem was solved graphically. The objective was to transport units of a product from the sources S_i to the destinations D_j, so as to satisfy all demands at minimum cost. The cost of shipping from each source to each destination was given in a table:

	D_1	D_2	D_3	*Capacity*
S_1	\$4	\$2	\$1	15
S_2	\$1	\$3	\$5	25
Requirement	10	20	10	(40)

We rule out shipments from destinations to sources as well as transhipments between two sources or two destinations; in other words, the *minimum* cost of shipping a unit of product from S_i to D_j is shown in the cell in row i, column j of the table. It is assumed that the cost of shipping is proportional to the amount shipped. The capacities at the S_i and the requirements at the D_j are listed in the borders of this table. The sum of the capacities equals the sum of the demands. A transportation problem which satisfies this condition is said to be *balanced.*

We saw in Chapter 2 that this problem can be represented algebraically by means of two unknowns. For example, x can represent the amount shipped from S_1 to D_1, and y the amount from S_1 to D_2. Then all other amounts shipped can be described in terms of x and y, as shown below.

	D_1	D_2	D_3	
S_1	x	y	$15 - x - y$	15
S_2	$10 - x$	$20 - y$	$-5 + x + y$	25
	10	20	10	(40)

Since each amount shipped must be nonnegative, the solution is subject to the following six constraints, one for each cell in the above table:

$$x \geq 0, \quad y \geq 0, \quad x + y \leq 15, \quad x \leq 10, \quad y \leq 20, \quad x + y \geq 5$$

The cost of shipping is clearly

$$\begin{aligned} C &= 4x + 2y + 1(15 - x - y) + 1(10 - x) + 3(20 - y) + 5(-5 + x + y) \\ &= 7x + 3y + 60 \end{aligned}$$

The objective is to minimize C subject to the six constraints.

The region of feasibility is shown in Figure 19. The coordinates of each extreme point and the corresponding value of C were found there to be

Feasible corner	*Value of* $C = 7x + 3y + 60$
(1) $x = 5$, $y = 0$	95
(2) $x = 10$, $y = 0$	130
(3) $x = 10$, $y = 5$	145
(4) $x = 0$, $y = 15$	105
(5) $x = 0$, $y = 5$	75

Each corner point corresponds to a shipping plan. These plans are shown below. Plan n corresponds to corner n:

Plan 1

	D_1	D_2	D_3	
S_1	5	0	10	15
S_2	5	20	0	25
	10	20	10	(40)

Plan 2

	D_1	D_2	D_3	
S_1	10	0	5	15
S_2	0	20	5	25
	10	20	10	(40)

Plan 3

	D_1	D_2	D_3	
S_1	10	5	0	15
S_2	0	15	10	25
	10	20	10	(40)

Plan 4

	D_1	D_2	D_3	
S_1	0	15	0	15
S_2	10	5	10	25
	10	20	10	(40)

Plan 5

	D_1	D_2	D_3	
S_1	0	5	10	15
S_2	10	15	0	25
	10	20	10	(40)

One of the goals of this section is to explore the exact correspondence between shipping plans and extreme points—under what conditions can we examine a table and decide whether or not it represents a basic feasible solution (that is, an extreme point) of the corresponding linear program? To answer this question it will be helpful to employ a more systematic notation. Accordingly, let x_{ij} denote the quantity shipped from S_i to D_j and c_{ij} the unit cost of shipping from S_i to D_j. In Problem 13.1, $c_{11} = 4$, $c_{12} = 2$, $c_{13} = 1$, and so on. Using this notation, our problem is to

$$\text{Minimize} \quad C = 4x_{11} + 2x_{12} + x_{13} + x_{21} + 3x_{22} + 5x_{23}$$

where all variables are nonnegative

$$\begin{aligned}
\text{and} \quad x_{11} + x_{12} + x_{13} \qquad\qquad\qquad\qquad &= 15 \qquad (r_1)\\
x_{21} + x_{22} + x_{23} &= 25 \qquad (r_2)\\
x_{11} \qquad\qquad + x_{21} \qquad\qquad &= 10 \qquad (k_1)\\
x_{12} \qquad\qquad + x_{22} \qquad &= 20 \qquad (k_2)\\
x_{13} \qquad\qquad + x_{23} &= 10 \qquad (k_3)
\end{aligned}$$

The equations numbered (r_i) and (k_j) correspond to the constraints associated with row i and column j of the following table:

	D_1	D_2	D_3	
S_1	x_{11}	x_{12}	x_{13}	15
S_2	x_{21}	x_{22}	x_{23}	25
	10	20	10	(40)

Thus, Eq. (r_2) states that the total of all units flowing out of S_2 must equal 25, and Eq. (k_3) states that the total flowing into D_3 is 10. The nontrivial constraints form a system of five equations in six unknowns. At first glance it might seem possible to solve for any five variables in terms of a sixth variable as parameter, but closer examination shows this to be false, because the five equations in question are dependent. Since Problem 13.1 is balanced, it follows that

$$\text{Eq. } (r_1) + \text{Eq. } (r_2) = \text{Eq. } (k_1) + \text{Eq. } (k_2) + \text{Eq. } (k_3)$$

and hence each equation can always be expressed in terms of the other four equations. This means that any one of the five equations can be disregarded. The remaining four can be shown to be independent. Often, rather than drop an equation, we retain them all for reasons of symmetry, merely keeping in mind that we are effectively dealing with four equations in six unknowns. In other words, we can actually solve for any four variables in terms of the other two.

To illustrate, consider plan 1. Clearly, the corner variables for this plan should be x_{12} and x_{23}. To construct the (x_{12}, x_{23}) dictionary note the special structure of Eqs. (k_2) and (k_3). Simply by transposing variables we can transform these equations into the first two lines of the desired simplex dictionary, after which the rest of the dictionary can be obtained by substitution. The complete dictionary will then read

$$\begin{aligned}
x_{22} &= 20 - x_{12} \\
x_{13} &= 10 \qquad\quad - x_{23} \\
x_{11} &= 5 - x_{12} + x_{23} \\
x_{21} &= 5 + x_{12} - x_{23} \\
C &= 95 - 4x_{12} + 7x_{23}
\end{aligned}$$

When we set the corner variables x_{12} and x_{23} equal to zero the solution shown in plan 1 is obtained.

It is also possible to use exchange algebra to construct the simplex tableau associated with plan 1. First note that the occupied cells in plan 1 stipulate the current basic variables and their values, so that the corresponding tableau must appear as follows:

	x_{11}	x_{12}	x_{13}	x_{21}	x_{22}	x_{23}	
x_{11}	1	?	0	0	0	?	5
x_{13}	0	?	1	0	0	?	10
x_{21}	0	?	0	1	0	?	5
x_{22}	0	?	0	0	1	?	20
	0	?	0	0	0	?	$C - 95$

To fill in the x_{12} column, we must determine *how feasibility can be maintained if we increase* x_{12} *and change the current basic variables.* The simplest way to see what happens is to visually consider the consequences of entering cell (1, 2). Turn to plan 1 and note that, in occupying cell (1, 2), we create a unique closed loop among the occupied cells shown in the following diagram:

(1, 1) −	+ (1, 2)
(2, 1) +	− (2, 2)

If we occupy cell (1, 2), in order to maintain feasibility, we must decrease the amounts shipped into cells (1, 1) and (2, 2) and then compensate by increasing the amount shipped into cell (2, 1). This pattern of adjustment is indicated by the alternating plus and minus signs in the above diagram and is equivalent to the exchange equation

$$x_{12} = x_{11} \ominus x_{21} \oplus x_{22}$$

Hence the x_{12} column of our simplex tableau must appear as

	C^j	
C_i	*Basic solution*	x_{12}
4	x_{11}	. . . 1 . . .
1	x_{13}	. . . 0 . . .
1	x_{21}	. . . −1 . . .
3	x_{22}	. . . 1 . . .

To find the objective coefficient of x_{12}, calculate the direct and indirect effects on cost of increasing x_{12} by 1 unit:

Direct increase in cost		= 2
Indirect increase in cost	= (4 − 1 + 3)	= 6
Net increase in cost	= 2 − 6	= −4

By similar reasoning, the unique exchange loop of occupied cells associated with entering cell (2, 3) is

(1, 1) +	− (1, 3)
(2, 1) −	+ (2, 3)

Thus, $x_{23} = x_{21} \ominus x_{11} \oplus x_{13}$, and the net effect on cost of entering cell (2, 3) is the difference between the direct effect 5 and the indirect effect −2, that is, $5 - (-2) = 7$.

The complete tableau now reads

	x_{11}	x_{12}	x_{13}	x_{21}	x_{22}	x_{23}		
x_{11}	1	①	0	0	0	−1	5	←(5)
x_{13}	0	0	1	0	0	1	10	
x_{21}	0	−1	0	1	0	1	5	
x_{22}	0	1	0	0	1	0	20	(20)
	0	−4 ↑	0	0	0	7	$C - 95$	

Cost can be further decreased by increasing x_{12}. The outgoing variable is x_{11}, and the maximum increase in x_{12} is 5. Note that this information can also be read directly from the exchange loop for occupying cell (1, 2). If we pivot as indicated, the new tableau reads

	x_{11}	x_{12}	x_{13}	x_{21}	x_{22}	x_{23}	
x_{12}	1	1	0	0	0	−1	5
x_{13}	0	0	1	0	0	1	10
x_{21}	1	0	0	1	0	0	10
x_{22}	−1	0	0	0	1	1	15
	4	0	0	0	0	3	$C - 75$

This tableau is terminal and clearly corresponds to plan 5.

The nonbasic columns of the terminal tableau specify that

$$x_{11} = x_{12} \ominus x_{22} \oplus x_{21}$$

$$x_{23} = x_{13} \ominus x_{12} \oplus x_{22}$$

and, as the reader can easily verify, these exchange equations describe exactly the unique exchange loops needed in plan 5 to adjust to the effect of entering cells (1, 1) and (2, 3), respectively.

Observe that we were able to determine visually the consequences of entering an empty cell in each of the above plans because we could associate with each empty cell a unique closed path which moved only vertically and horizontally via occupied cells back to the initial empty cell under consideration. This closed path is called the *path of adjustments*, the *exchange loop*, or the *stepping-stone loop*. This last name is suggestive of the way one traverses a stream. A stepping-stone loop contains the *empty cell to be occupied*, and all the other corners correspond to *occupied cells*. These occupied cells are like "stones," the unoccupied cells like "water squares." The analogy should now be clear—when crossing a stream we should place our feet on the stepping-stones and avoid the water squares. With each succeeding example the power of this analogy will be more evident—to illustrate, in specifying a stepping-stone loop we may skip over any cells, whether or not they are occupied.

The following table describes the *balanced transportation problem* in general terms:

	D_1	$\cdots$	D_j	$\cdots$	D_n	
S_1	c_{11} x_{11}	$\cdots$	c_{1j} x_{1j}	$\cdots$	c_{1n} x_{1n}	a_1
$\vdots$	$\vdots$		$\vdots$		$\vdots$	$\vdots$
S_i	c_{i1} x_{i1}	$\cdots$	c_{ij} x_{ij}	$\cdots$	c_{ni} x_{in}	a_i
$\vdots$	$\vdots$		$\vdots$		$\vdots$	$\vdots$
S_m	c_{m1} x_{m1}	$\cdots$	c_{mj} x_{mj}	$\cdots$	c_{mn} x_{mn}	a_m
	b_1	$\cdots$	b_j	$\cdots$	b_n	(Σ)

There are m sources S_i, each with capacity a_i, where $i = 1, \ldots, m$, and n destinations D_j, each with demand b_j, where $j = 1, \ldots, n$. Σ represents both total capacity and total demand. Each cell (i, j) above contains x_{ij}, the amount shipped from S_i to D_j. The per unit cost of shipping from S_i to D_j is c_{ij}, shown in the upper left corner of cell (i, j). We make no assumptions about the c_{ij}—a negative c_{ij} can be thought of as signifying that when a unit is shipped from S_i to D_j a *subsidy* is received. Unless the opposite is specifically indicated, we assumed that the a_i and b_j are positive integers greater than or equal to 1.

The objective is to minimize

$$C = c_{11}x_{11} + c_{12}x_{12} + \cdots + c_{ij}x_{ij} + \cdots + c_{mn}x_{mn}$$

subject to the trivial constraints $x_{ij} \geq 0$ and subject to $m + n$ nontrivial constraints, each of which is an equation corresponding to a row or column of the given table. The ith row equation reads

$$x_{i1} + x_{i2} + \cdots + x_{ij} + \cdots + x_{in} = a_i \qquad (r_i)$$

and the jth column equation reads

$$x_{1j} + x_{2j} + \cdots + x_{ij} + \cdots + x_{mj} = b_j \qquad (k_j)$$

These equations are not independent, because

$$\text{Eq. } (r_1) + \text{Eq. } (r_2) + \cdots + \text{Eq. } (r_m) = \text{Eq. } (k_1) + \text{Eq. } (k_2) + \cdots + \text{Eq. } (k_n)$$

Clearly we can solve for any of these equations in terms of the others.

Consider the dual to the balanced transportation problem as defined. Let $y_1, y_2, \ldots, y_m; y_{m+1}, y_{m+2}, \ldots, y_{m+n}$ denote the dual variables, the first m corresponding to the row constraints of the primal, in order, and the last n corresponding to the column constraints. Since the row and column constraints are equalities, the dual variables $y_1, \ldots, y_{m+n}$ are unrestricted as to sign. It is easy to show that the dual reads

$$\text{Maximize} \quad P = (a_1 y_1 + \cdots + a_m y_m) + (b_1 y_{m+1} + \cdots + b_n y_{m+n})$$

where the y_i are free variables

$$\begin{array}{llllll} \text{and} & y_1 & + y_{m+1} & & & \leq c_{11} \\ & \cdots & \cdots & & & \cdots \\ & y_i & & + y_{m+j} & & \leq c_{ij} \\ & \cdots & & \cdots & & \cdots \\ & y_m & & & + y_{m+n} & \leq c_{mn} \end{array}$$

Note that each dual inequality contains only two variables with nonzero coefficients, and that there is one such inequality for each value of i and j.

Let us now return to the primal, that is, the transportation problem. Soon we shall demonstrate the following fundamental result: A shipping plan which corresponds to an extreme point of the set of all possible solutions must contain at least $(m - 1)(n - 1)$ empty cells, or, equivalently, at most $(m + n - 1)$ nonempty cells.

As an illustration, consider

PROBLEM 13.2. Find the optimum shipping plan for the relevant data

	D_1	D_2	D_3	*Capacities*
S_1	\$3	\$17	\$11	4
S_2	\$12	\$18	\$10	12
S_3	\$10	\$25	\$24	8
Requirements	10	9	5	(24)

Let us seek an extreme solution by trying to construct a plan with at most $m + n - 1 = 5$ occupied cells. For example, consider the plan

Table 1

	D_1	D_2	D_3	
S_1	3 / 4	17	11	4
S_2	12	18 / 7	10 / 5	12
S_3	10 / 6	25 / 2	24	8
	10	9	5	(24)

This plan has five occupied cells and is clearly feasible. The variables corresponding to empty cells are x_{12}, x_{13}, x_{21}, and x_{33}, and the solution under consideration will certainly be a corner point if the system which remains when we set these variables equal to 0 possesses only one solution. Thus, consider the defining system

$$\begin{array}{llllllllllll} x_{11} + x_{12} + x_{13} & & & & & & & & & = 4 & & (r_1) \\ & & & x_{21} + x_{22} + x_{23} & & & & & & = 12 & & (r_2) \\ & & & & & & x_{31} + x_{32} + x_{33} & & & = 8 & & (r_3) \\ x_{11} & & & + x_{21} & & & + x_{31} & & & = 10 & & (k_1) \\ & x_{12} & & & + x_{22} & & & + x_{32} & & = 9 & & (k_2) \\ & & x_{13} & & & + x_{23} & & & + x_{33} & = 5 & & (k_3) \end{array}$$

Now set $x_{12} = x_{13} = x_{21} = x_{33} = 0$, and then solve for the remaining variables. Clearly, we obtain the unique feasible solution shown in Table 1. Since it is unique it must correspond to an extreme point. (The reader should also verify that the cost of shipping for Table 1 is \$298.)

Now, to test whether Table 1 can be improved, we must first determine the consequences of entering its four empty cells, namely, cells (1, 2), (1, 3), (2, 1), and (3, 3). Sketched below are the unique exchange loops that determine the adjustments attendant upon occupying each of these empty cells.

(1) *Occupying cell (1, 2).*

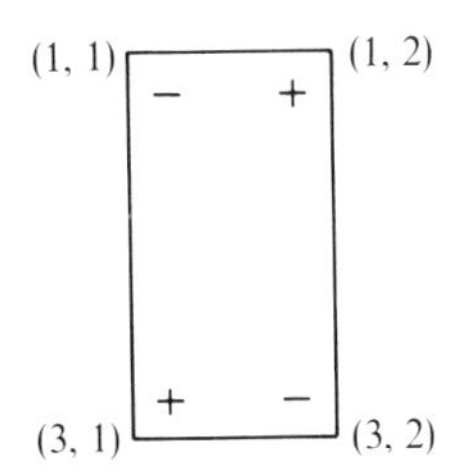

The corners of this loop designate those cells whose allocations must be readjusted when we occupy cell (1, 2). If a corner (i, j) carries a plus sign, we increase x_{ij}, but if it carries a minus sign, we decrease x_{ij}. Furthermore, the loop shows that the effect on cost of increasing x_{12} by 1 unit is to increase cost by

$$c_{12} - c_{32} + c_{31} - c_{11} = 17 - 25 + 10 - 3 = -1$$

This simple calculation is the exact equivalent of calculating the cost of entering cell (i, j) in terms of direct and indirect effect—because the cost relation

$$c_{12} - (c_{32} - c_{31} + c_{11})$$

mimics the direct-minus-indirect effect corresponding to the exchange relation

$$x_{12} \ominus (x_{32} \ominus x_{31} \oplus x_{11})$$

(2) *Occupying cell (1, 3).*

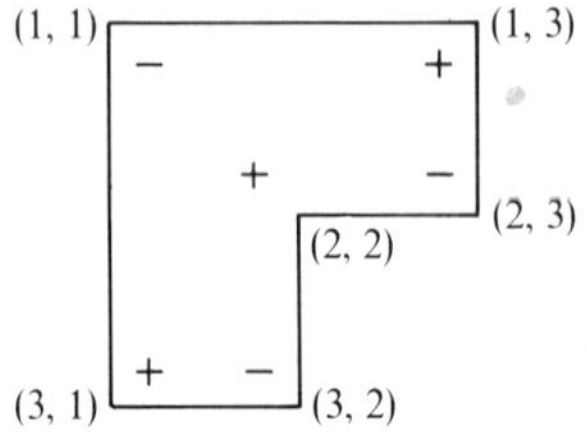

(1, 1) − (1, 3) +
+ (2, 2) − (2, 3)
(3, 1) + − (3, 2)

This loop shows that if we increase x_{13} by 1 unit, C will increase by

$$c_{13} - c_{23} + c_{22} - c_{32} + c_{31} - c_{11} = 11 - 10 + 18 - 25 + 10 - 3 = +1$$

(3) *Occupying cell (2, 1).*

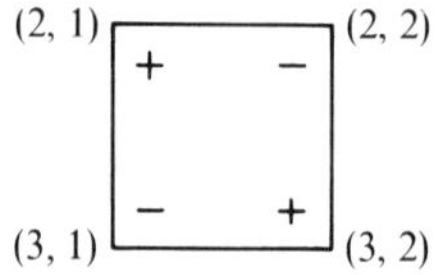

(2, 1) + − (2, 2)
(3, 1) − + (3, 2)

The increase in cost is clearly

$$c_{21} - c_{22} + c_{32} - c_{31} = 12 - 18 + 25 - 10 = +9$$

(4) *Occupying cell (3, 3).*

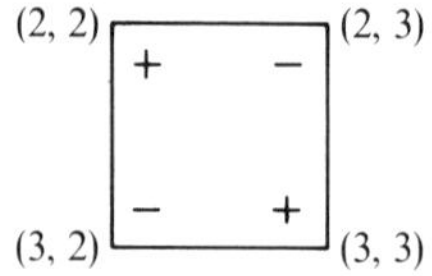

(2, 2) + − (2, 3)
(3, 2) − + (3, 3)

If we increase x_{33} by 1, then C will be increased by

$$c_{33} - c_{32} + c_{22} - c_{23} = 24 - 25 + 18 - 10 = +7$$

The examination of all empty cells is now complete and shows that we should increase x_{12} to the maximum possible extent. Moreover, examination of the appropriate loop indicates that the amount contained in cell (3, 2) puts a limit of 2 units on the allowable increase in x_{12}. The updated shipping plan is

Table 2

	D_1	D_2	D_3	
S_1	3; 2	17; 2	11	4
S_2	12	18; 7	10; 5	12
S_3	10; 8	25	24	8
	10	9	5	(24)

The table gives a cost of shipping equal to 296. It is left to the reader to verify that the increases in cost associated with occupying the empty cells of Table 2 are as follows:

Occupy	*Increase in cost*
Cell (1, 3)	+2
Cell (2, 1)	+8
Cell (3, 2)	+1
Cell (3, 3)	+8

This shows that Table 2 is optimal.

In Problems 13.1 and 13.2 each shipping plan with $m + n - 1$ occupied cells corresponds to an extreme point of the region of feasibility. But this is not always the case, as the following example will illustrate.

Given the 3 × 3 transportation table, with solution $x_{11} = x_{13} = x_{22} = x_{31} = x_{33} = 1$, as shown below:

	D_1	D_2	D_3	
S_1	1		1	2
S_2		1		1
S_3	1		1	2
	2	1	2	(5)

Clearly, this feasible solution contains $3 + 3 - 1 = 5$ occupied cells. To show that it is not extreme, we turn to the system of row and column equations, namely,

$$
\text{(I)}\quad \begin{cases}
x_{11} + x_{12} + x_{13} = 2 & (r_1)\\
x_{21} + x_{22} + x_{23} = 1 & (r_2)\\
x_{31} + x_{32} + x_{33} = 2 & (r_3)\\
x_{11} + x_{21} + x_{31} = 2 & (k_1)\\
x_{12} + x_{22} + x_{32} = 1 & (k_2)\\
x_{13} + x_{23} + x_{33} = 2 & (k_3)
\end{cases}
$$

In the above shipping plan, the potential corner variables are x_{12}, x_{21}, x_{23}, and x_{32}. When these variables are made equal to 0, the defining system reduces to the following fundamental system:

$$
\text{(II)}\quad \begin{cases}
x_{11} + x_{13} = 2\\
x_{22} = 1\\
x_{31} + x_{33} = 2\\
x_{11} + x_{31} = 2\\
x_{22} = 1\\
x_{13} + x_{33} = 2
\end{cases}
$$

System (II) implies that $x_{22} = 1$, but otherwise it can be satisfied in many ways, for example,

$$x_{11} = x_{33} = 0 \qquad x_{31} = x_{13} = 2 \tag{a}$$

$$x_{11} = x_{33} = 2 \qquad x_{13} = x_{31} = 0 \tag{b}$$

$$x_{11} = x_{13} = x_{31} = x_{33} = 1 \tag{c}$$

Solution (c) is the solution shown in the above table, with $(m + n - 1)$ occupied cells, but it cannot represent an extreme point because it is evidently the average of solutions (a) and (b). Solution (a), however, is extreme. It is the unique solution of the fundamental system that remains when, in system (I), we set x_{11}, x_{12}, x_{22}, and x_{23} equal to 0, as the reader can easily verify.

For another proof that the solution (a) is extreme, note that its nonzero coordinates are x_{13}, x_{22}, and x_{31}. The columns of system (I) corresponding to these nonzero variables are

$$\mathbf{x}_{13} = \begin{pmatrix} 1 \\ 0 \\ 0 \\ 0 \\ 0 \\ 1 \end{pmatrix} \qquad \mathbf{x}_{22} = \begin{pmatrix} 0 \\ 1 \\ 0 \\ 0 \\ 1 \\ 0 \end{pmatrix} \qquad \mathbf{x}_{31} = \begin{pmatrix} 0 \\ 0 \\ 1 \\ 1 \\ 0 \\ 0 \end{pmatrix}$$

Since no two of these columns has the unit 1 in the same row, they are clearly linearly independent; that is, the assumption that

$$\alpha \mathbf{x}_{13} + \beta \mathbf{x}_{22} + \gamma \mathbf{x}_{31} = \mathbf{0}$$

implies that $\alpha = \beta = \gamma = 0$.

Observe that the solution (c) contains a loop among its occupied cells, namely, the loop connecting cells (1, 1), (1, 3), (3, 3), and (3, 1). *It can be shown that if a shipping plan contains a loop among its occupied cells, it cannot correspond to an extreme point.* Furthermore, we cannot work with such solutions because they do not enable us to specify the exchange equations associated with entering an empty cell. For example, given solution (c), what are the consequences of entering cell (1, 2)?

Although solution (a) is extreme, we still have difficulty in determining the consequences of entering an empty cell. To illustrate, consider the plan corresponding to solution (a):

	D_1	D_2	D_3	
S_1	0	0	2	2
S_2	0	1	0	1
S_3	2	0	0	2
	2	1	2	(5)

A moment's thought will convince the reader that it is not possible to define clearly the cost of entering any unoccupied cell in the above table, because, despite the fact that the solution at hand is extreme, it is also *degenerate*. Since the table contains too many empty cells, it is visually difficult to identify the current degenerate basic variable (occupied cell, but containing zero items).

Exercises

1. Consider Problem 13.2.

(a) Is the following solution feasible: $x_{11} = \frac{40}{24}$, $x_{12} = \frac{36}{24}$, $x_{13} = \frac{20}{24}$; $x_{21} = \frac{120}{24}$, $x_{22} = \frac{108}{24}$, $x_{23} = \frac{60}{24}$; $x_{31} = \frac{80}{24}$, $x_{32} = \frac{72}{24}$, $x_{33} = \frac{40}{24}$?

(b) Does the following solution correspond to an extreme point: $x_{11} = 4$, $x_{21} = 1$, $x_{22} = 6$, $x_{23} = 5$, $x_{31} = 5$, $x_{32} = 3$, all other $x_{ij} = 0$?

2. Show that a transportation framework of size 50×250 corresponds to an unadjusted simplex tableau of size at least $300 \times 12{,}500$.

3. Consider the transportation problem

	D_1	D_2	D_3	
S_1	6	9	7	20
S_2	5	8	6	20
S_3	8	7	6	20
	30	15	15	

(a) The objective is the usual minimization of cost C. Express C as a function of the amount shipped x_{ij}, where $i = 1, 2, 3$, $j = 1, 2, 3$.

(b) Write the nontrivial constraints in equation form.

(c) Express all x_{ij} in terms of the four variables x_{11}, x_{12}, x_{21}, and x_{22}.

(d) Show that $C = 455 - 3x_{11} + x_{12} - 3x_{21} + x_{22}$. What conclusions can be drawn?

(e) Given the feasible plan $x_{11} = 10$, $x_{12} = 0$, $x_{13} = 10$; $x_{21} = 20$, $x_{22} = 0$, $x_{23} = 0$; $x_{31} = 0$, $x_{32} = 15$, $x_{33} = 5$. Does this plan correspond to an extreme point? If so, test for optimality.

13.2 FUNDAMENTAL THEOREMS

In this section we gather together all the definitions and theorems needed to clarify both the correspondence between patterns of occupied cells and extreme points, and also the conditions under which it is possible to test each currently unoccupied cell for improvement. As usual, a theorem is proved only when it is particularly instructive to do so.

DEFINITION 1. A *loop*, or *closed path*, in an $m \times n$ table is a specified finite sequence of cells such that:

(a) The first and last cells lie in the same row or column.

(b) Every successive pair of cells lies alternately in the same row or same column.

(c) No three consecutive cells are in the same row or same column.

We have already seen several examples of loops. For a somewhat less trivial example, consider the following diagram:

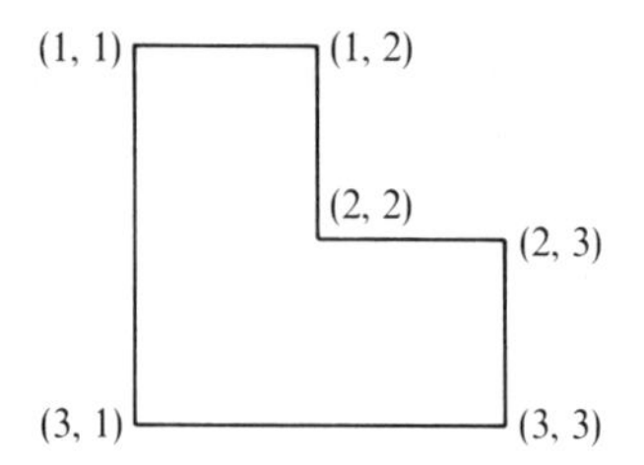

Note that if cell (3, 2) were included in the segment linking cells (3, 1) and (3, 3), then condition (c) would be violated.

DEFINITION 2. A set of occupied cells is in *independent position* if it does not contain any loops.

DEFINITION 3. A balanced transportation problem such that no partial sum of capacities can ever equal a partial sum of demands is *perfectly balanced.*

Note that the tables that define Problems 13.1 and 13.2 are perfectly balanced. For an example of a balanced, but not perfectly balanced, table, consider

	D_1	D_2	D_3	*Capacities*
S_1				$10 = a_1$
S_2				$15 = a_2$
S_3				$35 = a_3$
Demands	$5 = b_1$	$25 = b_2$	$30 = b_3$	(60)

Note that $a_1 + a_2 = b_2$, which shows lack of perfect balance.

Now let us examine in greater detail the system of equations which defines a transportation problem. For definiteness, consider the equations which correspond to the following 3×3 defining table:

	D_1	D_2	D_3	
S_1	x_{11}	x_{12}	x_{13}	a_1
S_2	x_{21}	x_{22}	x_{23}	a_2
S_3	x_{31}	x_{32}	x_{33}	a_3
	b_1	b_2	b_3	(Σ)

In tableau form, the defining equations appear as

x_{11}	x_{12}	x_{13}	x_{21}	x_{22}	x_{23}	x_{31}	x_{32}	x_{33}	
1	1	1	0	0	0	0	0	0	a_1
0	0	0	1	1	1	0	0	0	a_2
0	0	0	0	0	0	1	1	1	a_3
1	0	0	1	0	0	1	0	0	b_1
0	1	0	0	1	0	0	1	0	b_2
0	0	1	0	0	1	0	0	1	b_3

The broken partition line shows that every x_{ij} column contains only two 1's—in fact, each column is the juxtaposition of two unit vectors, one unit vector in the upper three rows corresponding to an a_i constraint and the other in the lower rows corresponding to a b_j constraint. In general, if a transportation problem has m sources and n destinations, then $\mathbf{x}_{ij}$ has a 1 in the ith and $(m + j)$th rows, and 0's everywhere else.

THEOREM 13.1. If there is a loop among cells, then the associated vectors are dependent.

To illustrate, consider a plan encountered earlier, namely,

	D_1	D_2	D_3	
S_1	1		1	2
S_2		1		1
S_3	1		1	2
	2	1	2	(5)

Clearly, there is a loop involving cells (1, 1), (1, 3), (3, 3), and (3, 1), with associated exchange equation

$$x_{11} = x_{13} \ominus x_{33} \ominus x_{31}$$

The dependency among the vectors $\mathbf{x}_{ij}$ is the vector analog of the above exchange equation, that is,

$$\mathbf{x}_{11} = \mathbf{x}_{13} - \mathbf{x}_{33} + \mathbf{x}_{31}$$

To verify this, observe that, by definition,

$$\mathbf{x}_{11} = \begin{pmatrix} 1 \\ 0 \\ 0 \\ 1 \\ 0 \\ 0 \end{pmatrix} \quad \mathbf{x}_{13} = \begin{pmatrix} 1 \\ 0 \\ 0 \\ 0 \\ 0 \\ 1 \end{pmatrix} \quad \mathbf{x}_{33} = \begin{pmatrix} 0 \\ 0 \\ 1 \\ 0 \\ 0 \\ 1 \end{pmatrix} \quad \mathbf{x}_{31} = \begin{pmatrix} 0 \\ 0 \\ 1 \\ 1 \\ 0 \\ 0 \end{pmatrix}$$

It is a demonstrated fact that any exchange equation corresponding to a pattern of occupied cells can be translated into the exactly analogous equation between the vectors associated with these cells. (For a sketch of the proof of Theorem 13.1, see Exercise 4.)

The converse of Theorem 13.1 is true, namely:

THEOREM 13.2. Given a transportation problem and its defining equations, if a subset of defining columns $\mathbf{x}_{ij}$ is linearly dependent, then the set of associated cells (i, j) contains a loop. (For a sketch of the proof of Theorem 13.2, see Exercise 5.)

THEOREM 13.3. If the occupied cells of a shipping plan contain a loop, this plan cannot represent an extreme point.

The reader will recall, from Chapter 10, that the columns belonging to the nonzero coordinates of an extreme point must be independent. Hence, Theorem 13.3 is an immediate consequence of Theorem 13.1.

THEOREM 13.4. Let B_1 be a set of $(m + n - 1)$ cells in independent position, and suppose cell (i, j) does not belong to B_1. Then:

(a) The set B_2 formed by adjoining cell (i, j) to B_1 contains one and only one loop.

(b) Suppose we form B_3 by deleting cell (i', j') from B_2, where cell $(i', j') \neq$ cell (i, j). Then the cells in B_3 are in independent position.
(For a sketch of the proof of Theorem 13.4(a), see Exercise 6.)

Theorems 13.1 to 13.4(a) establish a one-to-one correspondence between extreme points and feasible shipping plans having at most $(m + n - 1)$ occupied cells in independent position. Furthermore, Theorem 13.4 guarantees that no difficulties arise in testing such solutions for improvement.

The reader can hold the above theorems in memory by considering the following plan to be prototypic:

	D_1	D_2	·	·	·	D_n	
S_1	√	√	√	√	√	√	
S_2						√	
·						√	
·						√	
·						√	
S_m						√	

In this solution the occupied cells are $(1, 1), (1, 2), \ldots, (1, n), (2, n), (3, n), \ldots, (m, n)$. There are $(m + n - 1)$ occupied cells in independent position, and if we were to occupy one more cell, a unique closed loop would be created (the very stepping-stone loop needed to determine both the indirect cost of entering this cell, and also the adjustments needed to preserve feasibility).

Sometimes, when m and n are fairly large, theory may tell us that a loop exists, but just which cells do actually form the loop may not be apparent. To make the loop more apparent we can cross out those rows and columns that contain only one occupied cell; by definition, every row and column that occur in the loop must have at least two occupied cells. Having once completed this crossing-out process, we can start again on the reduced table and continue in this way until every row and every column in the remaining table has at least two occupied cells. At this stage the loop should be apparent.

It still remains to specify conditions under which an extreme point is characterized by precisely $(m + n - 1)$ occupied cells. The following sequence of theorems accomplishes this.

THEOREM 13.5. If a feasible shipping plan contains strictly less than $(m + n - 1)$ occupied cells, it cannot correspond to a nondegenerate corner of the region of feasibility.

Theorem 13.5 is a consequence of the fact that the simplex tableau of a transportation problem must contain $(m + n - 1)$ columns which are independent. (One such set is $\mathbf{x}_{11}, \mathbf{x}_{12}, \ldots, \mathbf{x}_{1n}, \mathbf{x}_{2n}, \mathbf{x}_{3n}, \ldots, \mathbf{x}_{mn}$, corresponding to the cells running across row 1 and column n of the transportation frame.)

THEOREM 13.6. If a feasible shipping plan contains no more than $(m + n - 1)$ occupied cells, it has a row or column containing only one occupied cell.

We suppose $m \leq n$; if this is not the case we may simply interchange the roles of m and n. Now assume that every column contains at least two occupied cells. Then the number N of occupied cells satisfies the inequalities

$$N \geq 2n \geq m + n > m + n - 1$$

But this is contrary to hypothesis, and hence there must be some column which contains only one occupied cell. It is this theorem of the "isolated occupied cell" that guarantees that every extreme point has all integer coordinates, and it is this theorem which enables us to solve fairly large transportation problems by hand. *Read with care the proof of the next theorem.*

THEOREM 13.7. If a feasible shipping plan contains no more than $(m + n - 1)$ occupied cells, then every x_{ij} is an integer. (We assume that each a_i and b_j is an integer.)

By Theorem 13.6, there is a column containing only one occupied cell. Call it column j, and suppose $x_{ij} \neq 0$. Then $x_{ij} = b_j$, clearly an integer. Now consider the table which remains after we delete column j and replace a_i by $a_i^{(1)} = a_i - b_j$. This new table has m rows and $(n - 1)$ columns; it has at most $m + n - 2 = m + (n - 1) - 1$ occupied cells, and all rim values are integers. Hence Theorem 13.6 can be applied to this reduced table; that is, at least one row or column of this table contains only one occupied cell, and so on. After k reductions, we are left with a table containing a total of $(m + n - k)$ rows and columns. The capacities $a_i^{(k)}$ and requirements $b_j^{(k)}$ of this reduced table are of the form

$$a_i^{(k)} = a_i + (\text{sum of } a_i\text{'s corresponding to certain previously deleted rows}) - (\text{sum of } b_j\text{'s corresponding to certain previously deleted columns})$$

$$b_j^{(k)} = b_j + (\text{sum of } b_j\text{'s corresponding to certain previously deleted columns}) - (\text{sum of } a_i\text{'s corresponding to certain previously deleted rows})$$

Clearly, as long as there exist reduced tables, the theorem of the isolated occupied cell can be applied to locate some row (or column) that contains no more than one basic x_{ij}, and the value of this basic variable is an $a_i^{(k)}$ or $b_j^{(k)}$. This shows that the amount contained in an occupied cell is the difference between a partial sum of row capacities and a partial sum of column demands.

To fix ideas, examine the following simple 3 × 3 table, with occupied cells marked by circles:

	D_1	D_2	D_3	
S_1	●	●		a_1
S_2		●	●	a_2
S_3			●	a_3
	b_1	b_2	b_3	(Σ)

The reader should deduce that $x_{11} = b_1$, $x_{12} = a_1 - b_1$, $x_{33} = a_3$, $x_{23} = b_3 - a_3$, and $x_{22} = a_2 + a_3 - b_3 = b_1 + b_2 - a_1$.

In general, the only way a basic variable can be 0 is for $a_i^{(k)}$ or $b_j^{(k)}$ to be 0. This proves that *degeneracy can occur in a transportation problem only when the sum of some proper subset of the a_i is equal to the sum of some proper subset of the b_j.*

Theorems 13.1 to 13.7 immediately imply Theorem 13.8, the fundamental result of this section.

THEOREM 13.8. For a perfectly balanced $m \times n$ transportation problem, a feasible shipping plan corresponds to an extreme point if and only if the plan contains precisely $(m + n - 1)$ nonempty cells.

To sum up Section 13.2, we repeat that, for a perfectly balanced $m \times n$ transportation problem:

(1) A feasible shipping pattern corresponds to an extreme point of the region of feasibility if and only if it contains precisely $(m + n - 1)$ nonempty cells.

(2) Such a plan cannot contain a loop, nor can it be degenerate.

(3) There is never any ambiguity in calculating the indirect cost of occupying an empty cell.

(4) Whenever we occupy a previously empty cell to the maximum allowable extent, the adjustment process will automatically take us to another pattern with exactly $(m + n - 1)$ nonempty cells.

Exercises

1. What can we say about the solutions shown below?

(a)

	D_1	D_2	
S_1	[2] 5	[7]	5
S_2	[8]	[3] 10	10
	5	10	

(b)

	D_1	D_2	D_3	D_4	
S_1	1			1	2
S_2		1	1		2
S_3	1			1	2
S_4		1			1
	2	2	1	2	

(c)

	D_1	D_2	D_3	
S_1	[2] 1	[2] 1	[2] 1	3
S_2	[2] 1	[2] 1	[2] 1	3
S_3	[2] 1	[2] 1	[2] 1	3
	3	3	3	

(d)

	1	2	3	4	5	6	7	8	9	
A	4	1								5
B	16	69	18		52		161			316
C				31		15		152	404	602
	20	70	18	31	52	15	161	152	404	(923)

2. *True or false?*

(a) If, for each i, we distribute the supply from source S_i in strict proportion to the requirements at all destinations, we always obtain an extreme solution. A feasible solution.

(b) The number of corners in a closed loop is always even.

(c) If the a_i and b_j are positive integers and a feasible solution contains non-integer entries x_{ij}, then the set of positive entries of this solution must contain a loop. What if the a_i and b_j are not required to be integers?

3. Show that we can calculate a lower bound to the minimum cost of shipping by ignoring the supply restrictions and supplying all requirements at least cost.

4. Fill in the details of the following proof of Theorem 13.1:

(a) A loop is defined by a sequence of neighboring cells.

(b) With each cell (i, j) in the loop associate $\mathbf{x}_{ij}$.

(c) Give an arbitrary algebraic sign to any such vector $\mathbf{x}_{ij}$ and specify that vectors corresponding to neighboring cells in the loop be given opposite signs.

(d) The number of cells in a loop is always even.

(e) Hence the algebraic sum of the $\mathbf{x}_{ij}$ in question is necessarily the null vector $\mathbf{0}$.

5. Fill in the details of the following proof of Theorem 13.2: Suppose the set of dependent column vectors is $\mathbf{x}_{i_1j_1}, \mathbf{x}_{i_2j_2}, \ldots$. Then, by the definition of dependence, we can write

$$\alpha_{i_1j_1}\mathbf{x}_{i_1j_1} + \alpha_{i_2j_2}\mathbf{x}_{i_2j_2} + \cdots + \alpha_{i_kj_k}\mathbf{x}_{i_kj_k} = \mathbf{0} \tag{1}$$

where not all the coefficients are 0. Suppose $\alpha_{i_1j_1} \neq 0$. Then there must be another summand in Eq. (1) with the row index i_1 and *nonzero coefficient*, say $\alpha_{i_1j_2}\mathbf{x}_{i_1j_2}$— because, otherwise, there could be no cancellation of the 1 in row i_1 of $\mathbf{x}_{i_1j_1}$. Again, by essentially the same argument, there must be a vector $\mathbf{x}_{i_2j_2}$ in Eq. (1) with $\alpha_{i_2j_2} \neq 0$, and so on. Since the set of vectors under study is finite, we must eventually return to the first vector.

6. Fill in the details for the following proof of Theorem 13.4(a): To prove (a), note that there are $(m + n)$ vectors $\mathbf{x}_{ij}$ in correspondence with the cells in B_2. These vectors must be dependent, and hence B_2 contains a loop. This loop is unique (except for direction) because there is one and only one way of expressing $\mathbf{x}_{ij}$ in terms of the $(m + n - 1)$ independent vectors associated with B_1.

7. Prove Theorem 13.4(b). (*Hint:* See the *exchange theorem*, that is, Theorem 10.6.)

13.3 THE STEPPING-STONE ALGORITHM PURSUED

In this section we complete the description of the stepping-stone method for solving $m \times n$ transportation problems. We first develop the theory under the assumption that the problem at hand is perfectly balanced. Then we show how to transform any balanced problem into an equivalent perfectly balanced problem.

In Section 13.2 we showed that, whenever a problem is perfectly balanced, a feasible plan is extreme if and only if it contains $(m + n - 1)$ nonempty cells. In our discussion of Problems 13.1 and 13.2 we indicated how to revise such an extreme plan. Now let us turn our attention to the problem of finding an initial plan with precisely $(m + n - 1)$ occupied cells.

One systematic method for finding such an initial solution is to apply the so-called *northwest-corner rule:* Begin at the upper left-hand (or northwest) cell of the table and assign the maximum possible shipment to this cell. Next assign the maximum possible shipment to either the vertical or horizontal neighbor of the cell just filled. (Because of perfect balance, no ambiguity can arise as to which neighbor is next.) Continue this procedure until all demands have been met.

To illustrate, we construct the northwest-corner solution to Problem 13.1 by first shipping 10 units from S_1 to D_1. This leaves 5 units at S_1 which are shipped to D_2. The requirement at D_2 is then completed by shipping 15 units from S_2. The rest of the supply at S_2 goes to D_3. Thus we obtain the solution:

Table 1

	D_1		D_2		D_3		
S_1	4	10	2	5	1	0	15
S_2	1	0	3	15	5	10	25
		10		20		10	(40)

The reader can verify that the cost of shipping according to this plan is \$145.

Although the word "corner" in the phrase "northwest corner" refers to a special "cell," that is, the cell in row 1, column 1, the northwest-corner rule does furnish a *corner*, or extreme-point solution in the sense of linear programming.

Phase II begins by specifying the consequences of entering each cell of Table 1. The same criterion used to improve a simplex tableau is now applied to improve a shipping table, namely:

(1) Revise the current table by occupying one empty cell at a time (this is equivalent to choosing one new incoming basic variable).

(2) Choose that empty cell which offers the greatest per unit improvement in the objective function.

(3) When entering an unoccupied cell, do so to the greatest extent feasible. (This guarantees that the updated table will contain the required number of empty cells.)

We explain again how to evaluate the effect on cost of entering an empty cell. Consider Table 1 and suppose 1 unit is shipped from S_1 to D_3. The currently empty cell (1, 3) will be part of a unique exchange loop whose other cells are occupied, namely,

(1, 2) −	+ (1, 3)
(2, 2) +	− (2, 3)

This loop shows that, in order to ship 1 unit from S_1 to D_3 and still maintain the rim conditions, we must ship one fewer unit from S_2 to D_3. But if we ship one fewer unit from S_2 to D_3 then one more unit must be shipped from S_2 to D_2. Finally, one fewer unit is shipped from S_1 to D_2, and all rim conditions are satisfied.

This set of instructions can be abbreviated:

$$x_{13} = x_{23} \ominus x_{22} \oplus x_{12}$$

or, equivalently,

$$x_{13} \ominus x_{23} \oplus x_{22} \ominus x_{12} = 0$$

This last exchange equation describes the net effect on the number of units shipped. It can be called the (closed) loop form of the exchange equation associated with occupying cell (1, 3) and should be read as follows: If we increase x_{13} by 1 unit, we must decrease x_{23} by 1 unit, increase x_{22} by 1 unit, and decrease x_{12} by 1 unit; then the net effect will be to ship the same number of items as previously.

The net benefit of occupying cell (1, 3) can be determined by noting the relevant cost changes at each corner of the relevant exchange loop:

	D_1	D_2	D_3	
S_1		2 −	1 +	
S_2		3 +	5 −	

At cells (1, 3) and (2, 2) we incur costs of \$1 and \$3, respectively, and at cells (1, 2) and (2, 3) savings of \$2 and \$5, respectively. Hence occupying cell (1, 3) leads to a net saving of \$3 per unit [a net cost of \$(−3) per unit].

We can come to this conclusion algebraically by proceeding as follows: Let c_{ij} denote the cost of occupying cell (i, j). First express the stepping-stone loop under consideration in equivalent exchange terms, in the closed-loop form in which all variables appear to the left of the equality sign. In the case at hand, this equation reads

$$x_{13} \ominus x_{23} \oplus x_{22} \ominus x_{12} = 0$$

Then *quantify* the left-hand side of the exchange equation by replacing x_{ij} with c_{ij} and each symbol of exchange with the analogous symbol of arithmetic, obtaining

$$c_{13} - c_{23} + c_{22} - c_{12} = 1 - 2 + 3 - 5 = -3$$

The value of this algebraic combination of costs is the *net increase in C* which occurs when x_{13} is increased by 1 unit, *given that cell (1, 3) is empty and that we compensate by adjusting only previously occupied cells.*

At this point we know that occupying cell (1, 3) is worthwhile, but the consequences of entering cell (2, 1) have yet to be evaluated. The exchange loop associated with occupying this cell reads

$$x_{21} \ominus x_{22} \oplus x_{12} \ominus x_{11} = 0$$

The attendant net increase in C is obviously

$$c_{21} - c_{22} + c_{12} - c_{11} = 1 - 3 + 2 - 4 = -4$$

Now revise Table 1 by occupying cell (2, 1) to the greatest extent possible, thereby providing the largest per unit reduction in cost. This procedure mimics the simplex algorithm exactly. Let θ denote the increase in x_{21}; then the loop is

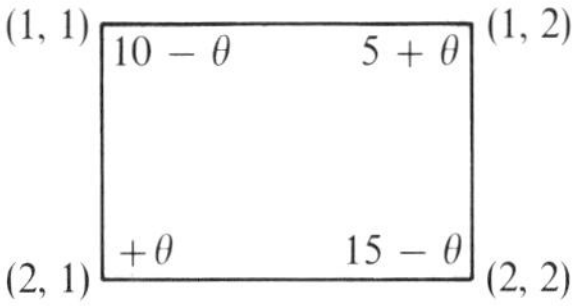

(1, 1)	$10 - \theta$	$5 + \theta$	(1, 2)
(2, 1)	$+\theta$	$15 - \theta$	(2, 2)

In the attempt to occupy cell (2, 1) to the maximum possible extent, the limiting factor will be the 10 units in cell (1, 1). If we occupy cell (2, 1), the number of units in cells (1, 1) and (2, 3) must be decreased, and, as the reader can easily see, it is cell (1, 1) which is emptied first. Thus, after 10 units are shipped into cell (2, 1), and, after adjusting, the new table will be

Table 2

	D_1	D_2	D_3	
S_1	[4] 0	[2] 15	[1] 0	15
S_2	[1] 10	[3] 5	[5] 10	25
	10	20	10	(40)

The cost of shipping is

$$C = (4)(0) + (2)(15) + (1)(0) + (1)(10) + (3)(5) + (5)(10) = \$105$$

This confirms our earlier observation that the net reduction in cost would be \$4 for each unit shipped into cell (2, 1).

Now we must test Table 2 for improvement possibilities. Note that in passing from Table 1 to Table 2 we emptied cell (1, 1); if we were to reoccupy this cell to the same extent that it was previously emptied we would necessarily return to Table 1. Hence, in studying Table 2, we need only determine the consequences of entering cell (1, 3). It is easy to see that this results in a net reduction in cost of \$3 per unit shipped. The maximum number of units that can be shipped is 10. The updated table reads

Table 3

	D_1	D_2	D_3	
S_1	[4] 0	[2] 5	[1] 10	15
S_2	[1] 10	[3] 15	[5] 0	25
	10	20	10	(40)

The cost of shipping is now \$75.

To test Table 3 we need only examine cell (1, 1). Entering this cell would be unprofitable, and hence Table 3 is optimal. Again, the current occupied cells correspond to the current basic variables; the unoccupied cells, to the corner variables. The "net reduction in cost" on entering the empty cell (i, j) equals the coefficient of x_{ij} in the current form of the objective function. If all these coefficients are nonnegative, the current plan is optimum.

There is nothing sacred about the northwest-corner rule. If the rows or columns had been arranged in another way, the northwest-corner solution would have been different. In fact, we can just as easily develop a northeast-corner rule. What is more important, the northwest-corner rule pays no heed to costs. Described below is the *least-cost rule*, a procedure for carrying out Phase I that does take costs into consideration.

To apply this rule we choose the smallest c_{ij} in the table and then allocate to cell (i, j) as large an amount x_{ij} as possible. To illustrate, consider the defining table of Problem 13.1. The smallest costs are c_{13} and c_{21}, both equal to 1. (Ties can be broken arbitrarily, for example, by referring to total rather than per unit reduction in cost.) Let us ship 10 units into cell (1, 3). This satisfies the total demand at D_3, and the remaining capacity at S_1 will be 5. If we readjust the rims to reflect demand that has been met and remaining capacity, the next allocation must be made in the reduced table

	D_1	D_2	
S_1	4	2	5
S_2	1	3	25
	10	20	(30)

Now, repeating the least-cost rule, we must take $x_{21} = 10$. After removal of this allocation, the next reduced table reads

	D_2	
S_1	2	5
S_2	3	15
	20	(20)

No choice remains; it follows that $x_{11} = 5$ and $x_{21} = 15$. Combining all the above allocations gives the following least-cost solution to Problem 13.1:

	D_1	D_2	D_3	
S_1		5	10	15
S_2	10	15		25
	10	20	10	(40)

This is the optimal shipping plan obtained earlier.

For another example of how to apply the least-cost rule, consider Problem 13.2 again. The cheapest route corresponds to cell (1, 1). We occupy this cell to the maximum—in other words, exhaust the capacity at S_1 by making $x_{11} = 4$. For the second decision, the field of choice corresponds to the following reduced set of data:

	D_1	D_2	D_3	
S_2	\$12	\$18	\$10	12
S_2	\$10	\$25	\$24	8
	6	9	5	(20)

The cheapest route in the reduced table is then determined and *occupied to the maximum*. Accordingly, we take $x_{31} = 6$. The next reduced table reads

	D_2	D_3	
S_2	\$18	\$10	12
S_3	\$25	\$24	2
	9	5	(14)

Now make $x_{23} = 5$, and the last reduced table is

	D_2	
S_2	\$18	7
S_3	\$25	2
	9	

Evidently, $x_{22} = 7$ and $x_{32} = 2$. Combining all these allocations gives the shipping table

	D_1	D_2	D_3	
S_1	3 4	17	11	4
S_2	12	18 7	10 5	12
S_3	10 6	25 2	24	8
	10	9	5	(24)

The least-cost rule leads to a shipping cost of \$298. As an exercise the reader should show that the northwest-corner rule leads to the following solution with cost of \$387. (Clearly, a low-cost initial solution can save much work.)

	D_1	D_2	D_3	
S_1	3 4	17	11	4
S_2	12 6	18 6	10	12
S_3	10	25 3	24 5	8
	10	9	5	(24)

A useful check on transportation calculations can be illustrated using the initial and terminal tables of Problem 13.1:

Initial Table 1

	D_1	D_2	D_3	
S_1	4 10	2 5	1 0	15
S_2	1 0	3 15	5 10	25
	10	20	10	(40)

Terminal Table 3

	D_1	D_2	D_3	
S_1	4 0	2 5	1 10	15
S_2	1 10	3 15	5 0	25
	10	20	10	(40)

In the initial table, the basic variables are $x_{11} = 10$, $x_{12} = 5$, $x_{22} = 15$, and $x_{23} = 10$. If we were to construct the simplex tableau corresponding to this table, the objective coefficients of these basic variables would be 0. In the terminal tableau, certain of the initial basic variables will have become nonbasic, and their objective coefficients in the terminal simplex tableau will be equal to the costs of occupying the corresponding vacant cells. Thus, of the original basic variables, x_{11} and x_{23} turn nonbasic in the terminal table, and the costs of entering the corresponding cells are easily seen to be 4 and 3, respectively.

The simplex multiplier theorem states that the inner product

$$[\text{Initial value of } x_{11}][\text{terminal cost of entering cell } (1, 1)] + [\text{initial value of } x_{23}][\text{terminal cost of entering cell } (2, 3)] = \Delta C$$

the resultant change in the objective function as we pass from the initial to the terminal table. The sign of ΔC is usually clear from the context at hand. Accordingly, in Problem 13.1 the initial values are $x_{11} = 10$ and $x_{23} = 10$, respectively, and hence

$$\Delta C = (4)(10) + (3)(10) = 70$$

When passing from the initial to the terminal table it is obvious that ΔC represents the *reduction* in C. By direct calculation, note that the cost of shipping according to the initial plan is 145, the cost of shipping for the terminal plan is 75, and hence ΔC is just as contended.

There is nothing special about which tables are called *initial* and *terminal*. *The multiplier checks hold for the passage between any two tables, in either direction.*

PROBLEM 13.3 (UPPER-BOUND RESTRICTION ON THE AMOUNT SHIPPED). Suppose the conditions of Problem 13.1 are modified by adding a new restriction limiting x_{22} to 12, that is, the amount shipped from S_2 to D_2 is not allowed to exceed 12 units. The optimum plan for Problem 13.1 does not satisfy the new condition. The new problem, with upper-bound restriction, will be called Problem 13.3. It is completely equivalent to the problem described in the following table:

Table 1

	D_1	D_2	D_2'	D_3	
S_1	\$4	\$2	\$2	\$1	15
S_2	\$1	\$3	\$$M$	\$5	25
	10	12	8	10	(40)

Destination D_2' of Table 1 serves to absorb the extra demand at D_2, beyond the upper-bound limit of 12 on x_{22}. The cell at the intersection of the S_2 row and D_2' column has a prohibitively high cost of entry, which ensures that the corresponding route will be priced out of solution.

Having already solved Problem 13.1, we can use its optimal solution to build a feasible solution to the new problem. The old optimal solution had $x_{12} = 5$ and $x_{22} = 15$. In order to meet the new restriction, we reduce x_{22} to 12. Then, in order to maintain feasibility, we send 3 units from S_2 to D_2' and 5 units from S_1 to D_2':

Table 2

	D_1		D_2		D_2'		D_3		
S_1	4	0	2	0	2	5	1	10	15
S_2	1	10	3	12	M	3	5	0	25
		10		12		8		10	(40)

The solution in Table 2 is a feasible solution to Problem 13.3. Furthermore, because exactly five cells are occupied, it corresponds to an extreme point. This is not an optimal solution, however, because 3 units are shipped over the high-cost route (S_2 to D_2'). It is left to the reader to show that it is worthwhile to occupy cell (2, 4); doing so to the maximum possible extent results in the following plan:

Table 3

	D_1	D_2	D_2'	D_3	
S_1	$\boxed{4}$ 0	$\boxed{2}$ 0	$\boxed{2}$ 8	$\boxed{1}$ 7	15
S_2	$\boxed{1}$ 10	$\boxed{3}$ 12	$\boxed{M}$ 0	$\boxed{5}$ 3	25
	10	12	8	10	(40)

Table 3 is optimal, with cost of shipping

$$C' = (2)(8) + (1)(7) + (1)(10) + (3)(12) + (5)(3) = 84$$

Let us impose on Table 1 of Problem 13.3 the additional restriction that x_{12} must turn out to be 0. Then Table 2 exhibits a solution which is no longer feasible. The extended version of Table 1 appropriate for solving the new problem, starting from the old optimal solution, is

	D_1	D_2	D_2'	D_3	
S_1	\$4	\$2	\$$M$	\$1	15
S_2	\$1	\$3	\$3	\$5	25
	10	0	20	10	(40)

This table effectively makes it nonoptimal to ship from S_1 to D_2, because, on the one hand, the demand at D_2 is 0, and, on the other hand, the cost of shipping from S_1 to D_2' is M, the prohibitively large constant. Moreover, since the demand at D_2 is 0, we can omit this column completely. Accordingly, we form an extreme solution to the new problem by modifying the old optimal solution:

	D_1	D_2'	D_3	
S_1	$\boxed{4}$ 0	$\boxed{M}$ 5	$\boxed{1}$ 10	15
S_2	$\boxed{1}$ 10	$\boxed{3}$ 15	$\boxed{5}$ 0	25
	10	20	10	(40)

The usual tests show that we are motivated to enter cell (1, 1) or cell (2, 3). Occupying cell (1, 1) to the maximum possible extent leads to the following solution:

	D_1	D_2'	D_3	
S_1	[4] 5	[M] 0	[1] 10	15
S_2	[1] 5	[3] 20	[5] 0	25
	10	20	10	(40)

This shipping plan is optimal, with cost of shipment

$$C = (4)(5) + (1)(10) + (1)(5) + (3)(20) = \$95$$

The next two problems are degenerate. When a problem is degenerate the stepping-stone algorithm may break down, but we can bypass this difficulty by working with a specially designed, perfectly balanced auxiliary problem. To illustrate, consider:

PROBLEM 13.4.

	D_1	D_2	D_3	
S_1	[4]	[2]	[1]	15
S_2	[1]	[3]	[5]	25
	10	5	25	(40)

This problem is balanced but not perfectly balanced. The northwest-corner rule furnishes the following initial solution:

Table 1

	D_1	D_2	D_3	
S_1	[4] 10	[2] 5	[1]	15
S_2	[1]	[3]	[5] 25	25
	10	5	25	(40)

Note that in occupying cell (1, 2) we satisfied both a row and a column requirement. Taking either $x_{13} = 0$, or $x_{22} = 0$, but not both, would really specify a northwest-corner solution with $(m + n - 1)$ occupied cells in independent position. In any case it is impractical to distinguish between empty cells and occupied cells containing a zero amount.

The consequences of entering cells (1, 3), (2, 1), and (2, 2) must now be determined. If we adjoin any one of these empty cells to the set of occupied cells, it is impossible to find an exchange loop among occupied cells, and therefore the stepping-stone algorithm cannot be used to revise Table 1. This algorithm has degenerated because one of the cells in Table 1 contains an amount shipped $x_{ij} = 0$, and it is not clear whether this cell is occupied or not.

Since Problem 13.4 is degenerate, it is instructive to obtain the graphical solution. Suppose the amounts shipped are as shown below:

	D_1	D_2	D_3	
S_1	4 | x	2 | y	1 | $15 - x - y$	15
S_2	1 | $10 - x$	3 | $5 - y$	5 | $10 + x + y$	25
	10	5	25	(40)

The cost of shipping is evidently

$$C = 4x + 2y + (15 - x - y) + (10 - x) + 3(5 - y) + 5(10 + x + y)$$
$$= 7x + 3y + 90$$

The objective is to minimize C subject to the six constraints

$$\begin{aligned} x \quad &\geq 0 \\ y &\geq 0 \\ x + y &\leq 15 \\ x \quad &\leq 10 \\ y &\leq 5 \\ x + y &\geq -10 \end{aligned}$$

The last constraint above is obviously superfluous.

The graphical solution to this problem is shown in Figure 39. Degeneracy occurs at the corner $D = (10, 5)$, which lies at the intersection of three limiting lines.

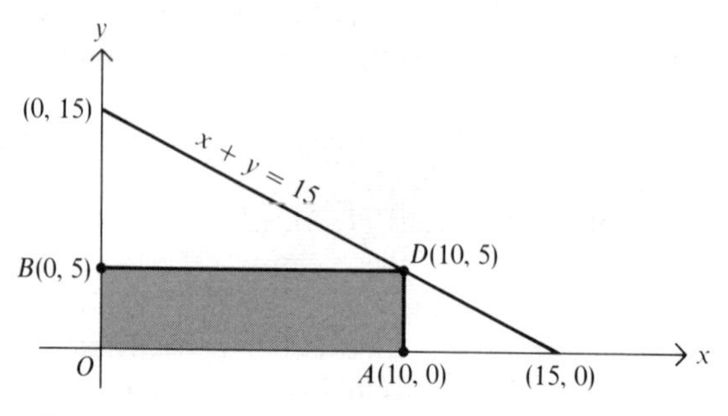

Figure 39

The optimum value of C is 90, at the corner $x = y = 0$. The optimum shipping schedule is clearly nondegenerate, and is

	S_1	S_2	S_3	
D_1	[4]	[2]	[1] 15	15
D_2	[1] 10	[3] 5	[5] 10	25
	10	5	25	(40)

As previously observed, the amount contained in an occupied cell can always be expressed as the difference between a partial sum of row capacities and a partial sum of column requirements. *Hence a cell can be occupied at zero level only if a problem is not perfectly balanced.* This is what occurs in Table 1. There are three empty cells because the sum of the requirements in columns 1 and 2 equals the capacity in row 1.

We cannot always remedy a degenerate table simply by perturbing some rim entries of the original problem. To illustrate, suppose we were to add ϵ to both the demand at D_1 and the supply at S_1, thereby obtaining

	D_1	D_2	D_3	
S_1				$15 + \epsilon = a_1$
S_2				$25 = a_2$
	$10 + \epsilon = b_1$	$5 = b_2$	$25 = b_3$	$(40 + \epsilon)$

But the northwest-corner solution to this modified problem is still degenerate. What we must do to avoid degeneracy is to systematically employ infinitesimal ϵ's in order to create a perfectly balanced table.

Dantzig showed that if a table contains m rows and n columns it is possible to create a perfectly balanced transportation problem by adding ϵ to each column requirement and $n\epsilon$ to any single row availability. In the auxiliary problem a partial sum over r columns will contain the infinitesimal term $r\epsilon$, where $r < n$. But any partial sum of rows will contain $n\epsilon$ or $0(\epsilon)$, depending upon whether the perturbed a_i is included in the partial sum. Hence the perturbed problem must be perfectly balanced.

By applying this profound ruse to Table 1, the following perturbed solution is obtained:

	D_1	D_2	D_3	
S_1	[4] $10 + \epsilon$	[2] $5 + \epsilon$	[1] ϵ	$15 + 3\epsilon$
S_2	[1]	[3]	[5] 25	25
	$10 + \epsilon$	$5 + \epsilon$	$25 + \epsilon$	$(40 + 3\epsilon)$

The above table is perfectly balanced, and we can now study the consequences of entering each empty cell. Thus

Occupy	*Increase in C*
Cell (2, 1)	−7
Cell (2, 2)	−3

By occupying cell (2, 1) to the maximum extent of $(10 + \epsilon)$ units, we obtain the solution

Table 2

	D_1	D_2	D_3	
S_1	[4]	[2] $5 + \epsilon$	[1] $10 + 2\epsilon$	$15 + 3\epsilon$
S_2	[1] $10 + \epsilon$	[3]	[5] $15 - \epsilon$	
	$10 + \epsilon$	$5 + \epsilon$	$25 + \epsilon$	$(40 + 3\epsilon)$

Testing empty cells of Table 2 gives the following information

Occupy	*Increase in C*
Cell (1, 1)	+7
Cell (2, 2)	−3

Now send $(5 + \epsilon)$ units into cell (2, 2) and obtain

Table 3

	D_1	D_2	D_3	
S_1	[4]	[2]	[1] $15 + 3\epsilon$	$15 + 3\epsilon$
S_2	[1] $10 + \epsilon$	[3] $5 + \epsilon$	[5] $10 - 2\epsilon$	25
	$10 + \epsilon$	$5 + \epsilon$	$25 + \epsilon$	$(40 + 3\epsilon)$

The reader should show that Table 3 is optimal. *To find the actual solution to Problem 13.4, we need only set $\epsilon = 0$ in Table 3.*

Rather than add ϵ to each column requirement we can add ϵ to each row availability and then add $m\epsilon$ to any single column requirement. The one precaution to observe is not to create degeneracy where no degeneracy previously existed; this can be guaranteed by taking ϵ less than $1/n$ (or $1/m$, as the case may be).

An example more substantial than Problem 13.4 will better illustrate the difficulties created by lack of perfect balance, and point up the cleverness of the $n\epsilon$ technique.

PROBLEM 13.5. The table which defines this problem is

	D_1	D_2	D_3	D_4	
S_1	[77]	[6]	[5]	[13]	12
S_2	[9]	[79]	[19]	[4]	12
S_3	[1]	[8]	[7]	[99]	5
	5	12	4	8	(29)

The northwest-corner rule leads to the following plan:

Table 1

	D_1	D_2	D_3	D_4	
S_1	5	7			12
S_2		5	4	3	12
S_3				5	5
	5	12	4	8	(29)

Note that, although Problem 13.5 is not perfectly balanced, Table 1 contains exactly $m + n - 1 = 6$ occupied cells. (All the extreme solutions of a degenerate problem are not necessarily degenerate solutions.) Now suppose we were motivated to occupy cell (3, 1) to the fullest. This revision would take us to the following solution:

Table 2

	D_1	D_2	D_3	D_4	
S_1		12			12
S_2			4	8	12
S_3	5				5
	5	12	4	8	(29)

Table 2 is "doubly degenerate" in that it contains four occupied cells rather than six. Clearly, it is not possible to construct any of the stepping-stone paths needed to evaluate the consequences of entering the empty cells of Table 2. Moreover, if we naively consider any two of the empty cells as occupied, it might still be impossible to construct all the needed stepping-stone paths. For example, if we place ϵ in cells (3, 3) and (3, 4) it will still not be possible to study the effects of entering cell (1, 1). *The $n\epsilon$ method of creating perfect balance avoids all difficulties; it does not allow us to fill empty cells with ϵ's in an indiscriminate fashion, but forces us to place them so that the occupied cells are always in independent position.*

Below is a complete $n\epsilon$ form of solution to Problem 13.5. Suppose the $n\epsilon$ version is formed by adding ϵ to each of the three capacities and 3ϵ to the requirement at D_4. Then the least-cost initial solution is:

	D_1	D_2	D_3	D_4	
S_1	[77]	[6] $8 + \epsilon$	[5] 4	[13]	$12 + \epsilon$
S_2	[9]	[79] $4 - 2\epsilon$	[19]	[4] $8 + 3\epsilon$	$12 + \epsilon$
S_3	[1] 5	[8] ϵ	[7]	[99]	$5 + \epsilon$
	5	12	4	$8 + 3\epsilon$	$(29 + 3\epsilon)$

By the stepping-stone algorithm, we are motivated to occupy cell (2, 1), obtaining

	D_1	D_2	D_3	D_4	
S_1		$8 + \epsilon$	4		$12 + \epsilon$
S_2	$4 - 2\epsilon$			$8 + 3\epsilon$	$12 + \epsilon$
S_3	$1 + 2\epsilon$	$4 - \epsilon$			$5 + \epsilon$
	5	12	4	$8 + 3\epsilon$	

This table is optimal. After discarding ϵ's, it is clear that the best plan for Problem 13.5 is $x_{12} = 8$, $x_{13} = 4$, $x_{21} = 4$, $x_{24} = 8$, $x_{31} = 1$, $x_{32} = 4$, all other $x_{ij} = 0$. The minimum cost is 169. Do alternative optimum solutions exist?

In summary, it should be stressed that:

(1) Balanced transportation problems do not occur naturally. Nevertheless, the balanced model is treated as a prototype because it is possible to develop a simple algorithm for solving it. (Later, we will see how easy it is to transform an unbalanced problem into an equivalent balanced one.)

(2) In this section we really worked only with *perfectly balanced* problems. (There is no need to test for perfect balance, because the $n\epsilon$ procedure applied mechanically to any balanced problem always gives an equivalent, perfectly balanced problem.)

Exercises

1. *True or false?*
   (a) A transportation problem can have only one northwest-corner solution.
   (b) A transportation problem can have only one least-cost solution.
   (c) A perfectly balanced problem can have only one least-cost solution.
2. Given the transportation problem

	D_1	D_2	D_3	D_4	
S_1	2	0	3	4	50
S_2	4	0	5	3	50
S_3	0	1	7	0	75
	40	55	60	20	(175)

Verify that the optimal plan is $x_{13} = 50$, $x_{22} = 40$, $x_{23} = 10$, $x_{31} = 40$, $x_{32} = 15$, $x_{34} = 20$; all other $x_{ij} = 0$.

3. *True or false?*

(a) A balanced transportation problem can have an initial northwest-corner solution which is degenerate.

(b) A balanced transportation problem which is not perfectly balanced must have a degenerate northwest-corner solution.

(c) A northwest-corner solution can contain a loop among occupied cells.

4. Given the following transportation problem:

	D_1	D_2	D_3	D_4	D_5	
S_1	$13	$ 5	$10	$ 6	$10	55
S_2	$ 8	$ 1	$ 6	$ 6	$ 7	46
S_3	$ 1	$12	$ 4	$ 7	$ 7	31
S_4	$11	$16	$ 7	$10	$ 2	51
	42	20	50	30	41	(183)

(a) Show that an optimal plan is given by $x_{12} = 12$, $x_{13} = 13$, $x_{14} = 30$, $x_{21} = 11$, $x_{22} = 8$, $x_{23} = 27$, $x_{31} = 31$, $x_{43} = 10$, $x_{45} = 41$, all other $x_{ij} = 0$.

(b) Are there any alternative optimum solutions?

5. Given the following transportation problem:

	D_1	D_2	D_3	D_4	D_5	D_6	
S_1	$10	$13	$10	$ 7	$10	$11	$a_1 = 5$
S_2	$ 4	$ 0	$ 4	$ 4	$ 2	$ 2	$a_2 = 18$
S_3	$ 6	$ 5	$ 9	$11	$ 3	$11	$a_3 = 3$
S_4	$ 7	$ 9	$12	$ 3	$ 3	$11	$a_4 = 30$
	$b_1 = 4$	$b_2 = 6$	$b_3 = 6$	$b_4 = 12$	$b_5 = 15$	$b_6 = 13$	(56)

(a) Show that an optimal plan is given by $x_{13} = 5$, $x_{22} = 4$, $x_{23} = 1$, $x_{26} = 13$, $x_{31} = 1$, $x_{32} = 2$, $x_{41} = 3$, $x_{44} = 12$, $x_{45} = 15$, all other $x_{ij} = 0$.

(b) Suppose a_2 and b_3 both increase by 2. Is the new optimum evident?

(c) Suppose a_2 and b_3 both decrease by 2. Is the new optimum evident?

6. Demonstrate that the optimal solution to a transportation problem remains optimal if the cost of entering an empty cell is increased. What if this cost is reduced?

7. Suppose that the following two tables are part of the sequence of tables needed to solve a given problem. Verify the simplex-multiplier check using these tables.

Table r

	D_1	D_2	D_3	D_4	
S_1	[6] 25	[5] 25	[3] 10	[4]	60
S_2	[2]	[4]	[6] 10	[5]	10
S_3	[2]	[3]	[2] 5	[3] 25	30
	25	25	25	25	(100)

Table s

	D_1	D_2	D_3	D_4	
S_1	[6]	[5] 25	[3] 25	[4] 10	60
S_2	[2] 10	[4]	[6]	[5]	10
S_3	[2] 15	[3]	[2]	[3] 15	30
	25	25	25	25	(100)

8. Discuss the statement: In degenerate transportation problems certain zero-valued variables x_{ij} should be considered basic. A straightforward method of determining these variables is to select them from among the basic variables that simultaneously fell to 0 in the immediately preceding iteration.

13.4 APPLICATIONS

In this section the versatility of the balanced transportation model is shown by extending it to cover several new situations. First, we show how to transform unbalanced problems into equivalent balanced ones. Second, we apply the transportation model to various problems having little to do with physical transportation but which, nevertheless, can be fitted into a transportation framework, for example, the problem of production scheduling.

The following theorems are useful in solving transportation problems by any method. By applying them in conjunction with the methods of the preceding sections, some of the stepping-stone calculations can be lightened appreciably. The reason we study these theorems, however, is simply that they are inherently important. Here they will help us in balancing unbalanced problems and, in Chapter 15, it will be shown how other algorithms for handling transportation problems can be built upon these very fundamental theorems.

THEOREM 13.9. Subtracting (or adding) a constant k from all the elements in any row or column of the cost matrix of a transportation problem does not affect the optimal shipping plan.

An an example, consider the table of costs, amounts available, and amounts demanded for Problem 13.1 of this chapter:

	D_1	D_2	D_3	*Capacity*
S_1	\$4	\$2	\$1	15
S_2	\$1	\$3	\$5	25
Requirement	10	20	10	(40)

As usual, let x_{ij} denote the quantity shipped from S_i to D_j. Then the total cost of shipping is

$$T = 4x_{11} + 2x_{12} + x_{13} + x_{21} + 3x_{22} + 5x_{23}$$

Now substract k from all the elements in row 1 of the original cost table. The reduced cost matrix

	D_1	D_2	D_3	
S_1	$4 - k$	$2 - k$	$1 - k$	
S_2	1	3	5	

is obtained. The total cost of shipping according to this reduced cost matrix is

$$\begin{aligned} T' &= (4 - k)x_{11} + (2 - k)x_{12} + (1 - k)x_{13} + x_{21} + 3x_{22} + 5x_{23} \\ &= (4x_{11} + 2x_{12} + x_{13} + x_{21} + 3x_{22} + 5x_{23}) - k(x_{11} + x_{12} + x_{13}) \end{aligned}$$

But $(x_{11} + x_{12} + x_{13})$ represents the quantity available at S_1, namely, 15. Hence

$$T' = T - 15k$$

and the same shipping plan which renders T' optimal must render T optimal, and vice versa.

To be more precise, if the matrix $X = (x_{ij})$ represents an optimal shipping plan for the transportation problem with costs $C = (c_{ij})$, then it is also an optimal plan for any transportation problem with the same amounts available and amounts demanded but with (reduced) cost matrix

$$C' = (c'_{ij}) = c_{ij} - r_i - k_j$$

where r_i denotes a constant subtracted from all entries in row i of C, and k_j denotes a constant subtracted from all entries in column j of C.

THEOREM 13.10. (a) Subtracting r_i from each of the costs of row i of a given transportation problem results in a reduced problem whose optimal cost is $r_i a_i$ less than that of the original problem, where a_i denotes the capacity of the ith supplier.

(b) If k_j is subtracted from each of the costs in column j of a transportation table, the cost of any (optimal) program is reduced by $k_j b_j$, where b_j denotes the demand at destination j.

(c) If $r_1, r_2, \ldots, r_m$ are subtracted from each of the costs in rows 1, 2, ..., m, respectively, and $k_1, \ldots, k_n$ from each of the costs in columns 1, 2, ..., n, respectively, of a transportation table, the cost of any (optimal) program is reduced by

$$S(r, k) = (r_1 a_1 + \cdots + r_m a_m) + (k_1 b_1 + \cdots + k_n b_n)$$

These theorems hold for any shipping plan and, hence, *a fortiori* for the optimal plan.

To prove Theorem 13.10, let $X = (x_{ij})$ denote the shipping plan in question, $C = (c_{ij})$ the given cost matrix, and $C' = (c'_{ij})$ the reduced cost matrix. Assume that $c'_{ij} = c_{ij} - r_i - k_j$, or

$$c_{ij} = c'_{ij} + r_i + k_j \tag{1}$$

In addition, let T denote the total cost of shipping for the given problem, and T' the total cost of shipping for the reduced problem. T can be written out, in detail, as

$$\begin{aligned} T = \quad & c_{11}x_{11} + c_{12}x_{12} + \cdots + c_{1n}x_{1n} \\ &+ c_{21}x_{21} + c_{22}x_{22} + \cdots + c_{2n}x_{2n} \\ &\cdot \\ &\cdot \\ &\cdot \\ &+ c_{m1}x_{m1} + c_{m2}x_{m2} + \cdots + c_{mn}x_{mn} \end{aligned}$$

After substituting Eq. (1) into this expression for T, we obtain

$$\begin{aligned} T = \quad & (c'_{11} + r_1 + k_1)x_{11} + (c'_{12} + r_1 + k_2)x_{12} + \cdots + (c'_{1n} + r_1 + k_n)x_{1n} \\ &+ (c'_{21} + r_2 + k_1)x_{21} + (c'_{22} + r_2 + k_2)x_{22} + \cdots + (c'_{2n} + r_2 + k_n)x_{2n} \\ &\cdot \\ &\cdot \\ &\cdot \\ &+ (c'_{m1} + r_m + k_1)x_{m1} + (c'_{m2} + r_m + k_2)x_{m2} + \cdots + (c'_{mn} + r_m + k_n)x_{mn} \end{aligned}$$

Upon regrouping of terms, T takes the form

$$\begin{aligned} T &= [(c'_{11}x_{11} + \cdots + c'_{12}x_n) + \cdots + (c'_{m1}x_{m1} + \cdots + c'_{mn}x_{mn})] \\ &\quad + r_1(x_{11} + x_{12} + \cdots + x_{1n}) + \cdots + r_m(x_{m1} + x_{m2} + \cdots + x_{mn}) \\ &\quad + k_1(x_{11} + x_{21} + \cdots + x_{m1}) + \cdots + k_n(x_{1n} + x_{2n} + \cdots + x_{mn}) \\ &= T' + [r_1a_1 + \cdots + r_ma_m] + (k_1b_1 + \cdots + k_nb_n) \\ &= T' + S(r, k) \end{aligned}$$

Thus, given any shipping plan $X = (x_{ij})$, the total cost T can be split into two parts, the first of which is dependent on the x_{ij} and the second of which, namely, $S(r, k)$, is independent of the x_{ij}.

We can verbalize Theorem 13.10 as follows: The *difference* between the original total cost and the reduced total cost can readily be calculated if all quantities subtracted from a row or column are multiplied by the corresponding rim values and if all products obtained are then added together.

In Chapter 15 we show how it is possible to construct a new algorithm for solving transportation problems by cleverly manipulating $S(r, k)$.

PROBLEM 13.6.

	D_1	D_2	D_3	*Capacity*
S_1	\$4	\$2	\$1	15
S_2	\$1	\$3	\$5	25
Requirement	10	20	9	≠

This problem is *unbalanced*—the sum of the quantities available is greater than the sum of the quantities demanded. Obviously, there exist many feasible shipping plans, but under any plan, one item will necessarily have to remain in storage, at S_1 or S_2. Of course, this problem can be formulated as a linear program and solved by the simplex algorithm, but our aim is to be able to solve it within the transportation framework previously studied. Remember that the stepping-stone method presupposes a balanced framework.

A way out of this difficulty is to use a fictitious destination, D_4, whose total demand will create exact balance. Clearly, in the given problem, the fictitious demand at D_4 should be for 1 unit. But what should be the cost of shipment from each supplier to the fictitious destination (remember, these costs really represent the "cost of overstocking" by 1 unit at S_1 and S_2)? We may not be able to quantify these costs, but often they can be assumed equal, that is, $c_{14} = c_{24} = k$. Then, the balanced equivalent of Problem 13.6 reads

	D_1	D_2	D_3	D_4	
S_1	\$4	\$2	\$1	\$$k$	1
S_2	\$1	\$3	\$5	\$$k$	25
	10	20	9	1	(40)

By Theorem 13.9 we can subtract a constant from all the entries of column 4 without changing the optimal shipping plan. Hence, after subtracting k from all the costs in column 4 of the above table, we obtain the equivalent reduced cost table:

	D_1	D_2	D_3	D_4	
S_1	\$4	\$2	\$1	\$0	15
S_2	\$1	\$3	\$5	\$0	25
	10	20	9	1	(40)

In practice, if we believe that the costs of storage or of oversupply at S_1 and S_2 are the *same*, we can always follow the rule of introducing fictitious demand with cost of shipping equal to zero.

The reader can now solve the given problem by any of the methods previously discussed. The optimal plan is

	D_1	D_2	D_3	D_4	
S_1	[4] 0	[2] 6	[1] 9	[0] 0	15
S_2	[1] 10	[3] 14	[5] 0	[0] 1	25
	10	20	9	1	(40)

This solution shows that S_2 should ship 1 unit to D_4; that is, the warehouse S_2 should retain 1 unit in stock.

PROBLEM 13.7.

	D_1	D_2	D_3	*Capacity*
S_1	\$4	\$2	\$1	14
S_2	\$1	\$3	\$5	25
Requirement	10	20	10	$\neq$

This problem is unbalanced, but, unlike Problem 13.6, the sum of the amounts required exceeds the total sum available for shipment. There is absolutely no way to satisfy the aggregate demand, that is, no feasible solution can be found, unless we agree that some destination will be undersupplied by one unit. Whenever requirements exceed capacity, it will be our aim to design a plan that utilizes the total capacity at the least possible "cost."

We can transform the present problem into a balanced one by using a fictitious supplier S_3, whose capacity is 1 unit. But what are the appropriate costs of shipment from S_3 to the three given destinations; that is, what are the "costs of leaving a customer (destination) unsatisfied?" Suppose these costs are the same for each customer. Then, as observed earlier, we may assume that all costs of shipping out of S_3 are 0, and the balanced equivalent of Problem 13.7 reads

	D_1	D_2	D_3	
S_1	\$4	\$2	\$1	14
S_2	\$1	\$3	\$5	25
S_3	\$0	\$0	\$0	1
	10	20	10	(40)

The rest is mechanics. The following plan is optimal

	D_1 (cost)	D_1	D_2 (cost)	D_2	D_3 (cost)	D_3	
S_1	4	0	2	4	1	10	14
S_2	1	10	3	15	5	0	25
S_3	0	0	0	1	0	0	1
		10		20		10	(40)

This solution identifies D_2 as the customer whose demand should be left unsatisfied.

PROBLEM 13.8. Suppose a manufacturer is confronted with the problem of planning production to meet a fixed schedule of future deliveries. To illustrate, suppose that 10, 12, and 15 units must be delivered over the next three time periods, which we call periods 1, 2, and 3. The production capacity per period is fixed at 13 units, but the costs of production are different from period to period, say 11, 12, and 13 monetary units per item produced during periods 1, 2, and 3, respectively. Assume that the excess of demand over capacity during period 3 can only be met by planned buildup of inventory. The storage cost per item produced is 0 during the period of production, but 2 monetary units over any other period.

An optimal production schedule can be determined by linear programming. Let x_i denote the amount to be produced in period i, for $i = 1, 2, 3$. Then total production cost is $(11x_1 + 12x_2 + 13x_3)$ and the combined cost of production and storage is evidently

$$\begin{aligned} C &= (11x_1 + 12x_2 + 13x_3) + 2(x_1 - 10) + 2(x_1 + x_2 - 10 - 12) \\ &= 15x_1 + 14x_2 + 13x_3 - 64 \end{aligned}$$

The objective is to:

$$\begin{array}{llr} \text{Minimize} & C & \\ \text{subject to} & x_1, \quad x_2, \quad x_3 \geq 0 & \\ \text{and} & x_1 \leq 13 & (1) \\ & x_2 \leq 13 & (2) \\ & x_3 \leq 13 & (3) \\ & x_1 \geq 10 & (4) \\ & x_1 + x_2 \geq 22 & (5) \\ & x_1 + x_2 + x_3 = 37 & (6) \end{array}$$

As an exercise the reader can solve this problem by the simplex algorithm. We show below how to solve this problem within the transportation framework, treating production periods as sources S_i and delivery periods as destinations D_j.

Accordingly, let x_{ij} denote the number of units to produce in period i for delivery in period j, and let c_{ij} be the unit cost of producing in period i for delivery in period j. Then

$$c_{ij} = k_i + 2(j - 1), \qquad \text{for } i \leq j$$

where k_i denotes the cost of production during period i and $2(j - i)$ denotes the storage cost. Thus, $c_{11} = 11$, $c_{12} = 11 + 2 = 13$, $c_{13} = 11 + 2 + 2 = 15$; $c_{22} = 12$, $c_{23} = 12 + 2 = 14$, $c_{33} = 13$.

We assume that it is impossible to produce in later periods for sale in prior periods. Accordingly, each of the variables x_{21}, x_{31}, and x_{32} must equal 0. The way to guarantee this is to price these variables out of solution by letting c_{21}, c_{31}, and c_{32} equal M, a prohibitively large cost.

Problem 13.8 can now be represented by the following table:

	D_1	D_2	D_3	D_4	
S_1	11	13	15	0	13
S_2	M	12	14	0	13
S_3	M	M	13	0	13
	10	12	13	2	(39)

Destination D_4, needed for balance, is fictitious. By taking 0 as the cost of shipment to D_4, we implicitly assume that the cost of underproducing is the same for all periods.

The reader can verify that the optimal plan is

	D_1	D_2	D_3	D_4	
S_1	[11] 10	[13] 1	[15] 0	[0] 2	13
S_2	[M] 0	[12] 11	[14] 2	[0] 0	13
S_3	[M] 0	[M] 0	[13] 13	[0] 0	13
	10	12	15	2	(39)

Are there any alternative optimal solutions?

PROBLEM 13.9 (VARIABLE PRODUCTION COSTS). In this example we extend the planning model of the previous example to include the possibility of *overtime production* (another way to meet excess of demand over regular production capacity). Suppose that maximum capacity under ordinary production is 15 units per period, while maximum excess capacity for overtime production is 10 units per period, so that total capacity per period can never exceed 25 units. In addition, assume that deliveries must be made according to the following fixed schedule: 24, 18, 32, 24 units in that order, over the next four periods.

Since the delivery requirement in period 3 exceeds the combined regular and overtime capacity of that period, planned inventory buildup will be necessary. Further, observe that the *cumulative* production capacity for any period is always greater than or equal to the cumulative delivery requirement in that period. If this were not the case, requirements could never be met solely by inventory buildup and overtime production. (For example, if the requirements were for 24, 19, 34, and 22 units, in that order, then the total requirement over the first three periods is 77, which is 2 units more than the maximum production capacity available.)

Assume that the cost of overtime production is higher than the cost of regular production and that these costs vary from period to period, according to the following cost schedule:

	Period			
	1	*2*	*3*	*4*
Regular production	5	4	6	6
Overtime production	6	7	7	9

The cost of storage has been estimated as 1 unit for every period of storage, except the period of production.

The production costs are seemingly nonlinear, but they can nevertheless be treated as linear if we imagine that the factory is producing two different products, one under regular conditions, the other under overtime conditions. As in the previous problem, to solve this scheduling problem within the transportation framework, we must focus on the relevant decision variables. Accordingly, let x_{ij} denote the number of units to produce in period i for delivery in period j, using regular facilities, and let y_{ij} be the number of units produced in period i for delivery in period j, using overtime facilities. In addition, let $p(i, j)$ be the regular per unit production cost and $c(i, j)$ the total unit cost of production, where we produce in period i for delivery in period j; and let $p(i', j)$ and $c(i', j)$ be the analogous overtime costs, again for producing in period i for delivery in period j. Then, by definition,

$$c(i, j) = p(i, j) + 1(j - i) \qquad \text{for } j \geq i$$
$$c(i', j) = p(i', j) + 1(j - i) \qquad \text{for } j \geq i$$

These $c(i, j)$ have been placed in the transportation table below—unprimed costs in the first four rows (which correspond to the period of regular production), primed costs (involving overtime production) in the last four rows. The first four columns represent the first four delivery periods, and column 5 is used to absorb the excess of supply over demand, balancing the table. We have assumed that the cost of not producing to total capacity is the same in all periods. This justifies the use of zero costs in column 5.

	1	2	3	4	5	
1	\$5	\$6	\$7	\$8	\$0	15
2	\$$M$	\$4	\$5	\$6	\$0	15
3	\$$M$	\$$M$	\$6	\$7	\$0	15
4	\$$M$	\$$M$	\$$M$	\$6	\$0	15
1′	\$6	\$7	\$8	\$9	\$0	10
2′	\$$M$	\$7	\$8	\$9	\$0	10
3′	\$$M$	\$$M$	\$7	\$8	\$0	10
4′	\$$M$	\$$M$	\$$M$	\$9	\$0	10
	24	18	32	24	2	(100)

The optimum solution is shown below:

	1	2	3	4	5	
1	[5] 15	[6]	[7]	[8]	[0]	15
2	[M]	[4] 15	[5]	[6]	[0]	15
3	[M]	[M]	[6] 15	[7]	[0]	15
4	[M]	[M]	[M]	[6] 15	[0]	15
1′	[6] 9	[7]	[8]	[9]	[0] 1	10
2′	[M]	[7] 3	[8] 7	[9]	[0]	10
3′	[M]	[M]	[7] 10	[8]	[0]	10
4′	[M]	[M]	[M]	[9] 9	[0] 1	10
	24	18	32	24	2	(100)

This table is degenerate, with one empty cell too many. To calculate the costs of entering empty cells we can fit an ϵ into one of the previously empty cells. Entering ϵ in cell (1, 5) will create a closed loop, as will entering ϵ in cell (4′, 1). It is quite legitimate to place an ϵ in cell (1, 2); doing so makes it possible to verify that the above solution is optimal.

PROBLEM 13.10 (TRANSHIPMENT). Consider the following transportation problem:

	D_1	D_2	*Amount available*
S_1	\$10	\$8	5
S_2	\$ 9	\$2	5
Amount required	4	6	(10)

The optimal solution is

	D_1	D_2	
S_1	[10] 4	[8] 1	5
S_2	[9]	[2] 5	5
	4	6	(10)

The minimum cost of shipment is

$$C = (10)(4) + (8)(1) + (2)(5) = \$58$$

Note that the table of cost data tacitly assumes that c_{ij} is the *minimum* cost of shipping a unit from S_i to D_j.

Now suppose it is possible to transfer items between D_1 and D_2, at a unit cost of \$1 in either direction. In such a case, the cost $c_{11} = 10$ no longer represents the cheapest way of shipping a unit from S_1 to D_1. Hence there must exist solutions whose total cost is less than \$58. For example, in the plan

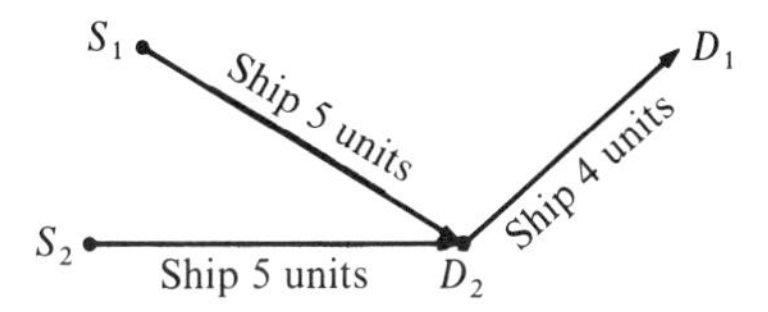

in which 5 units are shipped from S_1 to D_2 and 5 units are also shipped from S_2 to D_2, destination D_2 receives 4 more units than required and acts as an *intermediary*, or *point of transhipment*, by sending the 4 excess units to D_1. The reader should check that the total cost of shipping according to the above plan is \$54.

Given that transhipment is possible we can always find the overall least-cost solution by proceeding as follows: For every source S_i and destination D_j examine all possible ways of shipping from S_i to D_j, direct and indirect. Then use the *minimum* cost of shipping from S_i to D_j as the cost k_{ij} in the usual transportation table.

To illustrate, in the problem at hand, the minimum cost of shipping from S_1 to D_1 can be calculated from the following diagram:

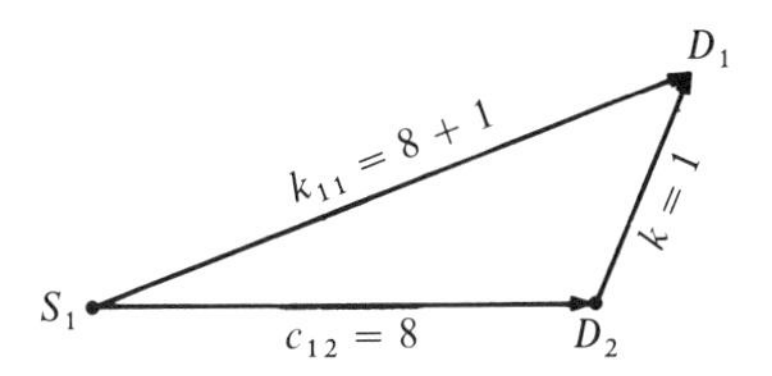

Here $c_{12} = 8$ represents the per unit cost over the direct route (1, 2), $k = 1$ represents the cost of transhipment between D_2 and D_1, and $k_{11} = 9$ represents the *minimum* cost of shipment from S_1 to D_1, indirect routes being allowed. The minimum cost k_{21}, from S_2 to D_1, is \$3 as the reader can verify.

The best overall shipping plan can now be found by using the cost data k_{ij}, where $k_{11} = 9$, $k_{21} = 3$, $k_{12} = c_{12} = 8$, and $k_{22} = c_{22} = 2$. In other words, we must solve the transportation problem

	D_1	D_2	*Excess*
S_1	\$9	\$8	5
S_2	\$3	\$2	5
Shortage	4	6	(10)

This problem has alternative optimal solutions, one of which is $x_{11} = 4$, $x_{12} = 1$, $x_{21} = 0$, $x_{22} = 5$. This is the very same set of values for the x_{ij} we found when transhipment was not allowed, but note the differences:

(1) The total cost of shipment using the k_{ij} as costs is

$$K = 9(4) + 8(1) + 2(5) = 54$$

(2) When we say that $x_{11} = 4$ in the transhipment problem we specify that 4 units should be shipped from S_1 to D_2 and then from D_2 to D_1, because x_{11} and all other x_{ij} represent amounts shipped over the cheapest possible route.

Exercises

1. Given the unbalanced transportation problem

	D_1	D_2	D_3	
S_1	10	5	8	5
S_2	6	9	3	7
S_3	8	6	4	5
	6	6	4	$\neq$

Show that there is an optimal plan where $C = 82$ and where S_3 will retain 1 unit in inventory.

2. Use Theorem 13.9 to show that the following problem has the same optimal shipping pattern as Problem 13.1:

	D_1	D_2	D_3	
S_1	\$3	\$0	\$0	15
S_2	\$0	\$1	\$4	25
	10	20	10	(40)

3. A company maintains three factories around the country and supplies four primary warehouses from them. The capacities of the factories are: S_1, 3000 gross/month; S_2, 4000 gross/month; S_3, 7000 gross/month. The warehouse requirements are D_1, 2000 gross/month; D_2, 3000 gross/month; D_3, 3500 gross/month; D_4, 4000 gross/month. Manufacturing costs are identical at all plants. Shipping costs (in dollars per gross) from factories to plants are given by the following table:

	Warehouses			
Factories	D_1	D_2	D_3	D_4
S_1	13	22	9	10
S_2	25	20	12	15
S_3	11	16	22	13

(a) Explain carefully why manufacturing costs do not influence the problem of finding the minimum-cost shipping schedule.

(b) Find the minimum-cost schedule and total shipping bill.

(c) Discuss the possibilities of transhipment.

4. Suppose that a firm has seven salesmen stationed at S_1, seven salesmen stationed at S_2, three salesmen stationed at S_3, and four salesmen stationed at S_4. These salesmen are to be sent to three territories, with territory T_1 requiring three salesmen, territory T_2 requiring eight salesmen, and territory T_3 requiring five salesmen. Because of their varying abilities and experience, the profit that the firm expects each salesman to contribute to operations will vary with the territory to which he is sent. However, the present groups are very homogeneous and it is assumed that all salesmen stationed at any S_i have the same profit potential in any given territory, as shown in the following table:

	T_1	T_2	T_3	
S_1	\$ 5,000	\$4,000	\$3,000	7
S_2	\$ 7,000	\$8,000	\$7,000	7
S_3	\$ 4,000	\$2,000	\$1,000	3
S_4	\$10,000	\$3,000	\$3.000	4
	3	8	5	$\neq$

Effect an assignment of salesmen to territories so as to maximize the total profit potential.

5. A company has three plants with the following variable production costs per unit and monthly production capacity:

	Plant		
	1	2	3
Production costs ($/unit)	42	40	42
Capacity (units/month)	60	70	70

The four company warehouses have the following requirements: $b_1 = 40$ units/month; $b_2 = 30$ units/month; $b_3 = 50$ units/month; $b_4 = 40$ units/month. The transportation costs (in dollars per unit) between the plants and warehouses are (S_i = plant and D_j = warehouse):

	D_1	D_2	D_3	D_4
S_1	18	13	18	14
S_2	12	12	18	12
S_3	15	12	14	18

(a) Show that the table of distribution costs (production + transportation) reads

	D_1	D_2	D_3	D_4
S_1	60	55	60	56
S_2	52	52	58	52
S_3	57	54	56	60

(b) Determine monthly production schedules for the plants and show how the units produced should be shipped.

6. Three plants, A, B, and C, can make two products, X and Y. The plants can work regular time and overtime, and two time periods are to be scheduled. Product X requires 2 hours per unit in any plant, while Y requires 3 hours per unit. Product requirements (in units) are

	January	*February*
X	100	135
Y	210	210

Plant capacities (in hours) are

	January		*February*	
	Regular time	*Overtime*	*Regular time*	*Overtime*
A	100	50	100	50
B	200	50	200	–
C	500	100	500	100

Costs per unit for the three plants are:

	A	*B*	*C*
X, regular time	10	12	14
X, overtime	15	16	15
Y, regular time	20	21	22
Y, overtime	25	24	26
X and Y, storage per unit of time	3	2	1

Formulate and solve the problem of satisfying product requirements as a transportation problem. (*Hint:* Let the sources of the transportation problem be bins containing the goods produced in the different months within regular time, with separate bins for goods produced in these months during overtime. Let the destinations be bins which must be filled with the different monthly requirements. It may also be useful to introduce a destination called "unused capacity" so that the total production capacity equals the total two-month demand.)

7. Four products, P_1, P_2, P_3, and P_4, can be made on any one of three machines, I, II, and III. Unit costs (in dollars) are

	P_1	P_2	P_3	P_4
I	0.45	0.35	0.30	0.32
II	0.30	0.31	0.18	0.20
III	0.25	0.33	0.20	0.24

Time (in minutes) required to produce each unit of product is

	P_1	P_2	P_3	P_4
I	10	5	15	20
II	8	4	12	16
III	6	3	9	12

Units of product required are

	P_1	P_2	P_3	P_4
Demand	200	400	100	150

Available time (in minutes) is

	I	*II*	*III*
Time available	4000	4000	4000

Formulate and solve the problem of assigning products to machines as a transportation problem.

8. Does Problem 13.9 contradict the statement that "linear programming assumes *constant returns to scale*, with respect to both output and profit"?

14

TRANSPORTATION PROBLEMS BY THE MULTIPLIER METHOD

The stepping-stone method is quite tedious because, in order to decide on which empty cell to occupy, the correct exchange loop corresponding to each empty cell must be sketched. This chapter describes a more efficient method for determining the best empty cell to occupy, one which will allow us to calculate the effect on cost of entering each empty cell without constructing the relevant exchange loop.

Consider, once again, Problem 13.1. As observed earlier, this problem can be reduced to the linear program

$$\text{Maximize} \quad (-C) = -4x_{11} - 2x_{12} - x_{13} - x_{21} - 3x_{22} - 5x_{23}$$
$$\text{where all} \quad x_{ij} \geq 0$$

$$\text{and} \quad \text{(I)} \quad \begin{cases} x_{11} + x_{12} + x_{13} = 15 & (r_1) \\ x_{21} + x_{22} + x_{23} = 25 & (r_2) \\ x_{11} + x_{21} = 10 & (k_1) \\ x_{12} + x_{22} = 20 & (k_2) \\ x_{13} + x_{23} = 10 & (k_3) \end{cases}$$

In what follows we retain all five equations in system (I) for reasons of symmetry.

The key to determining which empty cell to occupy—prior to constructing any exchange loop—is to employ undetermined simplex multipliers to construct the universal form of the objective function. To illustrate, here is the initial simplex tableau, in maximization form:

	x_{11}	x_{12}	x_{13}	x_{21}	x_{22}	x_{23}	
u_1	1	1	1	0	0	0	15
u_2	0	0	0	1	1	1	25
v_1	1	0	0	1	0	0	10
v_2	0	1	0	0	1	0	20
v_3	0	0	1	0	0	1	10
1	-4	-2	-1	-1	-3	-5	$-C$

The multipliers u_1 and u_2 correspond to Eqs. (r_1) and (r_2), respectively, and the multipliers v_1, v_2, and v_3 correspond to Eqs. (k_1), (k_2), and (k_3), respectively. The universal form of the objective function is

$$\begin{aligned}-C = {} & (u_1 + v_1 - 4)x_{11} + (u_1 + v_2 - 2)x_{12} + (u_1 + v_3 - 1)x_{13} \\ & + (u_2 + v_1 - 1)x_{21} + (u_2 + v_2 - 3)x_{22} + (u_2 + v_3 - 5)x_{23} \\ & - (15u_1 + 25u_2 + 10v_1 + 20v_2 + 10v_3)\end{aligned}$$

Now suppose we have at hand a transportation plan corresponding to a feasible extreme point. For example, the following is a northwest-corner solution:

Table 1

	D_1	D_2	D_3	
S_1	[4] 10	[2] 5	[1]	15
S_2	[1]	[3] 15	[5] 10	25
	10	20	10	(40)

As explained in Chapter 7, to determine the exact objective function at a given corner, we can eliminate the basic variables from the universal objective function by equating the relevant coefficients to 0. The occupied cells of Table 1 show that x_{11}, x_{12}, x_{22}, and x_{23} are basic variables, and hence the coefficients of these variables in the universal form of C should be 0. Thus we are led to consider the system of equations, where the notation "Eq. (i, j)" signifies that x_{ij} is basic:

$$(O_1)\quad \begin{cases} u_1 + v_1 = 4 & (1, 1) \\ u_1 + v_2 = 2 & (1, 2) \\ u_2 + v_2 = 3 & (2, 2) \\ u_2 + v_3 = 5 & (2, 3) \end{cases}$$

(Call this system (O_1); O for orthogonal, 1 for Table 1.)

With each such system (O_n), we associate a derived *system of numbers*, the (Q_n) system. The numbers in this system equal the coefficients of the current nonbasic variables appearing in the objective function. Thus

$$(Q_1) \qquad \begin{cases} q_{13} = c_{13} - (u_1 + v_3) = 1 - (u_1 + v_3) \\ q_{21} = c_{21} - (u_2 + v_1) = 1 - (u_2 + v_1) \end{cases}$$

Note that system (O_1) contains more unknowns than equations. It is never contradictory and always possesses an infinite number of solutions. But, fortunately, we need never know any particular u_i or v_j. It is only necessary to determine the values of the combinations of u_i's and v_j's which appear in system (Q_1). These sums turn out to be constant for any solution to system (O_1). For example, if we form the following combination of equations in system (O_1):

$$\text{Eq. } (1, 2) - \text{Eq. } (2, 2) + \text{Eq. } (2, 3)$$

we obtain the relation

$$u_1 + v_3 = 4$$

Similarly, the combination

$$\text{Eq. } (2, 2) - \text{Eq. } (1, 2) + \text{Eq. } (1, 1)$$

gives the relation

$$u_2 + v_1 = 5$$

These relations suggest that we determine the key sums $u_1 + v_3$ and $u_2 + v_1$ by first finding *any* numerical solution to system (O_1), and afterward using the numerical values obtained to calculate the values of the key sums. Accordingly, let $v_1 = 0$ in system (O_1). Then it follows that

$$\begin{aligned} v_1 &= 0 \\ u_1 &= 4 \\ v_2 &= -2 \qquad\qquad [v_1 = 0] \\ u_2 &= 5 \\ v_3 &= 0 \end{aligned}$$

On the other hand, if in system (O_1) we specify that $u_1 = 0$, the following solution is obtained:

$$\begin{aligned} u_1 &= 0 \\ v_1 &= 4 \\ v_2 &= 2 \qquad\qquad [u_1 = 0] \\ u_2 &= 1 \\ u_3 &= 4 \end{aligned}$$

Once the value of any unknown in system (O_1) is determined, the remaining 4×4 system is triangular and can be solved by sight, simply by substituting back from one equation to another. Whatever the numerical solution obtained it will always follow that

$$\begin{aligned} u_1 \qquad\quad + v_3 &= 4 \\ u_2 + v_1 \qquad\quad &= 5 \end{aligned}$$

From these two key sums the (Q_1) system of numbers can be determined immediately, as

$$(Q_1) \qquad \begin{cases} q_{13} = c_{13} - (u_1 + v_3) = 1 - 4 = -3 \\ q_{21} = c_{21} - (u_2 + v_1) = 1 - 5 = -4 \end{cases}$$

This shows that entering cell (2, 1) will decrease C at the fastest per-unit rate.

Now we can construct the unique exchange loop which describes the consequences of entering cell (2, 1). This loop shows that the maximum number of units that can be shipped into cell (2, 1) is 10. This takes us to the updated table

Table 2

	D_1	D_2	D_3	
S_1	[4] 0	[2] 15	[1] 0	15
S_2	[1] 10	[3] 5	[5] 10	25
	10	20	10	(40)

Now apply the multiplier method to Table 2, by writing down the system (O_2) of basic equations corresponding to its occupied cells. This gives

$$(O_2) \qquad \begin{cases} u_1 \qquad\qquad + v_2 \qquad = c_{12} = 2 \\ \qquad u_2 + v_1 \qquad\qquad = c_{21} = 1 \\ \qquad u_2 \qquad + v_2 \qquad = c_{22} = 3 \\ \qquad u_2 \qquad\qquad + v_3 = c_{23} = 5 \end{cases}$$

If $u_1 = 0$, then $v_2 = 2$, $u_2 = 1$, $v_1 = 0$, $v_3 = 4$. From this particular solution we obtain the (Q_2) system:

$$(Q_2) \qquad \begin{cases} q_{11} = c_{11} - (u_1 + v_1) = 4 - (0 + 0) = +4 \\ q_{13} = c_{13} - (u_1 + v_3) = 1 - (0 + 4) = -3 \end{cases}$$

Clearly, it pays to enter cell (1, 3). Only now do we construct the appropriate stepping-stone loop. By the usual procedures, the following updated table is obtained:

Table 3

	D_1	D_2	D_3	
S_1	[4]	[2] 5	[1] 10	15
S_2	[1] 10	[3] 15	[5]	25
	10	20	10	

The (O_3) system corresponding to Table 3 is

$$(O_3)\quad \begin{cases} u_1 + v_2 = c_{12} = 2 \\ u_1 + v_3 = c_{13} = 1 \\ u_2 + v_1 = c_{21} = 1 \\ u_2 + v_2 = c_{22} = 3 \end{cases}$$

Let $u_1 = 0$. Then a particular solution to (O_3) is $u_1 = 0$, $v_2 = 2$, $v_3 = 1$, $u_2 = 1$, $v_1 = 0$. This fixes the entries in the (Q_3) system, namely,

$$(Q_3)\quad \begin{cases} q_{11} = c_{11} - (u_1 + v_1) = 4 - 0 = +4 \\ q_{23} = c_{23} - (u_2 + v_3) = 5 - 2 = +3 \end{cases}$$

Since all q_{ij} in system (Q_3) are nonnegative, Table 3 must be optimal.

To sum up, the $(u_i + v_j)$ method is somewhat more straightforward than the stepping-stone procedure and is the exact equivalent of the simplex-multiplier method. It essentially amounts to examining the universal objective function and adjusting so as to eliminate all the basic variables from the objective function. It works so well because it is easy to determine the basic variables within the tabular framework. *Clearly, $(u_i + v_j)$ must equal z_{ij}, the indirect cost of bringing x_{ij} into solution.* As was stressed in Chapter 11, the multipliers u_i and v_j become feasible solutions to the *dual* of the transportation problem only when we have attained the optimum of the transportation problem.

For another illustration of the $(u_i + v_j)$ method, return to Problem 13.2. The least-cost rule gives the initial solution

Table 1

	D_1	D_2	D_3	
S_1	3 / 4	17	11	4
S_2	12	18 / 7	10 / 5	12
S_3	10 / 6	25 / 2	24	8
	10	9	5	(24)

The (O_1) system is

$$(O_1)\quad \begin{cases} u_1 + v_1 = 3 \\ u_2 + v_2 = 18 \\ u_2 + v_3 = 10 \\ u_3 + v_1 = 10 \\ u_3 + v_2 = 25 \end{cases}$$

By taking $u_1 = 0$, it follows in turn that $v_1 = 3$, $u_3 = 7$, $v_2 = 18$, $u_2 = 0$, $v_3 = 10$. The set (Q_1) of relevant q_{ij} is

$$(Q_1)\quad \begin{cases} q_{12} = c_{12} - u_1 - v_2 = 17 - 0 - 18 = -1 \\ q_{13} = c_{13} - u_1 - v_3 = 11 - 0 - 10 = +1 \\ q_{21} = c_{21} - u_2 - v_1 = 12 - 0 - \ \ 3 = +9 \\ q_{33} = c_{33} - u_3 - v_3 = 24 - 7 - 10 = +7 \end{cases}$$

This shows that it is most worthwhile to occupy cell (1, 2). The relevant exchange loop is

$$\begin{array}{ccc} (1,1)\ - & \longrightarrow & +\ (1,2) \\ | & & | \\ (3,1)\ + & \longrightarrow & -\ (3,2) \end{array}$$

Table 1 shows that the maximum increase in x_{12} is 2. Thus we obtain the improved shipping plan shown below:

Table 2

	D_1	D_2	D_3	
S_1	3 2	17 2	11	4
S_2	12	18 7	10 5	12
S_3	10 8	25	24	8
	10	9	5	(24)

Hence the (O_2) system is

$$(O_2)\quad \begin{cases} u_1 + v_1 = 3 \\ u_1 + v_2 = 17 \\ u_2 + v_2 = 18 \\ u_2 + v_3 = 10 \\ u_3 + v_1 = 10 \end{cases}$$

A particular solution of this system is $u_1 = 0$, $u_2 = 1$, $u_3 = 7$, $v_1 = 3$, $v_2 = 17$, $v_3 = 9$. The set (Q_2) is thus

$$(Q_2)\quad \begin{cases} q_{13} = c_{13} - u_1 - v_3 = 11 - 0 - 9 = +2 \\ q_{21} = c_{21} - u_2 - v_1 = 12 - 1 - 3 = +8 \\ q_{32} = c_{32} - u_3 - v_2 = 25 - 7 - 17 = +1 \\ q_{33} = c_{33} - u_3 - v_3 = 24 - 7 - 9 = +8 \end{cases}$$

These values show that Table 2 is optimal.

Note that we can employ the rims of the current shipping table to carry out the calculations needed to solve the corresponding (O) *system.* For example, suppose the table in question is 3×4 and that the occupied cells, with per-unit costs, are (1, 1), (1, 2), (2, 2), (2, 3), (3, 3), and (3, 4):

u_i \ v_j	$v_1 = 1$	$v_2 = 2$	$v_3 = 1$	$v_4 = 0$
$u_1 = 0$	1	2		
$u_2 = 1$		3	2	
$u_3 = 1$			2	1

With each row we can associate a value of u_i, with each column a value of v_j. The numerical values of these variables can be entered along the left rim and across the top of the table, as indicated.

Once we set $u_1 = 0$, then the v_j for the first two columns are determined, namely, $v_1 = 1$ and $v_2 = 2$. From $v_2 = 2$, it follows that $u_2 = 1$, and hence $v_3 = 1$; and from $v_3 = 1$ it follows that $u_3 = 1$, whence $v_4 = 0$. These values should be recorded as soon as determined. Once all u_i and v_j are known, the sums $u_i + v_j = z_{ij}$ corresponding to the empty cells are easy to calculate. [The z_{ij} can even be recorded somewhere in cell (i, j).] *If a problem is not perfectly balanced, the* $(u_i + v_j)$ *method can break down.* As an exercise the reader should reconsider Problems 13.4 and 13.5 by this method, first without the $n\epsilon$ adjustment and then with the adjustment. With regard to the behavior of the $(u_i + v_j)$ method in the face of degeneracy, it is particularly simple to find solutions to the (O_i) system in this case, but these solutions rarely lead to values of $(u_i + v_j)$ which correspond to z_{ij}, the sought-for indirect cost of occupying cell (i, j). To illustrate, consider Problem 13.4, whose northwest-corner solution is degenerate:

Table 1

	D_1	D_2	D_3	
S_1	4	10	2	5	1		15
S_2	1		3		5	25	25
	10	5	25	(40)

The (O) system determined by the occupied cells in Table 1 reads

$$\begin{aligned} u_1 \quad + v_1 \qquad\qquad &= 4 \\ u_1 \qquad\quad + v_2 \qquad &= 2 \\ u_2 \qquad\qquad + v_3 &= 5 \end{aligned}$$

This is a system of three equations in five unknowns, for which it is quite simple to find many particular solutions. For example, if $u_1 = 0$, then $v_1 = 4$ and $v_2 = 2$. The connection between equations is now broken, but we can continue to solve by taking $u_2 = 0$, from which it follows that $v_3 = 5$. Thus we obtain the following set of $z_{ij} = u_i + v_j$: $z_{11} = 4$, $z_{12} = 2$, $z_{13} = 5$, $z_{21} = 4$, $z_{22} = 2$, $z_{23} = 5$.

But, unfortunately, this set of z_{ij} cannot be interpreted as the indirect costs of entering the cells of Table 1. The only way to obtain a valid set of z_{ij} is to imagine one of the empty cells in Table 1 as containing an infinitesimal amount. There are three possibilities:

(1) Cell (1, 3) contains the amount ϵ_1. This assumption leads to the valid set of z_{ij}: $z_{11} = 4$, $z_{12} = 2$, $z_{13} = 1$, $z_{21} = 8$, $z_{22} = 6$, $z_{23} = 5$.

(2) Cell (2, 1) contains the amount ϵ_2. This leads to $z_{11} = 4$, $z_{12} = 2$, $z_{13} = 8$, $z_{21} = 1$, $z_{22} = -1$, $z_{23} = 5$.

(3) Cell (2, 2) contains the amount ϵ_3. This leads to $z_{11} = 4$, $z_{12} = 2$, $z_{13} = 4$, $z_{21} = 5$, $z_{22} = 3$, $z_{23} = 5$.

PROBLEM 14.1 (SENSITIVITY ANALYSIS ON RIM VALUES). Let us henceforth call Problem 13.2 the *old, or original, transportation problem.* For convenience, we repeat its optimal solution:

	D_1	D_2	D_3	
S_1	3 | 2	17 | 2	11 | 0	4
S_2	12 | 0	18 | 7	10 | 5	12
S_3	10 | 8	25 | 0	24 | 0	8
	10	9	5	(24)

The (O) system associated with the occupied cells of this table reads

$$(O)\quad \begin{cases} u_1 \qquad\quad + v_1 \qquad\qquad = 3 & (1) \\ u_1 \qquad\qquad\quad + v_2 \qquad = 17 & (2) \\ \quad u_2 \qquad\qquad + v_2 \qquad = 18 & (3) \\ \quad u_2 \qquad\qquad\qquad + v_3 = 10 & (4) \\ \qquad u_3 + v_1 \qquad\qquad = 10 & (5) \end{cases}$$

Note that the (O) system itself determines the values of five of the nine possible summands of the form $(u_i + v_j)$, for $i = 1, 2, 3$ and $j = 1, 2, 3$. The reader can verify that the values of the four other sums are

$$\begin{array}{llllllll} u_1 & & & & & + v_3 & = & 9 \\ & u_2 & & + v_1 & & & = & 4 \\ & & u_3 & & + v_2 & & = & 24 \\ & & u_3 & & & + v_3 & = & 16 \end{array}$$

We remind the reader that the existence of many solutions to the (O) system shows that the value of any individual u_i or v_j cannot be considered as automatically indicating the implicit value of a change in capacity or demand alone. Great care must be exercised, as the following remarks should make clear.

Suppose the old problem (that is, Problem 13.2) changes so that one fewer unit is demanded at D_2, and one fewer unit is available at S_1. *Call this new problem, Problem 14.1.* It is clearly balanced (although no longer perfectly balanced). In the optimal solution to the old problem, cell (1, 2) is occupied, and, hence, to find the optimal solution to Problem 14.1 we need only remove 1 unit from cell (1, 2) in the old solution, that is, reduce x_{12} by 1. The new solution obtained is necessarily optimal, because it has the very same costs and pattern of empty cells as the old optimal solution. This new solution is

	D_1	D_2	D_3	
S_1	3 2	17 1	11 0	3
S_2	12 0	18 7	10 5	12
S_3	10 8	25 0	24 0	8
	10	8	5	(23)

The cost of shipping is now $C = \$279$. Note that the difference between the new optimal cost and the old optimal cost is \$17, which represents the actual cost of shipping a unit from S_1 to D_2. Furthermore, since cell (1, 2) was occupied in the old optimal solution, the value of the summand $(u_1 + v_2)$ must be 17.

In general, if an availability a_i and a requirement b_j are both reduced (or increased) by 1 unit and cell (i, j) was occupied in the old optimal solution, we can pass to the new optimal solution by decreasing (or increasing) x_{ij} by 1 unit. The cost of the new shipping plan goes down (or up) by c_{ij}, where c_{ij} is the direct unit cost of shipping into cell (i, j). Since this cell is occupied, c_{ij} equals $(u_i + v_j)$. Furthermore, if we decrease a_i and b_j by k units, then optimum cost will decrease by kc_{ij}, provided only that $k \leq x_{ij}$. And for equal increases in a_i and b_j, there is no upper limit.

Problem 14.2 below furnishes a somewhat more complex illustration of modified rim values.

PROBLEM 14.2.

	D_1	D_2	D_3	
S_1	\$ 3	\$17	\$11	4
S_2	\$12	\$18	\$10	12
S_3	\$10	\$25	\$24	9
	10	9	6	(25)

Problem 14.2, like Problem 14.1, is a modification of Problem 13.2. In order to transform Problem 13.2 into the current problem we have made one more unit available at S_3 and required one more unit at D_3. Note that in the old optimal solution, cell (3, 3) is unoccupied. Keep the old occupied cells non-empty, and, at the same time, ship 1 unit into cell (3, 3). This is possible and leads to the following feasible solution:

Table 1

	D_1		D_2		D_3		
S_1	3	2	17	2	11		4
S_2	12		18	7	10	5	12
S_3	10	8	25		24	1	9
	10		9		6		(25)

The shipping plan in Table 1 does not correspond to an extreme point. There is now a closed loop among the occupied cells, but this can be eliminated by emptying cell (3, 3) and making the compensatory adjustments needed to maintain feasibility. Thus we obtain the extreme solution:

Table 2

	D_1		D_2		D_3		
S_1	3	1	17	3	11		4
S_2	12		18	6	10	6	12
S_3	10	9	25		24		9
	10		9		6		(25)

This solution has the same pattern of occupied cells as the optimal solution to the old problem. Since both problems have the same costs, the solution in Table 2 must also be optimal, with

$$C = (3)(1) + (17)(3) + (18)(6) + (10)(6) + (10)(9) = \$312$$

The difference between the new and old optimal costs is

$$\Delta C = 312 - 296 = \$16$$

The value of ΔC could have been predicted *a priori* from the (O) system of the old problem, since this system shows that $u_3 + v_3 = 16$.

PROBLEM 14.3. Modify Problem 13.2 by increasing the supply at S_1 by 1 unit, but now make no increase in demand anywhere. The new problem is then specified by the following table, where D_4 denotes a *fictitious destination* needed to restore balance. Call this *Problem 14.3*.

	D_1	D_2	D_3	D_4	
S_1	\$ 3	\$17	\$11	\$0	5
S_2	\$12	\$18	\$10	\$0	12
S_3	\$10	\$25	\$24	\$0	8
	10	9	5	1	(25)

By incorporating the old optimal solution into the framework of the new problem, the following feasible solution is obtained.

Table 1

	D_1	D_2	D_3	D_4	
S_1	3 / 2	17 / 2	11	0 / 1	5
S_2	12	18 / 7	10 / 5	0	12
S_3	10 / 8	25	24	0	8
	10	9	5	1	(25)

This solution corresponds to an extreme point, but it is not optimal since the cost of entering cell (3, 4) is -7. Occupying cell (3, 4) to the maximum possible extent leads to the following revised solution.

Table 2

	D_1		D_2		D_3		D_4		
S_1	3	3	17	2	11		0		5
S_2	12		18	7	10	5	0		12
S_3	10	7	25		24		0	1	8
		10		9		5		1	(25)

It is easy to verify that this solution is optimal, with

$$C = (3)(3) + (17)(2) + (18)(7) + (10)(5) + (10)(7) = 289$$

The reader might ask whether there is any information contained in the optimal $(u_i + v_j)$ equations of the old problem, from which we can immediately deduce the exact reduction in cost attendant upon increasing S_1 capacity by 1 unit? The answer is "no." Although we know beforehand that the new optimum cost cannot be larger than the old optimum, there is not enough information in the old optimum $(u_i + v_j)$ equations to enable us to calculate quickly the exact decrease in cost achieved when we increase S_1 capacity alone. This is because Problem 13.2 is balanced and the optimum u_i cannot be separated from the optimum v_j—all we know is the value of the sum $(u_i + v_j)$.

The optimal table for Problem 14.3 designates S_3 as the source with excess capacity. By the theorem on complementary slackness, the optimal value of the dual variable u_3 must necessarily be zero. (In less sophisticated terms, suppose we were to decrease the capacity of S_3 by 1 unit. We know beforehand that total shipping cost will not change, because not all of S_3's capacity can be utilized. Hence $u_3 = 0$.)

Furthermore, the (O) system associated with the optimal solution of Problem 14.3 determines a set of dual variables satisfying the following relations:

$$\begin{array}{lllllllll}
u_1 & & & + v_1 & & & & = & 3 \\
u_1 & & & & + v_2 & & & = & 17 \\
& u_2 & & & + v_2 & & & = & 18 \\
& u_2 & & & & + v_3 & & = & 10 \\
& & u_3 & + v_1 & & & & = & 10 \\
& & u_3 & & & & + v_4 & = & 10
\end{array}$$

This is a system of six equations in seven unknowns, but, as we have just observed, $u_3 = 0$. This precise specification allows us to discover the values of all *individual* u_i and v_j, as follows: $u_1 = -7$, $u_2 = -6$, $u_3 = 0$; $v_1 = 10$, $v_2 = 24$, $v_3 = 16$, $v_4 = 10$.

Note carefully that, although we cannot predict the optimal cost of Problem 14.3 solely from the information contained in the optimal table of Problem 13.2, we can proceed in reverse order and actually predict the optimal cost of Problem 13.2 solely from information contained in the optimal table of Problem 14.3. We have already determined the exact values of the dual variables associated with Problem 14.3; in particular, the dual variable associated with S_1 capacity is $u_1 = -7$. In passing from Problem 14.3 to Problem 13.2 we have to decrease the capacity at S_1 by 1 unit, and the optimal value of C must increase by $-(-7) = +7$ dollars.

For another example, consider the unbalanced problem defined below.

PROBLEM 14.4.

	D_1	D_2	D_3	
S_1	\$ 3	\$17	\$11	5
S_2	\$12	\$18	\$10	13
S_3	\$10	\$25	\$24	8
	10	9	5	$\neq$

Think of Problem 14.4 as a modification of Problem 14.3, obtained by increasing the capacity at S_2 by 1 unit. The optimal solution of Problem 14.3 indicated that

$$u_2 = -6$$

Hence the optimal solution to Problem 14.4 will be \$6 less than that obtained in Problem 14.3—*provided, of course, that the passage from the old problem to the new one does not alter the old optimal set of occupied cells.*

We now verify this prediction. The balanced defining table for Problem 14.4 reads

	D_1	D_2	D_3	D_4	
S_1	\$ 3	\$17	\$11	\$0	5
S_2	\$12	\$18	\$10	\$0	13
S_3	\$10	\$25	\$24	\$0	8
	10	9	5	2	(26)

A feasible solution to Problem 14.4 is shown in Table 1 below. This solution was obtained by slightly modifying the optimal solution to Problem 14.3, namely, by shipping 1 unit into cell (2, 4):

Table 1

	D_1	D_2	D_3	D_4	
S_1	[3] 3	[17] 2	[11]	[0]	5
S_2	[12]	[18] 7	[10] 5	[0] 1	13
S_3	[10] 7	[25]	[24]	[0] 1	8
	10	9	5	2	(26)

Table 1 contains seven occupied cells, instead of the desired six. A moment's thought shows that we should immediately seek to empty cell (2, 4) because this cell was empty in the *optimal* table of Problem 14.3. Remember that when a table is optimal, then *empty* and *unprofitable* are synonyms.

Upon emptying cell (2, 4), we obtain the following solution:

Table 2

	D_1	D_2	D_3	D_4	
S_1	[3] 4	[17] 1	[11]	[0]	5
S_2	[12]	[18] 8	[10] 5	[0]	13
S_3	[10] 6	[25]	[24]	[0] 2	8
	10	9	5	2	(26)

This program is optimal, with

$$C = (3)(4) + (17)(1) + (18)(8) + (10)(5) + (10)(6) = 283$$

as predicted.

Exercises

1. Solve by the $(u_i + v_j)$ method:

	D_1	D_2	D_3	
S_1	0	14	8	4
S_2	2	8	0	12
S_3	0	15	14	5
	7	9	5	(21)

2. *True or false?* If a problem is not perfectly balanced, its optimum solution is degenerate. Explain.

3. Given the transportation problem

	D_1	D_2	D_3	
S_1	100	114	108	4
S_2	2	8	0	12
S_3	0	15	14	5
	7	9	5	(21)

verify that the optimum plan is $x_{11} = 2$, $x_{12} = 2$, $x_{13} = 0$, $x_{21} = 0$, $x_{22} = 7$, $x_{23} = 5$, $x_{31} = 5$, $x_{32} = 0$, $x_{33} = 0$.

4. Given the transportation problem

	D_1	D_2	D_3	D_4	
S_1	\$ 23	\$ 8	\$ 0	\$28	11
S_2	\$ 10	\$20	\$25	\$15	17
S_3	\$100	\$65	\$ 0	\$ 2	20
	5	15	13	15	(48)

(a) Verify that the following solution is optimal:

	D_1	D_2	D_3	D_4	
S_1		3	8		11
S_2	5	12			17
S_3			5	15	20
	5	15	13	15	(48)

(b) Do any alternative optima exist?

5. *True or false?* Suppose there is a reduction in unit cost of g dollars upon entering cell (i, j) of Table n and that Table (n + 1) is formed by occupying this cell to the maximum extent possible. Then the unit costs of entering an empty cell for Tables n and (n + 1) are either equal, or differ by $\pm g$ dollars.

6. Given the problem

	D_1	D_2	
S_1	\$1	\$2	6
S_2	\$3	\$4	10
	4	11	≠

(a) Verify that the following two solutions are optimal.

Solution 1

	D_1	D_2	D_3	
S_1	[1] 4	[2] 2	[0]	6
S_2	[3]	[4] 9	[0] 1	10
	4	11	1	(16)

Solution 2

	D_1	D_2	D_3	
S_1	[1] 0	[2] 6	[0]	6
S_2	[3] 4	[4] 5	[0] 1	10
	4	11	1	(16)

(b) Show that the (O) systems corresponding to these solutions can be specified uniquely, and that they give the same solution, namely, $u_1 = -2$, $u_2 = 0$; $v_1 = 3$, $v_2 = 4$, $v_3 = 0$.

(c) Examine the validity of the following sensitivity predictions on the given problem:

(i) Increase capacity at S_1 by 1 unit, and optimal cost will drop by \$2.
(ii) Increase demand at D_1 by 1 unit, and optimal cost will rise by \$3.
(iii) Increase demand at D_2 by 1 unit, and cost will rise by \$4.

(iv) Increase demand at D_3 by 1 unit, and cost will remain unchanged.
(v) Increase capacity at S_1 by 1 unit and simultaneously increase demand at D_1 by 1 unit, and cost will rise by $\$(-2 + 3) = \$1$.

7. Construct an example to show that when the capacity at some S_i and the demand at some D_j are both increased by one, it is possible for the minimum cost of shipment to decrease.

15

AN INTRODUCTION TO GRAPHS AND NETWORKS

The aim of this last chapter is to introduce some topics usually treated in a second course in linear programming. We begin by sketching the essential ideas behind a graph-theoretic, or "network," algorithm for solving transportation problems. One of the several interesting features of this algorithm is that it completely avoids all matters of degeneracy, a great advantage in solving the assignment problem, the most degenerate of all transportation problems. After a brief discussion of the assignment problem we show how to apply this model to questions as varied as the existence of codes and the determination of shortest paths in a network.

We recall a fundamental theorem of Chapter 13, which we here refer to as Theorem 15.1.

THEOREM 15.1. Subtracting (or adding) a constant k from all the elements in any row or column of the cost matrix of a balanced transportation problem does not affect the optimal shipping plan.

In this chapter we sketch a complete theory around Theorem 15.1. Let us first pause to develop a clear conception of a form of the cost matrix from which the optimal shipping plan is self-evident. Consider the following table:

	D_1	D_2	D_3	
S_1	\$0	\$0	\$7	15
S_2	\$5	\$0	\$0	25
	10	20	10	(40)

This table describes a perfectly balanced transportation problem with zero cost of shipping from S_1 to both D_1 and D_2, as well as zero cost of shipping from S_2 to both D_2 and D_3. The optimal shipping plan is almost self-evident, namely, $x_{11} = 10$, $x_{12} = 5$, $x_{13} = 0$, $x_{21} = 0$, $x_{22} = 15$, and $x_{23} = 10$:

	D_1	D_2	D_3	
S_1	[0] 10	[0] 5	[7] 0	15
S_2	[5] 0	[0] 15	[0] 10	25
	10	20	10	(40)

This plan meets all requirements, at a total shipping cost of 0. It is clearly optimal, whether or not we understand simplex theory, simply because all individual costs are nonnegative.

The cost table above deserves the name *canonical* because it leads almost immediately to an optimal plan. The reasons for this are:

(1) There are sufficient zero costs of shipping to find a feasible shipping plan at total cost of zero.

(2) No cost is negative, and hence no plan could cost less than 0.

Whenever $c_{ij} = 0$ we will say that cell (i, j) is *cost free*, or equivalently, route (i, j) is *free*. A cost table with all $c_{ij} \geq 0$ and enough free routes to find a feasible plan at total cost of 0 will be called *canonical*.

In this section we will show how to transform a noncanonical cost table into an equivalent canonical table. For a first indication of things to come, consider the problem defined by the following data:

PROBLEM 15.1.

Table 1

	D_1	D_2	D_3	
S_1	\$4	\$2	\$1	15
S_2	\$1	\$3	\$5	25
	10	20	10	(40)

By subtracting the minimum cost in each row of Table 1 from all the costs in that row, we obtain the following equivalent, but reduced, cost table:

Table 2

	D_1	D_2	D_3	
S_1	3	1	0	15
S_2	0	2	4	25
	10	20	10	(40)

By our very choice of subtrahends, the reduced costs in Table 2 are nonnegative, and the smallest cost in each row is 0. Now subtract the smallest cost in each column of Table 2 from all entries in that column. This gives

Table 3

	D_1	D_2	D_3	
S_1	3	0	0	15
S_2	0	1	4	25
	10	20	10	(40)

Table 3 is not canonical, since it is impossible to find a feasible shipping plan at total cost of 0. However, by reducing by 1 each cost in row 2 of Table 3, the following equivalent table is obtained:

Table 4

	D_1	D_2	D_3	
S_1	3	0	0	15
S_2	-1	0	3	25
	10	20	10	(40)

Since Table 4 contains a negative cost it is obviously not canonical. But note that if all costs in column 1 of this table are increased by 1, we obtain the equivalent table

Table 5

	D_1	D_2	D_3	
S_1	4	0	0	15
S_2	0	0	3	25
	10	20	10	(40)

Table 5 enables us to specify the following optimal shipping plan:

	D_1	D_2	D_3	
S_1	4 | 0	0 | 5	0 | 10	15
S_2	0 | 10	0 | 15	3 | 0	25
	10	20	10	(40)

Note carefully the logic of the procedure—by Theorem 15.1, Tables 1 to 5 are equivalent in the sense that any plan which is optimal with respect to the costs in one table is optimal for the costs in any other table; and it was not difficult to specify an optimal plan for Table 5.

The solution above was not obtained in a systematic fashion—somehow we scanned Table 3 and had an insight as to how to proceed to Tables 4 and 5. In order to solve this problem and others like it, in a systematic fashion, we need the following consequences of Theorem 15.1, the proof of which can be found in Chapter 13.

THEOREM 15.2. (a) Subtracting r_i from each of the costs of row i of a given transportation problem gives a reduced problem whose optimal cost is $r_i a_i$ less than that of the original problem, where a_i denotes the capacity of the ith supplier.

(b) Subtracting k_j from each of the costs in column j of a transportation table reduces the cost of any (optimal) program by $k_j b_j$, where b_j denotes the demand at destination j.

(c) Subtracting $r_1, r_2, \ldots, r_m$ from each of the costs in row $1, 2, \ldots, m$, respectively, and $k_1, \ldots, k_n$ from each of the costs in columns $1, 2, \ldots, n$, respectively, reduces the cost of any (optimal) program for a given transportation model by

$$S(r, k) = [r_1 a_1 + \cdots + r_m a_m] + (k_1 b_1 + \cdots + k_n b_n)$$

Recall that we have employed c_{ij} to denote the original cost of shipping a unit from S_i to D_j, and c'_{ij} to denote the corresponding reduced cost. In addition, if T denotes the total cost of shipping for some definite plan, using the original c_{ij}, and T' denotes the total cost for this plan using the reduced c'_{ij}, then $T = T' + S(r, k)$. The following theorems are immediate consequences of Theorem 15.2, and are fundamental to the algorithm of this section.

THEOREM 15.3. Suppose all reduced costs are nonnegative; that is, for all i and j, $c'_{ij} \geq 0$. Then $S(r, k)$ serves as a *lower bound* for T, the total cost of shipping:

$$S(r, k) \leq T$$

The proof is immediate, for $c'_{ij} \geq 0$ implies that $T' \geq 0$ and, from $T = T' + S(r, k)$ and $T' \geq 0$, it follows that $T \geq S(r, k)$.

THEOREM 15.4. If the matrix C' is canonical, then the value T of the optimal shipping plan is given by $T = S(r, k)$.

The proof again follows directly from the equation

$$T = T' + S(r, k)$$

because if C' is canonical then $T' = 0$, by definition.

THEOREM 15.5. The cost table of a transportation problem can always be reduced to canonical form. More precisely, it is possible to find subtrahends r_i and k_j so that $S(r, k) = T$, and hence $T' = 0$.

As observed earlier, all subtrahends r_i and k_j such that $(r_i + k_j) \leq c_{ij}$ must be feasible solutions to the dual of the transportation problem. Remember that we have written the primal in a form in which all nontrivial constraints are equalities, and hence the dual variables are unrestricted. Thus, Theorem 15.5 is an immediate consequence of the fundamental theorem of duality. However, it is possible to give a proof of Theorem 15.5 which is independent of the theory of duality in linear programming. Such a proof is implicitly contained in the constructive algorithm for obtaining a canonical cost table, which follows.

We return to Problem 15.1, solving it in detail, *and keeping a record of the current row and column subtrahends.* The original cost matrix is

Table 1

	D_1	D_2	D_3	
S_1	\$4	\$2	\$1	15
S_2	\$1	\$3	\$5	25
	10	20	10	(40)

To reduce Table 1, subtract the minimum cost in each row, thereby forming

Table 2

	$k_1 = 0$	$k_2 = 0$	$k_3 = 0$	
$r_1 = 1$	3	1	0	15
$r_2 = 1$	0	2	4	25
	10	20	10	(40)

In the left rim and in the upper border of Table 2 we record the current values of the r_i and k_j which have been subtracted from the *initial* table to form the *current* table.

Now reduce Table 2 by subtracting the minimum element in each column of the body of this table from every element in that column, obtaining

Table 3

	$k_1 = 0$	$k_2 = 1$	$k_3 = 0$	
$r_1 = 1$	3	0	0	15
$r_2 = 1$	0	1	4	25
	10	20	10	(40)

The numbers in the margins of Table 3 now indicate how to pass directly, in one giant step, from Table 1 to Table 3—simply take as subtrahends $r_1 = 1$, $r_2 = 1$, $k_1 = 0$, $k_2 = 1$, $k_3 = 0$.

As observed earlier, Table 3 is not optimal. One way of showing that it is impossible to find a feasible solution over the free routes of Table 3 is to observe that only one free route starts at S_2, namely, the route (2, 1), but using this route to the maximum extent possible will not exhaust all the capacity at S_2.

We can get greater insight into the deficiencies of Table 3 and, at the same time, discover how to remove these deficiencies, by drawing a *graph* connecting all supply and demand points which can be joined by a free route. Thus we obtain the picture

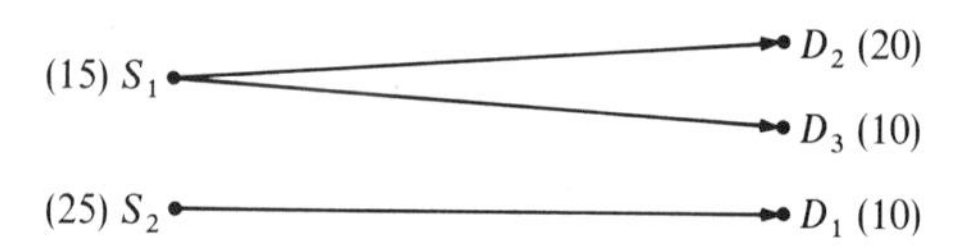

Each *node* in the graph is labeled S_i, or D_j, depending on whether it represents a supply or demand point. The capacity at S_i and the requirement at D_j are recorded in parenthesis. S_i and D_j are connected by a *directed edge*, carrying an arrowhead pointing into D_j, if and only if there exists a free route between S_i and D_j.

Observe that this graph splits into two subgraphs:

(1) A subgraph connecting S_1 to D_2 and D_3. S_1 can only ship at zero cost to D_2 and D_3, neither of which can be supplied free from any other origin.

(2) A subgraph connecting S_2 to D_1. S_2 can ship free to D_1, but D_1 can be supplied free from no other origin.

Either of these subgraphs, taken alone, suffices to show that Table 3 is not canonical: Subgraph (1) shows that it is impossible to meet the aggregate demand using only free routes, while subgraph (2) shows that although the demand at D_1 can be satisfied there will necessarily be excess supply left over at S_2. We can, however, use the information contained in either subgraph (1) or (2) to find revised subtrahends, that is, new and better r_i and k_j which will *increase* the value of $S(r, k)$ and thereby take us closer to optimality.

For example, observe from case (2) that if we increase r_2 and *decrease* k_1 by the same integral amount θ, keeping all other subtrahends as they are, then route (2, 1) will remain free, but the contribution of the summand $(r_2a_2 + k_1b_1)$ will increase $S(r, k)$ by $\theta(a_2 - b_1) = \theta(25 - 10) = 15\theta$. How large θ can actually become is governed by the stipulation that no cost become negative in the revised table. A moment's thought shows that we can only take $\theta = 1$, that is, increase r_2 by 1, at which point the cost in cell (2, 2) of Table 3 will drop to 0. A more mechanical way of finding the maximum allowable θ is to draw lines through row 2 and column 1. Since we have decided to increase r_2 the costs along the dotted line in row 2 will necessarily decrease. The smallest cost in this row whose decrease cannot be counteracted by the adjustments in column 1 is the entry in column 2, as shown below:

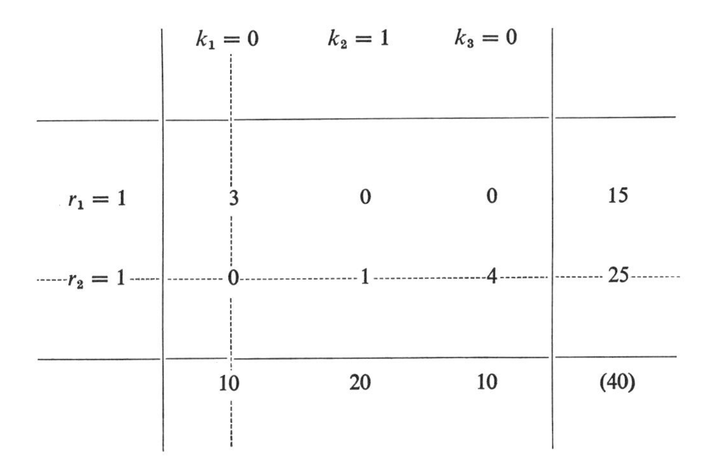

	$k_1 = 0$	$k_2 = 1$	$k_3 = 0$	
$r_1 = 1$	3	0	0	15
$r_2 = 1$	0	1	4	25
	10	20	10	(40)

Hence we increase r_2 by 1 and decrease k_1 by 1 and obtain the following table:

Table 4

	$k_1 = -1$	$k_2 = 1$	$k_3 = 0$	
$r_1 = 1$	4	0	0	15
$r_2 = 2$	0	0	3	25
	10	20	10	(40)

This table is in canonical form (it was Table 5 in the earlier, nonsystematic solution).

The next example is intended to clarify the intuitive content of the graph of free-routes technique.

PROBLEM 15.2.

Table 1

	D_1	D_2	D_3	
S_1	\$ 3	\$17	\$11	4
S_2	\$12	\$18	\$10	12
S_3	\$10	\$25	\$24	8
	10	9	5	(24)

First subtract the minimum cost in each row of the cost matrix from every cost in that row. Then subtract the minimum cost in each of the resulting columns from every cost in that column, obtaining

Table 2

	$k_1 = 0$	$k_2 = 8$	$k_3 = 0$	
$r_1 = 3$	0	6	8	4
$r_2 = 10$	2	0	0	12
$r_3 = 10$	0	7	14	8
	10	9	5	(24)

The graph of free routes for Table 2 is

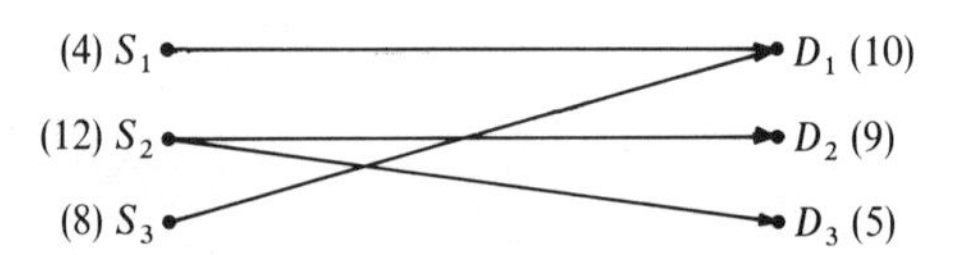

Evidently there is imbalance in the sector connecting S_2 to D_2 and D_3, which means that $S(r, k)$ can be increased by increasing k_2 and k_3 and decreasing r_2, all by the same amount θ. By drawing lines down columns 2 and 3 and across row 2 of Table 2 we see that the critical value is $\theta = 6$. After decreasing r_2 by 6 and increasing k_2 and k_3 by 6, the following table is obtained:

Table 3

	$k_1 = 0$	$k_2 = 14$	$k_3 = 6$	
$r_2 = 3$	0	0	2	4
$r_2 = 4$	8	0	0	12
$r_3 = 10$	0	1	8	8
	10	9	5	(24)

The graph of free routes belonging to Table 3 is

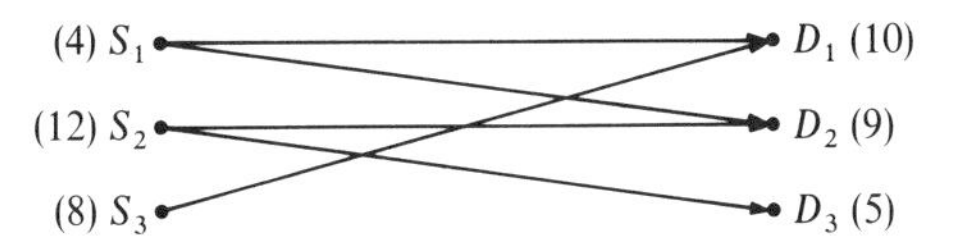

There are no apparent imbalances in this graph, which suggests that a feasible solution can be found using the free routes in Table 3. Indeed, there is only one such solution, namely, $x_{31} = 8$, $x_{11} = 2$, $x_{12} = 2$, $x_{22} = 7$, $x_{23} = 5$.

The optimal plan for the given problem, with the original cost data taken from Table 1, is

	D_1	D_2	D_3	
S_1	[3] 2	[17] 2	[11] 0	4
S_2	[12] 0	[18] 7	[10] 5	12
S_3	[10] 8	[25] 0	[24] 0	8
	10	9	5	(24)

The total cost T for this plan is

$$T = (3)(2) + (17)(2) + (11)(0) + (12)(0) + (18)(7) + (10)(5) + (10)(8) + (25)(0) = 296$$

As an independent check on optimality we can calculate the value of $S(r, k)$ associated with Table 3:

$$S(r, k) = (3)(4) + (4)(12) + (10)(8) + (0)(10) + (14)(9) + (6)(5) = 296$$

The reader may have noticed that Problem 15.3 is not perfectly balanced. Degeneracy might have manifested itself, but no matter, because the graph of free-routes technique of this chapter performs just as smoothly in the degenerate case as in the nondegenerate case. This should become quite clear from the exercises.

We began the solution to Problem 15.2 by first subtracting the minimum cost in each row from all other costs, and then we did the same thing for the resulting columns. It is also possible to begin with columns and then operate on rows. Each of these procedures leads to a reduced cost table with 0 as the smallest cost in any row or column, but the reduced cost tables obtained are generally not the same.

There exist several algorithms for determining whether the current graph of free routes corresponds to a canonical cost table and, if not, for determining a better choice of r_i's and k_j's. One such procedure, sufficient for our purposes, can be based upon the following remarkable *supply–demand theorem:*

The graph of free routes corresponds to a canonical cost table if and only if, for *every* subset D of demand nodes, the sum of the associated b_j is less than or equal to the sum of the a_i at the origins of the edges pointing into D.

To illustrate, consider the graph of free routes corresponding to Problem 15.2, Table 3. To test whether this table is canonical, we can examine all the different subsets of demand nodes:

(1) D_1
(2) D_2
(3) D_3
(4) D_1, D_2
(5) D_1, D_3
(6) D_2, D_3
(7) D_1, D_2, D_3

In case (1), there is only one b_j, namely $b_1 = 10$, and the sum of the a_i at the other end of the edges pointing into D_1 is $a_1 + a_3 = 4 + 8 \geq 10$, and so on; in case (5), the sum of the b_j is $b_1 + b_3 = 10 + 5$ and the sum of the a_i is $a_1 + a_2 + a_3 = 24$, and so on. The reader can easily verify that Table 3 satisfies all the conditions of the supply–demand theorem.

Clearly, if the theorem were not satisfied, there would exist an unbalanced sector, and we could use this sector to find a better set of r_i and k_j, as indicated in the above examples. Knowing that a table is canonical is not the same thing as constructing the optimal shipping plan. For canonical tables of small dimensionality the optimal plan can usually be found by inspection or trial and error; with tables of high dimensionality a formal algorithm is needed.*

PROBLEM 15.3. Given three jobs, J_1, J_2, J_3, and three candidates, P_1, P_2, P_3. Each candidate is more or less unqualified for each of the jobs. The table below gives the *incompatibility ratings* for each individual with respect to each job.

	J_1	J_2	J_3
P_1	8	4	7
P_2	5	3	4
P_3	4	1	5

The objective is to assign each individual to one job so as to minimize the total sum of incompatibilities.

* The interested reader can find such an algorithm, as well as the proof of the supply–demand theorem, in L. Ford and D.R. Fulkerson, *Flows in Networks* (Princeton University Press, Princeton, N.J., 1962).

Think of the individuals as sources, each of capacity equal to 1, *and the jobs as destinations, each with demand equal to* 1. Then this "assignment problem" is clearly equivalent to the following transportation problem:

Table 1

	J_1	J_2	J_3	*Capacity*
P_1	8	4	7	1
P_2	5	3	4	1
P_3	4	1	5	1
Requirement	1	1	1	(3)

Obviously, any amount shipped x_{ij} can only be 0 or 1. As a consequence, no row or column can contain more than one nonzero amount shipped. Note further that an assignment problem, when viewed as a transportation problem, is never perfectly balanced. This suggests that we attack the assignment problem by the flow-graph techniques of this chapter rather than by the methods of the previous chapters. By reducing "costs" in the usual way, first in each row, then in each column, we obtain

Table 2

	$k_1 = 2$	$k_2 = 0$	$k_3 = 1$	
$r_1 = 4$	2	0	2	1
$r_2 = 3$	0	0	0	1
$r_3 = 1$	1	0	3	1
	1	1	1	(3)

The graph of free routes associated with Table 2 is

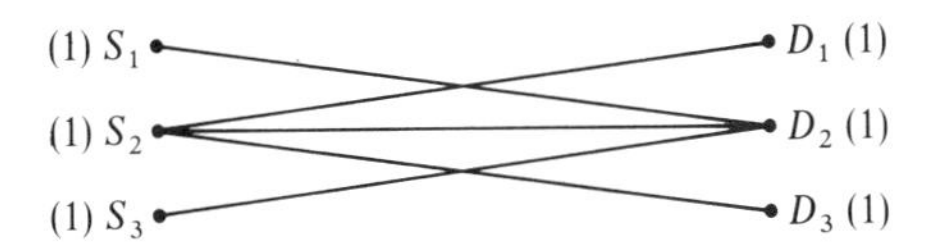

In discussing assignment problems, instead of saying that there is a free route between S_i and D_j, we will say that "individual i and job j can be *matched*." Since this relation is symmetric, the routes, or edges, will not be given a direction.

The graph shows that job 3 can only be matched with person 2. Once this is accomplished, it is clear that no individual remains who can be matched with job 1, and hence Table 2 is not canonical. But the very bottleneck involving person 2 and jobs 1 and 3 can be used to find a better solution:

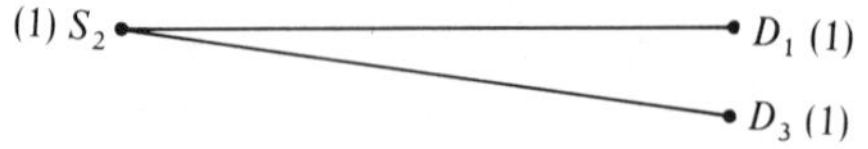

(1) S_2 — D_1 (1)
(1) S_2 — D_3 (1)

Thus, to increase $S(r, k)$ we should increase k_1 and k_3 and decrease r_2. By the usual techniques, it is seen that the maximum allowable change in these subtrahends is $\theta = 1$. Thus, by increasing k_1 and k_3 by 1 and decreasing r_2 by 1, the following reduced table and graph are obtained:

Table 3

	$k_1 = 3$	$k_2 = 0$	$k_3 = 2$	
$r_1 = 4$	1	0	1	1
$r_2 = 2$	0	1	0	1
$r_3 = 1$	0	0	2	1
	1	1	1	(3)

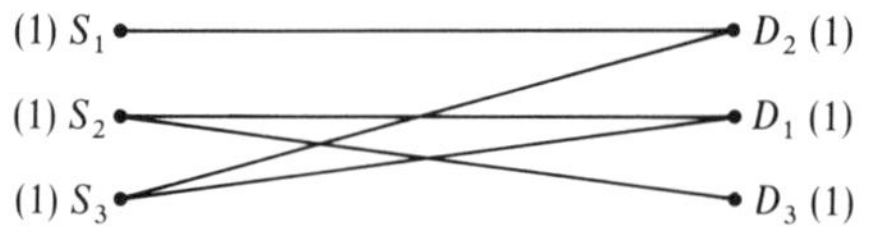

(1) S_1 — D_2 (1)
(1) S_2 — D_1 (1)
(1) S_2 — D_3 (1)
(1) S_3 — D_2 (1)
(1) S_3 — D_1 (1)

Evidently, at least one complete matching is possible: S_1 with D_2, S_2 with D_3, S_3 with D_1; in other words, the optimal assignment is $x_{12} = 1$, $x_{23} = 1$, $x_{31} = 1$. This leads to a total rating of

$$T = (4)(1) + (4)(1) + (4)(1) = 12$$

As an independent check, note that the value of $S(r, k)$ associated with Table 3 equals the value of T:

$$S(r, k) = (4)(1) + (2)(1) + (1)(1) + (3)(1) + (0)(1) + (2)(1) = 12$$

We repeat that it was relatively easy to find a bottleneck, or unbalanced sector, associated with Table 2 above. In more complex problems, a *systematic* (not necessarily efficient) way of seeking an unbalanced sector is to apply the supply–demand test.

The example below illustrates how to transform an assignment, or transportation problem, of maximization type into an equivalent minimization problem.

PROBLEM 15.4. Consider the following rating table:

Table 1

	D_1	D_2	D_3	D_4	D_5	
S_1	15	19	20	12	13	1
S_2	20	18	10	13	17	1
S_3	18	21	15	21	20	1
S_4	20	21	13	21	17	1
S_5	18	21	16	17	21	1
	1	1	1	1	1	(5)

The S_i stand for salesmen (sources), the D_j for territories (destinations). The entries in the body of the table are ratings or measures of the profit potentiality of each salesman working in each territory. The objective is to assign one and only one salesman to each territory, so as to *maximize* the sum of the corresponding ratings.

We can easily change this problem into the usual minimization problem by multiplying all ratings by -1. The most negative rating will then be -21. By adding $+21$ to all the negative ratings we obtain a table with nonnegative ratings, namely,

Table 1′

	D_1	D_2	D_3	D_4	D_5	
S_1	6	2	1	9	8	1
S_2	1	3	11	8	4	1
S_3	3	0	6	0	1	1
S_4	1	0	8	0	4	1
S_5	3	0	5	4	0	1
	1	1	1	1	1	(5)

The objective now is to find an assignment which will *minimize* "total cost" according to Table 1′. Reducing costs in the usual way, first across rows and then down columns, leads to the following reduced table and graph:

Table 2

	$k_1 = 0$	$k_2 = 0$	$k_3 = 0$	$k_4 = 0$	$k_5 = 0$	
$r_1 = 1$	5	1	0	8	7	1
$r_2 = 1$	0	2	10	7	3	1
$r_3 = 0$	3	0	6	0	1	1
$r_4 = 0$	1	0	8	0	4	1
$r_5 = 0$	3	0	5	4	0	1
	1	1	1	1	1	(5)

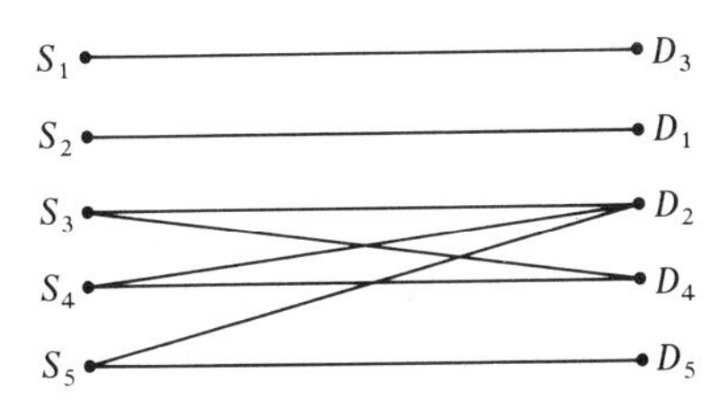

The graph shows that D_3 and S_1 can only join each other, likewise for D_1 and S_2, and, furthermore, D_5 can only be matched with S_5. This leaves only D_2

and D_4, on the one hand, and S_3 and S_4, on the other, for further consideration. There exist alternative ways of matching this subsector of S's and D's—for example, match S_3 with D_2 and S_4 with D_4. This gives an optimal assignment, namely, $x_{13} = 1$, $x_{21} = 1$, $x_{32} = 1$, $x_{44} = 1$, $x_{55} = 1$, all other $x_{ij} = 0$. Accordingly, the maximum profit potential is

$$20 + 20 + 21 + 21 + 21 = 103$$

Observe that the supply–demand theorem takes the following form when applied to an $n \times n$ assignment problem: *Take any subset of the jobs and examine all the men who can be "matched" to some job in that subset. If this number is at least as great as the number of jobs in the subset under scrutiny, and if this is true for all subsets of jobs, then it is indeed possible to match men and jobs completely.*

The problem of matching can be generalized somewhat, as is shown below. The following definitions will serve our purpose:

DEFINITION 1. A *graph* is a set of nodes V, some of which are joined in pairs by interconnecting lines, or *edges*. Two nodes are said to be *adjacent* if they are at the ends of a connecting edge.

DEFINITION 2. A *bipartite graph* is a graph whose node set V can be divided into two nonempty, nonoverlapping subsets, V_1 and V_2, such that no node in either subset is adjacent to another node in the same subset. In other words, every edge goes from V_1 to V_2. (So far, every graph drawn in this chapter is bipartite.)

DEFINITION 3. Let V_1 and V_2 denote the two disjoint sets of nodes in a bipartite graph. (In general, V_1 and V_2 contain unequal numbers of nodes.) A *matching* of V_1 into V_2 is defined as a selection of edges such that no two edges chosen pass through the same node. A matching is said to be *complete* if there is an edge passing through every node of V_1.

To illustrate, consider the following problem: The members of a club S are u, v, w, x, y, z. The following committees are in existence:

$$\begin{aligned} C_1 &= \{v, x, z\} \\ C_2 &= \{y\} \\ C_3 &= \{u, x, y, z\} \\ C_4 &= \{u, w\} \\ C_5 &= \{u, v, w\} \end{aligned}$$

Can we choose a representative from each committee in such a way that no representative will stand for more than one committee? In graph-theoretical terms, this problem reduces to studying the existence of a complete matching of V_1 into V_2 in a bipartite graph, as follows.

Let V_1 be the collection of all committees, that is, $V_1 = \{C_1, C_2, C_3, C_4, C_5\}$, and let V_2 be the set of members, that is, $V_2 = \{u, v, w, x, y, z\}$. Connect a node in V_2 to a node in V_1 if and only if the node in V_2 is a member of the committee represented by the node in V_1:

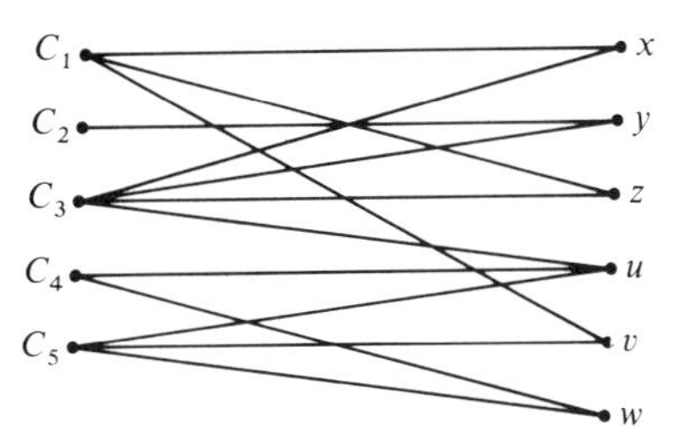

The reader may find it interesting to interpret the problem of committees as a problem in *coding*. To this end, think of $C_1, C_2, \ldots, C_5$ as words to be transmitted. The existence of a complete matching is equivalent to a way of representing each word by one and only one of its letters. If such a matching exists we can represent the word by the letter and transmit the letter rather than the whole word.

The following result is a direct consequence of the supply–demand theorem. It is not a necessary condition for the existence of a complete matching but merely a sufficient condition. As such, it is sometimes useful, always charming.

THEOREM 15.6. Let X and Y be the two disjoint sets of nodes in a bipartite graph. There exists a complete matching of X into Y whenever we can determine a number k such that each node in X is adjacent to k or more nodes in Y and each node in Y is adjacent to k or less nodes in X.

To prove this, let $S_X = \{x_1, \ldots, x_n\}$ be a set of n nodes in X. By hypothesis, each x_i in S can be joined to at least k nodes in Y. Hence there are at least k times n edges leading out of S_X. Let S_Y denote the nodes in Y reached by these edges. Clearly, there are at least k times n edges leading into S_Y. Furthermore, since by hypothesis every node in Y is adjacent to at most k nodes in X, it follows that there must be at least $(kn)/k = n$ nodes in S_Y. Thus, for any set of n nodes S_X in X, the set of possible matches in Y contains at least n nodes and hence, by the supply–demand theorem, there must exist a complete matching of X into Y.

This survey of some basic concepts and results in matching theory and their application to the assignment and transportation problem has omitted some detail and has not even mentioned much famous work in this field. Our only aim is to motivate the reader to study further.*

We terminate the chapter with an example of problem formulation, namely the reduction of the problem of finding the shortest path in a network to an assignment problem. Given a directed network in which each edge (i, j) has associated with it a positive number d_{ij}, which may be thought of as the cost of traversing the arc, or as the length of the arc. To determine the *shortest path* in

* Besides the earlier reference to Ford and Fulkerson, we recommend D. Gale, *The Theory of Linear Economic Models* (McGraw-Hill, New York, 1960).

such a network between nodes P_i and P_j, consider the following simple example:

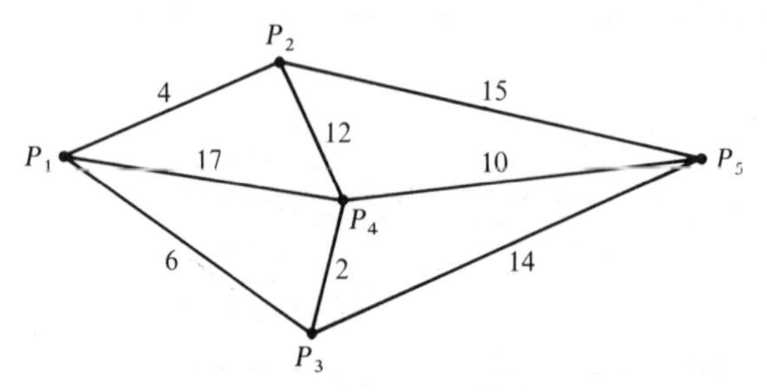

The numbers shown along the above edges are the d_{ij}. Assume that it is possible to traverse each edge in both directions and that distance between nodes is symmetric. The distances can be exhibited in a table:

Table 1

	P_1	P_2	P_3	P_4	P_5
P_1	0	4	6	17	M
P_2	4	0	M	12	15
P_3	6	M	0	2	14
P_4	17	12	2	0	10
P_5	M	15	14	10	0

If no arc connects nodes P_i and P_j, then $d_{ij} = M$, a preponderantly large positive number.

Suppose that the objective is to find the shortest path in this network between P_1 and P_5, a purely combinatorial problem. In the case at hand the shortest route is

$$P_1 \xrightarrow{6} P_3 \xrightarrow{2} P_4 \xrightarrow{10} P_5$$

for a total distance of 18.

Evidently, when many nodes are involved, the problem can become quite complicated. There exist many fairly efficient algorithms for treating this problem, one of which is to transform the shortest-path problem into an equivalent assignment problem. To this end, define

$$\begin{aligned} x_{ij} &= 1 \quad \text{if the arc from } P_i \text{ to } P_j \text{ is in the solution} \\ x_{ij} &= 0 \quad \text{if the arc from } P_i \text{ to } P_j \text{ is not in the solution} \end{aligned}$$

Observe that $x_{11} = x_{22} = x_{33} = x_{44} = x_{55} = 1$ would give an assignment of cost equal to 0 if Table 1 were the relevant rating table—but, obviously, this cannot be the case.

The rating table which works is Table 1′, where the distance from P_5 to P_1 has been changed to $-M^2$:

Table 1′

	P_1	P_2	P_3	P_4	P_5
P_1	0	4	6	17	M
P_2	4	0	M	12	15
P_3	6	M	0	2	14
P_4	17	12	2	0	10
P_5	$-M^2$	15	14	10	0

By using $d_{51} = -M^2$ we are, in effect, *forcing* the imaginary route P_5P_1 into solution. This means that if we solve the assignment problem of Table 1′ then $x_{51} = 1$, and hence $x_{55} \neq 1$, $x_{11} \neq 1$. But there must be some x_{1j} which does equal 1 and this will determine the optimum branch out of P_1, terminating in P_j. As above, this means that $x_{jj} \neq 1$ and therefore some $x_{jk} = 1$, for $j \neq k$. In this way all the necessary intermediate branches will be forced into solution and the complete shortest route from P_1 to P_5 will be determined. The actual mechanics is left to the reader.

In general to determine the shortest route from P_α to P_γ we must force the reverse arc, from P_γ to P_α, into solution. This can be accomplished by taking distance $d_{\gamma\alpha} = -M^2$, where M is a very large positive number. Then, when we solve the resulting assignment problem not all $x_{ii} = 1$; but if $x_{jj} = 1$ for some j, then P_j will not be a node along the optimum path. Obviously, the reverse branch from P_γ to P_α is not part of the solution—it is merely the catalyst.

Exercises

1. Consider the following transportation problem:

	D_1	D_2	
S_1	\$1	\$7	5
S_2	\$4	\$2	10
	5	10	(15)

(a) Is the optimal shipping plan intuitively obvious?
(b) Solve this problem by the methods of Chapters 13 or 14.
(c) Solve this problem by the free-routes technique.

2. Consider the following problem:

	D_1	D_2	
S_1	\$7	\$1	5
S_2	\$2	\$4	10
	5	10	(15)

(a) Is the solution obvious?
(b) Solve this problem by the methods of Chapters 13 or 14.
(c) Solve this problem by the free-routes technique.

3. Consider the following transportation problem (Problem 13.3).

	D_1	D_2	D_3	
S_1	\$4	\$2	\$1	15
S_2	\$1	\$3	\$5	25
	10	5	25	(40)

(a) Verify that the tables below give an optimal shipping plan:

Reduced Cost Table

	$k_1 = -1$	$k_2 = 1$	$k_3 = 3$
$r_1 = -2$	7	3	0
$r_2 = 2$	0	0	0

Shipping Plan

	D_1	D_2	D_3	
S_1	4 0	2 0	1 15	15
S_2	1 10	3 5	5 10	25
	10	5	25	(40)

(b) Solve this problem from the beginning, by the free-routes technique.

4. Given the following transportation problem:

	D_1	D_2	D_3	D_4	
S_1	\$7	\$6	\$ 5	\$13	12
S_2	\$9	\$7	\$14	\$10	12
S_3	\$1	\$8	\$ 7	\$ 9	5
	5	12	4	8	(29)

(a) Verify that the tables below furnish an optimal solution:

Reduced Cost Table

	$k_1 = 0$	$k_2 = 0$	$k_3 = -1$	$k_4 = 3$
$r_1 = 6$	1	0	0	4
$r_2 = 7$	2	0	8	0
$r_3 = 1$	0	7	7	5

Shipping Plan

	D_1	D_2	D_3	D_4	
S_1	7 0	6 8	5 4	13 0	12
S_2	9 0	7 4	14 0	10 8	12
S_3	1 5	8 0	7 0	4 0	5
	5	12	4	8	(29)

(b) Solve this problem from the beginning, by the free-routes technique.

5. What happens to the optimal value of the objective function of a transportation problem if all c_{ij} are replaced by $c_{ij} + k$, where k is a constant?

6. A steamship company has four ships scheduled to arrive on the same day. They can be unloaded at any of four different docks. The following table gives unloading time in hours:

	Dock			
Ship	D_1	D_2	D_3	D_4
S_1	8	7	9	6
S_2	10	9	7	9
S_3	10	9	7	10
S_4	8	8	7	8

(a) Assign ships to docks so as to minimize the total unloading time.
(b) Assign ships to docks so as to maximize the total unloading time.

7. Express the following transportation problem as a 5 × 5 assignment problem:

	D_1	D_2	
S_1	c_{11}	c_{12}	2
S_2	c_{21}	c_{22}	3
	1	4	(5)

8. A car rental service has a surplus of one car in each of five cities, A, B, C, D, and E; it has a shortage of two cars in city F and a shortage of one car in cities G, H, and I. The following table gives the minimum cost of shifting a car from a surplus city to a shortage city:

	F	*G*	*H*	*I*
A	13	12	14	11
B	11	14	12	13
C	14	13	11	14
D	10	10	10	11
E	11	11	13	13

Find an optimal plan for shifting cars from surplus areas to shortage areas.

9. Use the methods of this chapter to solve the assignment problems discussed in illustrative Problems 9.17 and 9.18.

10. Why do we face an inherently degenerate situation when we try to solve an assignment problem by the simplex algorithm?

11. Explain why the supply–demand theorem, when applied to assignment problems, is sometimes called the *diversity condition*.

12. *True or false?*

(a) To turn the problem of degeneracy and solve the assignment problem by means of the stepping-stone algorithm, it suffices to take any assignment and add ϵ to cell (i, j) whenever $x_{i+1,j} = 1$.

(b) In any optimal solution to the assignment problem there is always at least one person assigned to the job he does best and at least one job handled by the individual best qualified.

(c) Each girl at the dance has been previously introduced to precisely k boys, where $k \geq 1$, and each boy has been introduced to precisely k girls. Given that there are n girls and n boys at the dance, there is at least one way of pairing off boys and girls into dance partners so that every girl dances with a partner to whom she has already been introduced.

(d) In the statement of (c) above, assume that each of the n girls has been introduced to *at least* k boys, and that each boy has been introduced to *at least* k girls. Then a pairing into dance partners who have been previously introduced does not necessarily exist.

13. Use the assignment model to determine the shortest route from node 1 to node 5 in the following network:

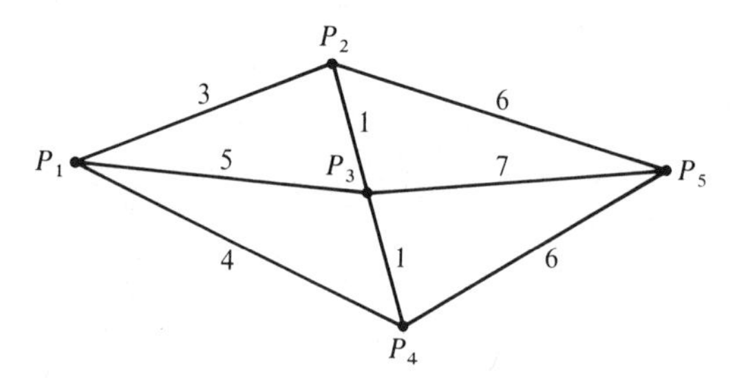

INDEX